Graduate Texts in Mathematics 78

Stanley Burris
H. P. Sankappanavar

# A Course in Universal Algebra

With 36 Illustrations

Springer-Verlag
New York Heidelberg Berlin

Stanley Burris

Department of Pure Mathematics
Faculty of Mathematics
University of Waterloo
Waterloo, Ontario N2L 3G1
CANADA

H. P. Sankappanavar

Instituto de Matemática
Universidade Federal da Bahia
Salvador, Bahia 40000
BRAZIL



AMS Subject Classification (1980): 08-01, 08AXX, 03C05

Library of Congress Cataloging in Publication Data
Burris, S.
A course in universal algebra.
(Graduate texts in mathematics; 78)
Bibliography: p.
Includes index.
1. Algebra, Universal. I. Sankappanavar, H. P.
II. Title. III. Series.
QA251.B87 512 81-1619
AACR2



Printed in the United States of America.

9 8 7 6 5 4 3 2 1

ISBN 0-387-90578-2 Springer-Verlag New York Heidelberg Berlin
ISBN 3-540-90578-2 Springer-Verlag Berlin Heidelberg New York

*This book is dedicated to our children*

*Kurosh Phillip Burris*
*Veena and Geeta Sankappanavar*

# Acknowledgments

First we would like to express gratitude to our colleagues who have added so much vitality to the subject of Universal Algebra during the past twenty years. One of the original reasons for writing this book was to make readily available the beautiful work on sheaves and discriminator varieties which we had learned from, and later developed with H. Werner. Recent work of, and with, R. McKenzie on structure and decidability theory has added to our excitement, and conviction, concerning the directions emphasized in this book.

In the late stages of polishing the manuscript we received valuable suggestions from M. Valeriote, W. Taylor, and the reviewer for Springer-Verlag. For help in proof-reading we also thank A. Adamson, M. Albert, D. Higgs, H. Kommel, G. Krishnan, and H. Riedel. A great deal of credit for the existence of the final product goes to Sandy Tamowski whose enthusiastic typing has been a constant inspiration. The Natural Sciences and Engineering Research Council of Canada has generously funded both the research which has provided much of the impetus for writing this book as well as the preparation of the manuscript through NSERC Grant No. A7256. Also thanks go to the Pure Mathematics Department of the University of Waterloo for their kind hospitality during the several visits of the second author, and to the Institute of Mathematics, Federal University of Bahia, for their generous cooperation in this venture.

The second author would most of all like to express his affectionate gratitude and appreciation to his wife—Nalinaxi—who, over the past four years has patiently endured the many trips between South and North America which were necessary for this project. For her understanding and encouragement he will always be indebted.

And finally we are delighted that Springer-Verlag agreed to our request to have this book appear in their GTM series.

# Preface

Universal algebra has enjoyed a particularly explosive growth in the last twenty years, and a student entering the subject now will find a bewildering amount of material to digest.

This text is not intended to be encyclopedic; rather, a few themes central to universal algebra have been developed sufficiently to bring the reader to the brink of current research. The choice of topics most certainly reflects the authors' interests.

Chapter I contains a brief but substantial introduction to lattices, and to the close connection between complete lattices and closure operators. In particular, everything necessary for the subsequent study of congruence lattices is included.

Chapter II develops the most general and fundamental notions of universal algebra—these include the results that apply to all types of algebras, such as the homomorphism and isomorphism theorems. Free algebras are discussed in great detail—we use them to derive the existence of simple algebras, the rules of equational logic, and the important Mal'cev conditions. We introduce the notion of classifying a variety by properties of (the lattices of) congruences on members of the variety. Also, the center of an algebra is defined and used to characterize modules (up to polynomial equivalence).

In Chapter III we show how neatly two famous results—the refutation of Euler's conjecture on orthogonal Latin squares and Kleene's characterization of languages accepted by finite automata—can be presented using universal algebra. We predict that such "applied universal algebra" will become much more prominent.

Chapter IV starts with a careful development of Boolean algebras, including Stone duality, which is subsequently used in our study of Boolean sheaf representations; however, the cumbersome formulation of general sheaf theory has been replaced by the considerably simpler definition of a Boolean product. First we look at Boolean powers, a beautiful tool for transferring results about Boolean algebras to other varieties as well as for providing a structure theory for certain varieties. The highlight of the chapter is the study of discriminator varieties. These varieties have played a remarkable role in the study of spectra, model companions, decidability, and Boolean product representations. Probably no other class of varieties is so well-behaved yet so fascinating.

The final chapter gives the reader a leisurely introduction to some basic concepts, tools, and results of model theory. In particular, we use the ultraproduct construction to derive the compactness theorem and to prove fundamental preservation theorems. Principal congruence formulas are a favorite model-theoretic tool of universal algebraists, and we use them in the study of the sizes of subdirectly irreducible algebras. Next we prove three general results on the existence of a finite basis for an equational theory. The last topic is semantic embeddings, a popular technique for proving undecidability results. This technique is essentially algebraic in nature, requiring no familiarity whatsoever with the theory of algorithms. (The study of decidability has given surprisingly deep insight into the limitations of Boolean product representations.)

At the end of several sections the reader will find selected references to source material plus state of the art texts or papers relevant to that section, and at the end of the book one finds a brief survey of recent developments and several outstanding problems.

The material in this book divides naturally into two parts. One part can be described as "what every mathematician (or at least every algebraist) should know about universal algebra." It would form a short introductory course to universal algebra, and would consist of Chapter I; Chapter II except for §4, §12,§13, and the last parts of §11, §14; Chapter IV §1–4; and Chapter V §1 and the part of §2 leading to the compactness theorem. The remaining material is more specialized and more intimately connected with current research in universal algebra.

Chapters are numbered by Roman numerals I through V, the sections in a chapter are given by Arabic numerals, §1, §2, etc. Thus II§6.18 refers to item 18, which happens to be a theorem, in Section 6 of Chapter II. A citation within Chapter II would simply refer to this item as 6.18. For the exercises we use numbering such as II§5 Ex. 4, meaning the fourth exercise in §5 of Chapter II. The bibliography is divided into two parts, the first containing books and survey articles, and the second research papers. The books and survey articles are referred to by number, e.g., G. Birkhoff [3], and the research papers by year, e.g., R. McKenzie [1978].

## Diagram of Prerequisites

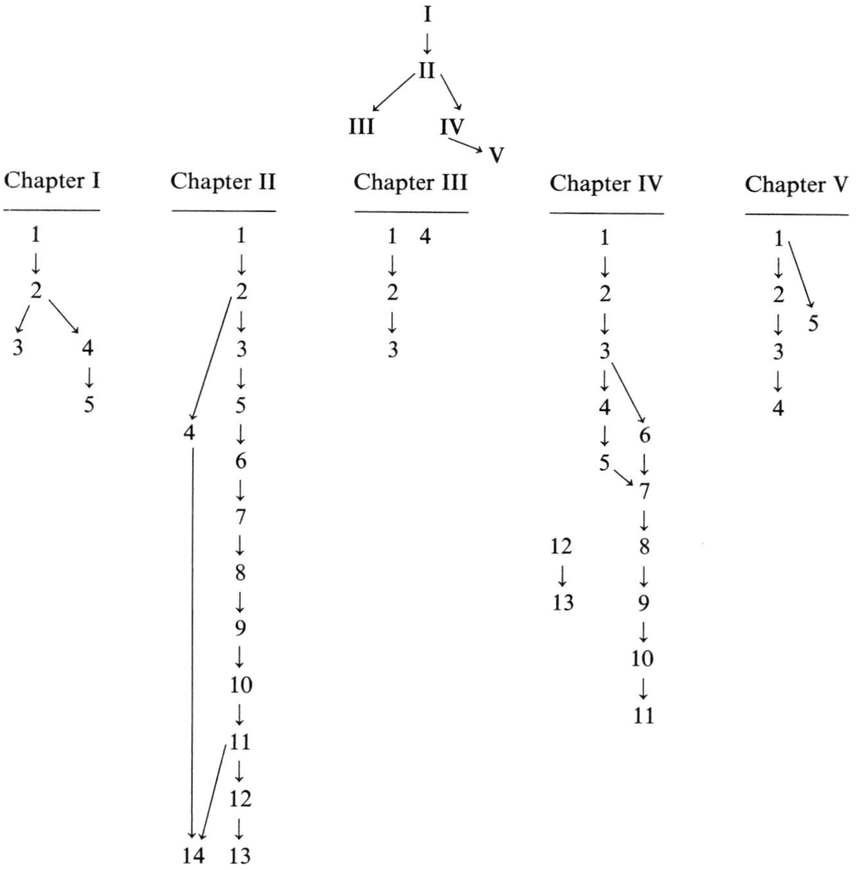

# Contents

# Special Notation

# Preliminaries

We have attempted to keep our notation and conventions in agreement with those of the closely related subject of model theory, especially as presented in Chang and Keisler's *Model Theory* [8]. The reader needs only a modest exposure to classical algebra; for example he should know what groups and rings are.

We will assume a familiarity with the most basic notions of *set theory*. Actually, we use *classes* as well as *sets*. A class of sets is frequently called a *family* of sets. The notations $A_i$, $i \in I$, and $(A_i)_{i \in I}$ refer to a *family of sets indexed by a set I*. A naive theory of sets and classes is sufficient for our purposes. We assume the reader is familiar with *membership* ($\in$), *set-builder notation* ($\{—:—\}$), *subset* ($\subseteq$), *union* ($\cup$), *intersection* ($\cap$), *difference* ($-$), *ordered n-tuples* ($\langle x_1, \ldots, x_n \rangle$), *(direct) products* of sets ($A \times B$, $\prod_{i \in I} A_i$), and *(direct) powers* of sets ($A^I$). Also, it is most useful to know that

(a) concerning relations:
  - (i) an *n-ary relation* on a set $A$ is a subset of $A^n$;
  - (ii) if $n = 2$ it is called a *binary* relation on $A$;
  - (iii) the *inverse* $r^{\smile}$ of a binary relation $r$ on $A$ is specified by $\langle a,b \rangle \in r^{\smile}$ iff $\langle b,a \rangle \in r$;
  - (iv) the *relational product* $r \circ s$ of two binary relations $r$, $s$ on $A$ is given by: $\langle a,b \rangle \in r \circ s$ iff for some $c$, $\langle a,c \rangle \in r$, $\langle c,b \rangle \in s$;

(b) concerning functions:
  - (i) a *function* $f$ from a set $A$ to a set $B$, written $f: A \to B$, is a subset of $A \times B$ such that for each $a \in A$ there is exactly one $b \in B$ with $\langle a,b \rangle \in f$; in this case we write $f(a) = b$ or $f: a \mapsto b$;
  - (ii) the set of all functions from $A$ to $B$ is denoted by $B^A$;
  - (iii) the function $f \in B^A$ is *injective* (or *one-to-one*) if $f(a_1) = f(a_2) \Rightarrow a_1 = a_2$;
  - (iv) the function $f \in B^A$ is *surjective* (or *onto*) if for every $b \in B$ there is an $a \in A$ with $f(a) = b$;

(v) the function $f \in B^A$ is *bijective* if it is both injective and surjective;
(vi) for $f \in B^A$ and $X \subseteq A$, $f(X) = \{b \in B : f(a) = b \text{ for some } a \in X\}$;
(vii) for $f \in B^A$ and $Y \subseteq B$, $f^{-1}(Y) = \{a \in A : f(a) \in Y\}$;
(viii) for $f : A \to B$ and $g : B \to C$, let $g \circ f : A \to C$ be the function defined by $(g \circ f)(a) = g(f(a))$. [This does not agree with the relational product defined above—but the ambiguity causes no problem in practice.];

(c) given a family $F$ of sets, the *union* of $F$, $\bigcup F$, is defined by $a \in \bigcup F$ iff $a \in A$ for some $A \in F$ (define the *intersection* of $F$, $\bigcap F$, dually);
(d) a *chain* of sets $C$ is a family of sets such that for each $A,B \in C$ either $A \subseteq B$ or $B \subseteq A$;
(e) *Zorn's lemma* says that if $F$ is a nonempty family of sets such that for each chain $C$ of members of $F$ there is a member of $F$ containing $\bigcup C$ (i.e., $C$ has an *upper bound* in $F$) then $F$ has a *maximal* member $M$ (i.e., $M \in F$ and $M \subseteq A \in F$ implies $M = A$);
(f) concerning ordinals:
(i) the *ordinals* are generated from the empty set $\varnothing$ using the operations of *successor* ($x^+ = x \cup \{x\}$) and *union*;
(ii) $0 = \varnothing$, $1 = 0^+$, $2 = 1^+$, etc.; the *finite ordinals* are $0,1,\ldots$; and $n = \{0,1,\ldots,n-1\}$; the *natural numbers* are $1,2,3,\ldots$, the nonzero finite ordinals;
(iii) the first *infinite ordinal* is $\omega = \{0,1,2,\ldots\}$;
(iv) the ordinals are *well-ordered* by the relation $\in$, also called $<$;
(g) concerning cardinality:
(i) two sets $A$ and $B$ have the *same cardinality* if there is a bijection from $A$ to $B$;
(ii) the *cardinals* are those ordinals $\kappa$ such that no earlier ordinal has the same cardinality as $\kappa$. The *finite cardinals* are $0,1,2,\ldots$; and $\omega$ is the smallest *infinite cardinal*;
(iii) the *cardinality* of a set $A$, written $|A|$, is that (unique) cardinal $\kappa$ such that $A$ and $\kappa$ have the same cardinality;
(iv) $|A| \cdot |B| = |A \times B|$ [$= \max(|A|,|B|)$ if either is infinite and $A,B \neq \varnothing$]. $A \cap B = \varnothing \Rightarrow |A| + |B| = |A \cup B|$ [$= \max(|A|,|B|)$ if either is infinite];
(h) one usually recognizes that a *class* is not a set by noting that it is *too big* to be put in one-to-one-correspondence with a cardinal (for example, the class of all groups).

In Chapter IV the reader needs to know the basic definitions from point set topology, namely what a *topological space*, a *closed* (*open*) *set*, a *subbasis* (*basis*) for a topological space, a closed (open) *neighborhood* of a point, a *Hausdorff* space, a *continuous* function, etc., are.

The symbol "$=$" is used to express the fact that both sides name the same object, whereas "$\approx$" is used to build equations which may or may not be true of particular elements. (A careful study of $\approx$ is given in Chapter II.)

CHAPTER I
# Lattices

In the study of the properties common to all algebraic structures (such as groups, rings, etc.) and even some of the properties that distinguish one class of algebras from another, lattices enter in an essential and natural way. In particular, congruence lattices play an important role. Furthermore, lattices, like groups or rings, are an important class of algebras in their own right, and in fact one of the most beautiful theorems in universal algebra, Baker's finite basis theorem, was inspired by McKenzie's finite basis theorem for lattices. In view of this dual role of lattices in relation to universal algebra, it is appropriate that we start with a brief study of them. In this chapter the reader is acquainted with those concepts and results from lattice theory which are important in later chapters. Our notation in this chapter is less formal than that used in subsequent chapters. We would like the reader to have a casual introduction to the subject of lattice theory.

The origin of the lattice concept can be traced back to Boole's analysis of thought and Dedekind's study of divisibility. Schroeder and Peirce were also pioneers at the end of the last century. The subject started to gain momentum in the 1930's, and was greatly promoted by Birkhoff's book *Lattice Theory* in the 1940's.

## §1. Definitions of Lattices

There are two standard ways of defining lattices—one puts them on the same (algebraic) footing as groups or rings, and the other, based on the notion of order, offers geometric insight.

**Definition 1.1.** A nonempty set $L$ together with two binary operations $\vee$ and $\wedge$ (read "*join*" and "*meet*" respectively) on $L$ is called a *lattice* if it satisfies the following identities:

L1: (a) $x \vee y \approx y \vee x$
(b) $x \wedge y \approx y \wedge x$ (commutative laws)
L2: (a) $x \vee (y \vee z) \approx (x \vee y) \vee z$
(b) $x \wedge (y \wedge z) \approx (x \wedge y) \wedge z$ (associative laws)
L3: (a) $x \vee x \approx x$
(b) $x \wedge x \approx x$ (idempotent laws)
L4: (a) $x \approx x \vee (x \wedge y)$
(b) $x \approx x \wedge (x \vee y)$ (absorption laws).

EXAMPLE. Let $L$ be the set of propositions, let $\vee$ denote the connective "or" and $\wedge$ denote the connective "and". Then L1 to L4 are well-known properties from propositional logic.

EXAMPLE. Let $L$ be the set of natural numbers, let $\vee$ denote the least common multiple and $\wedge$ denote the greatest common divisor. Then properties L1 to L4 are easily verifiable.

Before introducing the second definition of a lattice we need the notion of a partial order on a set.

**Definition 1.2.** A binary relation $\leq$ defined on a set $A$ is a *partial order* on the set $A$ if the following conditions hold identically in $A$:

(i) $a \leq a$ (reflexivity)
(ii) $a \leq b$ and $b \leq a$ imply $a = b$ (antisymmetry)
(iii) $a \leq b$ and $b \leq c$ imply $a \leq c$ (transitivity).

If, in addition, for every $a,b$ in $A$

(iv) $a \leq b$ or $b \leq a$

then we say $\leq$ is a *total order* on $A$. A nonempty set with a partial order on it is called a *partially ordered set*, or more briefly a *poset*, and if the relation is a total order then we speak of a *totally ordered set*, or a *linearly ordered set*, or simply a *chain*. In a poset $A$ we use the expression $a < b$ to mean $a \leq b$ but $a \neq b$.

EXAMPLES. (1) Let $\mathrm{Su}(A)$ denote the *power set* of $A$, i.e., the set of all subsets of $A$. Then $\subseteq$ is a partial order on $\mathrm{Su}(A)$.

(2) Let $A$ be the set of natural numbers and let $\leq$ be the relation "divides." Then $\leq$ is a partial order on $A$.

(3) Let $A$ be the set of real numbers and let $\leq$ be the usual ordering. Then $\leq$ is a total order on $A$.

Most of the concepts developed for the real numbers which involve only the notion of order can be easily generalized to partially ordered sets.

**Definition 1.3.** Let $A$ be a subset of a poset $P$. An element $p$ in $P$ is an *upper bound* for $A$ if $a \le p$ for every $a$ in $A$. An element $p$ in $P$ is the *least upper bound* of $A$ (l.u.b. of $A$), or *supremum* of $A$ (sup $A$) if $p$ is an upper bound of $A$, and $a \le b$ for every $a$ in $A$ implies $p \le b$ (i.e., $p$ is the smallest among the upper bounds of $A$). Similarly we can define what it means for $p$ to be a *lower bound* of $A$, and for $p$ to be the *greatest lower bound* of $A$ (g.l.b. of $A$), also called the *infimum* of $A$ (inf $A$). For $a,b$ in $P$ we say *$b$ covers $a$*, or *$a$ is covered by $b$*, if $a < b$, and whenever $a \le c \le b$ it follows that $a = c$ or $c = b$. We use the notation $a \prec b$ to denote $a$ is covered by $b$. The *closed interval* $[a,b]$ is defined to be the set of $c$ in $P$ such that $a \le c \le b$, and the *open interval* $(a,b)$ is the set of $c$ in $P$ such that $a < c < b$.

Posets have the delightful characteristic that we can draw pictures of them. Let us describe in detail the method of associating a diagram, the so-called *Hasse diagram*, with a finite poset $P$. Let us represent each element of $P$ by a small circle "$\circ$". If $a \prec b$ then we draw the circle for $b$ above the circle for $a$, joining the two circles with a line segment. From this diagram we can recapture the relation $\le$ by noting that $a < b$ holds iff for some finite sequence of elements $c_1, \ldots, c_n$ from $P$ we have $a = c_1 \prec c_2 \cdots c_{n-1} \prec c_n = b$. We have drawn some examples in Figure 1. It is not so clear how one would draw

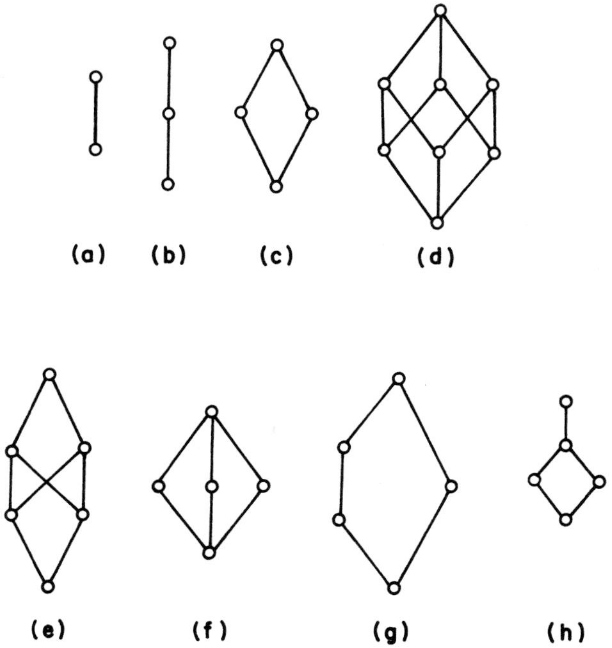

Figure 1 Examples of Hasse diagrams

Figure 2 Drawing the poset of the integers

an infinite poset. For example, the real line with the usual ordering has no covering relations, but it is quite common to visualize it as a vertical line. Unfortunately, the rational line would have the same picture. However, for those infinite posets for which the ordering is determined by the covering relation it is often possible to draw diagrams which do completely convey the order relation to the viewer; for example, consider the diagram in Figure 2 for the integers under the usual ordering.

Now let us look at the second approach to lattices.

**Definition 1.4.** A poset $L$ is a *lattice* iff for every $a,b$ in $L$ both $\sup\{a,b\}$ and $\inf\{a,b\}$ exist (in $L$).

The reader should verify that for each of the diagrams in Figure 1 the corresponding poset is a lattice, with the exception of (e). The poset corresponding to diagram (e) does have the interesting property that every pair of elements has an upper bound and a lower bound.

We will now show that the two definitions of a lattice are equivalent in the following sense: if $L$ is a lattice by one of the two definitions then we can construct in a simple and uniform fashion on the same set $L$ a lattice by the other definition, and the two constructions (converting from one definition to the other) are inverses. First we describe the constructions:

(A) If $L$ is a lattice by the first definition, then define $\leq$ on $L$ by $a \leq b$ iff $a = a \wedge b$;

(B) If $L$ is a lattice by the second definition, then define the operations $\vee$ and $\wedge$ by $a \vee b = \sup\{a,b\}$, and $a \wedge b = \inf\{a,b\}$.

Suppose that $L$ is a lattice by the first definition and $\leq$ is defined as in (A). From $a \wedge a = a$ follows $a \leq a$. If $a \leq b$ and $b \leq a$ then $a = a \wedge b$ and $b = b \wedge a$; hence $a = b$. Also if $a \leq b$ and $b \leq c$ then $a = a \wedge b$ and $b = b \wedge c$, so $a = a \wedge b = a \wedge (b \wedge c) = (a \wedge b) \wedge c = a \wedge c$; hence $a \leq c$. This shows $\leq$ is a partial order on $L$. From $a = a \wedge (a \vee b)$ and $b = b \wedge (a \vee b)$ follow $a \leq a \vee b$ and $b \leq a \vee b$, so $a \vee b$ is an upper bound of both $a$ and $b$. Now if $a \leq u$ and $b \leq u$ then $a \vee u = (a \wedge u) \vee u = u$, and likewise $b \vee u = u$, so $(a \vee u) \vee (b \vee u) = u \vee u = u$; hence $(a \vee b) \vee u = u$, giving $(a \vee b) \wedge u = (a \vee b) \wedge [(a \vee b) \vee u] = a \vee b$ (by the absorption law), and this says $a \vee b \leq u$. Thus $a \vee b = \sup\{a,b\}$. Similarly, $a \wedge b = \inf\{a,b\}$.

If, on the other hand, we are given a lattice $L$ by the second definition, then the reader should not find it too difficult to verify that the operations $\vee$ and $\wedge$ as defined in (B) satisfy the requirements L1 to L4, for example the absorption law L4(a) becomes $a = \sup\{a, \inf\{a,b\}\}$, which is clearly true as $\inf\{a,b\} \le a$.

The fact that these two constructions (A) and (B) are inverses is now an easy matter to check. Throughout the text we will be using the word lattice to mean lattice by the first definition (with the two operations join and meet), but it will often be convenient to freely make use of the corresponding partial order.

## References


1. R. Balbes and P. Dwinger [1]
2. G. Birkhoff [3]
3. P. Crawley and R. P. Dilworth [10]
4. G. Grätzer [15]


## Exercises §1

1. Verify that Su($X$) with the partial order $\subseteq$ is a lattice. What are the operations $\vee$ and $\wedge$?
2. Verify L1–L4 for $\vee$, $\wedge$ as defined in (B) below Definition 1.4.
3. Show that the idempotent laws L3 of lattices follow from L1, L2, and L4.
4. Let $C[0,1]$ be the set of continuous functions from $[0,1]$ to the reals. Define $\le$ on $C[0,1]$ by $f \le g$ iff $f(a) \le g(a)$ for all $a \in [0,1]$. Show that $\le$ is a partial order which makes $C[0,1]$ into a lattice.
5. If $L$ is a lattice with operations $\vee$ and $\wedge$, show that interchanging $\vee$ and $\wedge$ still gives a lattice, called the *dual* of $L$. (For contrast, note that interchanging $+$ and $\cdot$ in a ring usually does not give another ring.) Note that dualization turns the Hasse diagram upside down.
6. If $G$ is a group, show that the set of subgroups $S(G)$ of $G$ with the partial ordering $\subseteq$ forms a lattice. Describe all groups $G$ whose lattices of subgroups look like (b) of Figure 1.
7. If $G$ is a group, let $N(G)$ be the set of normal subgroups of $G$. Define $\vee$ and $\wedge$ on $N(G)$ by $N_1 \wedge N_2 = N_1 \cap N_2$, and $N_1 \vee N_2 = N_1N_2 = \{n_1n_2 : n_1 \in N_1,\ n_2 \in N_2\}$. Show that under these operations $N(G)$ is a lattice.
8. If $R$ is a ring, let $I(R)$ be the set of ideals of $R$. Define $\vee$ and $\wedge$ on $I(R)$ by $I_1 \wedge I_2 = I_1 \cap I_2$, $I_1 \vee I_2 = \{i_1 + i_2 : i_1 \in I_1, i_2 \in I_2\}$. Show that under these operations $I(R)$ is a lattice.
9. If $\le$ is a partial order on a set $A$, show that there is a total order $\le^*$ on $A$ such that $a \le b$ implies $a \le^* b$. (Hint: Use Zorn's lemma.)
10. If $L$ is a lattice we say that an element $a \in L$ is *join irreducible* if $a = b \vee c$ implies $a = b$ or $a = c$. If $L$ is a finite lattice show that every element is of the form $a_1 \vee \cdots \vee a_n$, where each $a_i$ is join irreducible.

## §2. Isomorphic Lattices, and Sublattices

The word isomorphism is used to signify that two structures are the same except for the nature of their elements (for example, if the elements of a group are painted blue one still has essentially the same group). The following definition is a special case of II§2.1.

**Definition 2.1.** Two lattices $L_1$ and $L_2$ are *isomorphic* if there is a bijection $\alpha$ from $L_1$ to $L_2$ such that for every $a,b$ in $L_1$ the following two equations hold: $\alpha(a \vee b) = \alpha(a) \vee \alpha(b)$ and $\alpha(a \wedge b) = \alpha(a) \wedge \alpha(b)$. Such an $\alpha$ is called an *isomorphism.*

It is useful to note that if $\alpha$ is an isomorphism from $L_1$ to $L_2$ then $\alpha^{-1}$ is an isomorphism from $L_2$ to $L_1$, and if $\beta$ is an isomorphism from $L_2$ to $L_3$ then $\beta \circ \alpha$ is an isomorphism from $L_1$ to $L_3$. One can reformulate the definition of isomorphism in terms of the corresponding order relations.

**Definition 2.2.** If $P_1$ and $P_2$ are two posets and $\alpha$ is a map from $P_1$ to $P_2$, then we say $\alpha$ is *order-preserving* if $\alpha(a) \leq \alpha(b)$ holds in $P_2$ whenever $a \leq b$ holds in $P_1$.

**Theorem 2.3.** *Two lattices $L_1$ and $L_2$ are isomorphic iff there is a bijection $\alpha$ from $L_1$ to $L_2$ such that both $\alpha$ and $\alpha^{-1}$ are order-preserving.*

PROOF. If $\alpha$ is an isomorphism from $L_1$ to $L_2$ and $a \leq b$ holds in $L_1$ then $a = a \wedge b$, so $\alpha(a) = \alpha(a \wedge b) = \alpha(a) \wedge \alpha(b)$, hence $\alpha(a) \leq \alpha(b)$, and thus $\alpha$ is order-preserving. As $\alpha^{-1}$ is an isomorphism, it is also order-preserving.

Conversely, let $\alpha$ be a bijection from $L_1$ to $L_2$ such that both $\alpha$ and $\alpha^{-1}$ are order-preserving. For $a,b$ in $L_1$ we have $a \leq a \vee b$ and $b \leq a \vee b$, so $\alpha(a) \leq \alpha(a \vee b)$ and $\alpha(b) \leq \alpha(a \vee b)$, hence $\alpha(a) \vee \alpha(b) \leq \alpha(a \vee b)$. Furthermore, if $\alpha(a) \vee \alpha(b) \leq u$ then $\alpha(a) \leq u$ and $\alpha(b) \leq u$, hence $a \leq \alpha^{-1}(u)$ and $b \leq \alpha^{-1}(u)$, so $a \vee b \leq \alpha^{-1}(u)$, and thus $\alpha(a \vee b) \leq u$. This implies that $\alpha(a) \vee \alpha(b) = \alpha(a \vee b)$. Similarly, it can be argued that $\alpha(a) \wedge \alpha(b) = \alpha(a \wedge b)$. □

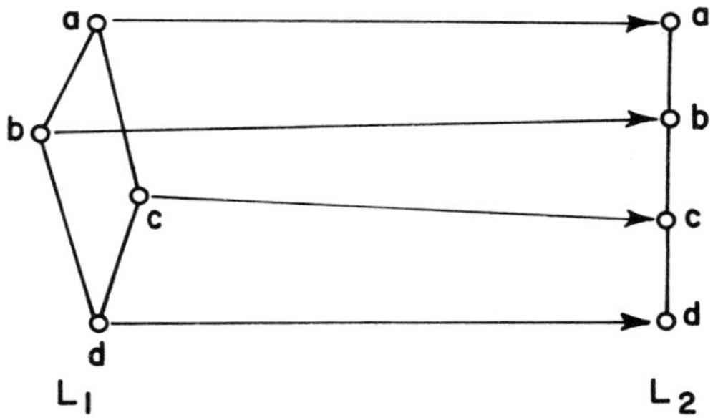


Figure 3 An order-preserving bijection

It is easy to give examples of bijections $\alpha$ between lattices which are order-preserving but are not isomorphisms; for example, consider the map $\alpha(a) = a, \ldots, \alpha(d) = d$ where $L_1$ and $L_2$ are the two lattices in Figure 3.

A sublattice of a lattice $L$ is a subset of $L$ which is a lattice in its own right, using the same operations.

**Definition 2.4.** If $L$ is a lattice and $L' \neq \emptyset$ is a subset of $L$ such that for every pair of elements $a,b$ in $L'$ both $a \vee b$ and $a \wedge b$ are in $L'$, where $\vee$ and $\wedge$ are the lattice operations of $L$, then we say that $L'$ with the same operations (restricted to $L'$) is a *sublattice* of $L$.

If $L'$ is a sublattice of $L$ then for $a,b$ in $L'$ we will of course have $a \leq b$ in $L'$ iff $a \leq b$ in $L$. It is interesting to note that given a lattice $L$ one can often find subsets which as posets (using the same order relation) are lattices, but which do not qualify as sublattices as the operations $\vee$ and $\wedge$ do not agree with those of the original lattice $L$. The example in Figure 4 illustrates this, for note that $P = \{a,c,d,e\}$ as a poset is indeed a lattice, but $P$ is not a sublattice of the lattice $\{a,b,c,d,e\}$.

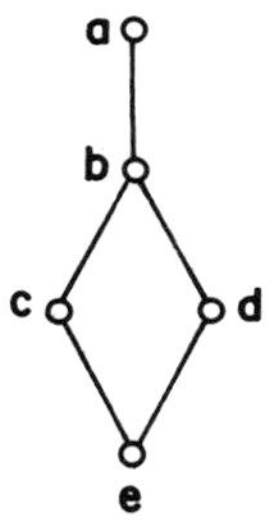


Figure 4

**Definition 2.5.** A lattice $L_1$ can be *embedded* into a lattice $L_2$ if there is a sublattice of $L_2$ isomorphic to $L_1$; in this case we also say $L_2$ *contains a copy of* $L_1$ *as a sublattice.*

## Exercises §2

1. If $(X,T)$ is a topological space show that the closed subsets, as well as the open subsets, form a lattice using $\subseteq$ as the partial order. Show that the lattice of open subsets is isomorphic to the dual (see §1, Exercise 5) of the lattice of closed subsets.

2. If $P$ and $Q$ are posets, let $Q^P$ be the poset of order-preserving maps from $P$ to $Q$, where for $f, g \in Q^P$ we define $f \leq g$ iff $f(a) \leq g(a)$ for all $a \in P$. If $Q$ is a lattice show that $Q^P$ is also a lattice.

3. If $G$ is a group, is $N(G)$ a sublattice of $S(G)$ (see §1, Exercises 6,7)?

4. If $\leq$ is a partial order on $P$ then a *lower segment* of $P$ is a subset $S$ of $P$ such that if $s \in S$, $p \in P$, and $p \leq s$ then $p \in S$. Show that the lower segments of $P$ form a lattice with the operations $\cup$, $\cap$. If $P$ has a least element show that the set $L(P)$ of nonempty lower segments of $P$ forms a lattice.

5. If $L$ is a lattice, then an *ideal* $I$ of $L$ is a nonempty lower segment closed under $\vee$. Show that the set of ideals $I(L)$ of $L$ forms a lattice under $\subseteq$.

6. Given a lattice $L$, an ideal $I$ of $L$ is called a *principal ideal* if it is of the form $\{b \in L : b \leq a\}$, for some $a \in L$. (Note that such subsets are indeed ideals.) Show that the principal ideals of $L$ form a sublattice of $I(L)$ isomorphic to $L$.

## §3. Distributive and Modular Lattices

The most thoroughly studied classes of lattices are distributive lattices and modular lattices.

**Definition 3.1.** A *distributive lattice* is a lattice which satisfies either (and hence, as we shall see, both) of the *distributive laws*,

D1: $x \wedge (y \vee z) \approx (x \wedge y) \vee (x \wedge z)$
D2: $x \vee (y \wedge z) \approx (x \vee y) \wedge (x \vee z)$.

**Theorem 3.2.** *A lattice $L$ satisfies* D1 *iff it satisfies* D2.

PROOF. Suppose D1 holds. Then

$$\begin{aligned}
x \vee (y \wedge z) &\approx (x \vee (x \wedge z)) \vee (y \wedge z) && \text{(by L4(a))}\\
&\approx x \vee ((x \wedge z) \vee (y \wedge z)) && \text{(by L2(a))}\\
&\approx x \vee ((z \wedge x) \vee (z \wedge y)) && \text{(by L1(b))}\\
&\approx x \vee (z \wedge (x \vee y)) && \text{(by D1)}\\
&\approx x \vee ((x \vee y) \wedge z) && \text{(by L1(b))}\\
&\approx (x \wedge (x \vee y)) \vee ((x \vee y) \wedge z) && \text{(by L4(b))}\\
&\approx ((x \vee y) \wedge x) \vee ((x \vee y) \wedge z) && \text{(by L1(b))}\\
&\approx (x \vee y) \wedge (x \vee z) && \text{(by D1).}
\end{aligned}$$

Thus D2 also holds. A similar proof shows that if D2 holds then so does D1. □

Actually every lattice satisfies both of the inequalities $(x \wedge y) \vee (x \wedge z) \leq x \wedge (y \vee z)$ and $x \vee (y \wedge z) \leq (x \vee y) \wedge (x \vee z)$. To see this, note for example that $x \wedge y \leq x$ and $x \wedge y \leq y \vee z$; hence $x \wedge y \leq x \wedge (y \vee z)$, etc. Thus to verify the distributive laws in a lattice it suffices to check either of the following inequalities:

$$x \wedge (y \vee z) \leq (x \wedge y) \vee (x \wedge z)$$
$$(x \vee y) \wedge (x \vee z) \leq x \vee (y \wedge z).$$

**Definition 3.3.** A *modular lattice* is any lattice which satisfies the *modular law*

M: $x \leq y \rightarrow x \vee (y \wedge z) \approx y \wedge (x \vee z)$.

The modular law is obviously equivalent (for lattices) to the identity

$$(x \wedge y) \vee (y \wedge z) \approx y \wedge ((x \wedge y) \vee z)$$

since $a \leq b$ holds iff $a = a \wedge b$. Also it is not difficult to see that every lattice satisfies

$$x \leq y \rightarrow x \vee (y \wedge z) \leq y \wedge (x \vee z),$$

so to verify the modular law it suffices to check the implication

$$x \leq y \rightarrow y \wedge (x \vee z) \leq x \vee (y \wedge z).$$

**Theorem 3.4.** *Every distributive lattice is a modular lattice.*

PROOF. Just use D2, noting that $a \vee b = b$ whenever $a \leq b$. □

The next two theorems give a fascinating characterization of modular and distributive lattices in terms of two five-element lattices called $M_5$ and $N_5$ depicted in Figure 5. In neither case is $a \vee (b \wedge c) = (a \vee b) \wedge (a \vee c)$, so neither $M_5$ nor $N_5$ is a distributive lattice. For $N_5$ we also see that $a \leq b$ but $a \vee (b \wedge c) \neq b \wedge (a \vee c)$, so $N_5$ is not modular. With a small amount of effort one can verify that $M_5$ does satisfy the modular law, however.

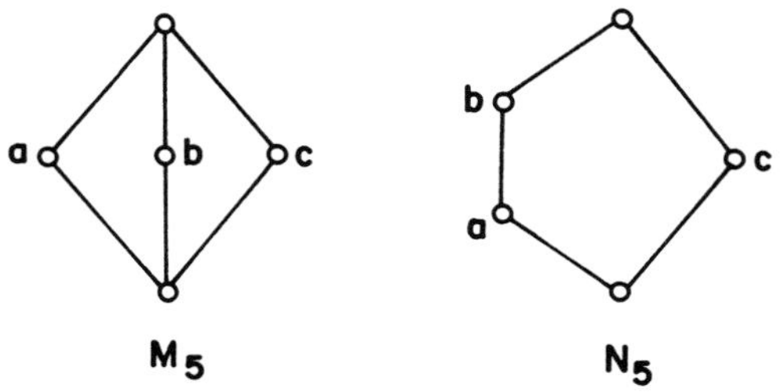


Figure 5

**Theorem 3.5** (Dedekind). *L is a nonmodular lattice iff $N_5$ can be embedded into L.*

PROOF. From the remarks above it is clear that if $N_5$ can be embedded into $L$, then $L$ does not satisfy the modular law. For the converse, suppose that $L$ does not satisfy the modular law. Then for some $a,b,c$ in $L$ we have $a \leq b$

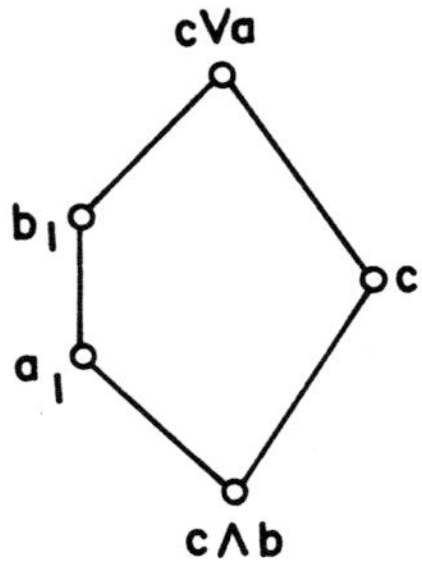


Figure 6

but $a \vee (b \wedge c) < b \wedge (a \vee c)$. Let $a_1 = a \vee (b \wedge c)$ and $b_1 = b \wedge (a \vee c)$. Then

$$\begin{aligned} c \wedge b_1 &= c \wedge [b \wedge (a \vee c)] \\ &= [c \wedge (c \vee a)] \wedge b && \text{(by L1(a), L1(b), L2(b))} \\ &= c \wedge b && \text{(by L4(b))} \end{aligned}$$

and

$$\begin{aligned} c \vee a_1 &= c \vee [a \vee (b \wedge c)] \\ &= [c \vee (c \wedge b)] \vee a && \text{(by L1(a), L1(b), L2(a))} \\ &= c \vee a && \text{(by L4(a)).} \end{aligned}$$

Now as $c \wedge b \leq a_1 \leq b_1$ we have $c \wedge b \leq c \wedge a_1 \leq c \wedge b_1 = c \wedge b$, hence $c \wedge a_1 = c \wedge b_1 = c \wedge b$. Likewise $c \vee b_1 = c \vee a_1 = c \vee a$.

Now it is straightforward to verify that the diagram in Figure 6 gives the desired copy of $N_5$ in $L$. □

**Theorem 3.6** (Birkhoff). *$L$ is a nondistributive lattice iff $M_5$ or $N_5$ can be embedded into $L$.*

PROOF. If either $M_5$ or $N_5$ can be embedded into $L$, then it is clear from earlier remarks that $L$ cannot be distributive. For the converse, let us suppose that $L$ is a nondistributive lattice and that $L$ does not contain a copy of $N_5$ as a sublattice. Thus $L$ is modular by 3.5. Since the distributive laws do not hold in $L$, there must be elements $a,b,c$ from $L$ such that $(a \wedge b) \vee (a \wedge c) < a \wedge (b \vee c)$. Let us define

$$\begin{aligned} d &= (a \wedge b) \vee (a \wedge c) \vee (b \wedge c) \\ e &= (a \vee b) \wedge (a \vee c) \wedge (b \vee c) \\ a_1 &= (a \wedge e) \vee d \\ b_1 &= (b \wedge e) \vee d \\ c_1 &= (c \wedge e) \vee d. \end{aligned}$$

Then it is easily seen that $d \leq a_1, b_1, c_1 \leq e$. Now from

$$a \wedge e = a \wedge (b \vee c) \qquad \text{(by L4(b))}$$

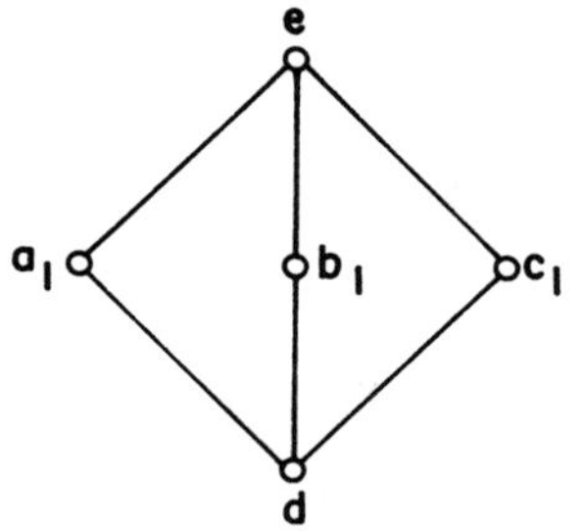


Figure 7

and (applying the modular law to switch the underlined terms)

$$\begin{aligned} a \wedge d &= \underline{a} \wedge (\underline{(a \wedge b) \vee (a \wedge c)} \vee (b \wedge c)) \\ &= ((a \wedge b) \vee (a \wedge c)) \vee (a \wedge (b \wedge c)) \qquad \text{(by M)} \\ &= (a \wedge b) \vee (a \wedge c) \end{aligned}$$

it follows that $d < e$.

We now wish to show that the diagram in Figure 7 is a copy of $M_5$ in $L$. To do this it suffices to show that $a_1 \wedge b_1 = a_1 \wedge c_1 = b_1 \wedge c_1 = d$ and $a_1 \vee b_1 = a_1 \vee c_1 = b_1 \vee c_1 = e$. We will verify one case only and the others require similar arguments (in the following we do not explicitly state several steps involving commutativity and associativity; the terms to be interchanged when the modular law is applied have been underlined):

$$\begin{aligned} a_1 \wedge b_1 &= ((a \wedge e) \vee \underline{d}) \wedge (\underline{(b \wedge e) \vee d}) \\ &= ((a \wedge e) \wedge ((b \wedge \underline{e}) \vee \underline{d})) \vee d && \text{(by M)} \\ &= ((a \wedge e) \wedge ((b \vee d) \wedge e)) \vee d && \text{(by M)} \\ &= ((a \wedge e) \wedge e \wedge (b \vee d)) \vee d \\ &= ((a \wedge e) \wedge (b \vee d)) \vee d && \text{(by L3(b))} \\ &= (a \wedge (\underline{b \vee c}) \wedge (\underline{b} \vee (a \wedge c))) \vee d && \text{(by L4)} \\ &= (a \wedge (b \vee ((b \vee c) \wedge (a \wedge c)))) \vee d && \text{(by M)} \\ &= (\underline{a} \wedge (b \vee (\underline{a \wedge c}))) \vee d && (a \wedge c \leq b \vee c) \\ &= (a \wedge c) \vee (b \wedge a) \vee d && \text{(by M)} \\ &= d. \end{aligned}$$

□

EXERCISES §3

1. If we are given a set $X$, a sublattice of $\mathrm{Su}(X)$ under $\subseteq$ is called a *ring of sets* (following the terminology used by lattice theorists). Show that every ring of sets is a distributive lattice.

2. If $L$ is a distributive lattice, show that the set of ideals $I(L)$ of $L$ (see §2 Exercise 5) forms a distributive lattice.

3. Let $(X, T)$ be a topological space. A subset of $X$ is *regular open* if it is the interior of its closure. Show that the family of regular open subsets of $X$ with the partial order $\subseteq$ is a distributive lattice.

4. If $L$ is a finite lattice let $J(L)$ be the poset of join irreducible elements of $L$ (see §1 Exercise 10), where $a \leq b$ in $J(L)$ means $a \leq b$ in $L$. Show that if $L$ is a finite distributive lattice then $L$ is isomorphic to $L(J(L))$ (see §2 Exercise 4), the lattice of nonempty lower segments of $J(L)$. Hence a finite lattice is distributive iff it is isomorphic to some $L(P)$, for $P$ a finite poset with least element. (This will be used in V§5 to show the theory of distributive lattices is undecidable.)

5. If $G$ is a group show that $N(G)$, the lattice of normal subgroups of $G$ (see §1 Exercise 7), is a modular lattice. Is the same true of $S(G)$? Describe $N(Z_2 \times Z_2)$.

6. If $R$ is a ring show that $I(R)$, the lattice of ideals of $R$ (see §1 Exercise 8), is a modular lattice.

7. If $M$ is a left module over a ring $R$ show that the submodules of $M$ under the partial order $\subseteq$ form a modular lattice.

## §4. Complete Lattices, Equivalence Relations, and Algebraic Lattices

In the 1930's Birkhoff introduced the class of complete lattices to study the combinations of subalgebras.

**Definition 4.1.** A poset $P$ is *complete* if for every subset $A$ of $P$ both sup $A$ and inf $A$ exist (in $P$). The elements sup $A$ and inf $A$ will be denoted by $\bigvee A$ and $\bigwedge A$, respectively. All complete posets are lattices, and a lattice $L$ which is complete as a poset is a *complete lattice*.

**Theorem 4.2.** *Let $P$ be a poset such that $\bigwedge A$ exists for every subset $A$, or such that $\bigvee A$ exists for every subset $A$. Then $P$ is a complete lattice.*

Proof. Suppose $\bigwedge A$ exists for every $A \subseteq P$. Then letting $A^u$ be the set of upper bounds of $A$ in $P$, it is routine to verify that $\bigwedge A^u$ is indeed $\bigvee A$. The other half of the theorem is proved similarly. □

In the above theorem the existence of $\bigwedge \varnothing$ guarantees a largest element in $P$, and likewise the existence of $\bigvee \varnothing$ guarantees a smallest element in $P$. So an equivalent formulation of Theorem 4.2 would be to say that $P$ is complete if it has a largest element and the inf of every nonempty subset exists, or if it has a smallest element and the sup of every nonempty subset exists.

Examples. (1) The set of extended reals with the usual ordering is a complete lattice.

(2) The open subsets of a topological space with the ordering $\subseteq$ form a complete lattice.

(3) $\mathrm{Su}(I)$ with the usual ordering $\subseteq$ is a complete lattice.

A complete lattice may, of course, have sublattices which are incomplete (for example, consider the reals as a sublattice of the extended reals). It is also possible for a sublattice of a complete lattice to be complete, but the sups and infs of the sublattice not to agree with those of the original lattice (for example look at the sublattice of the extended reals consisting of those numbers whose absolute value is less than one together with the numbers $-2, +2$).

**Definition 4.3.** A sublattice $L'$ of a complete lattice $L$ is called a *complete* sublattice of $L$ if for every subset $A$ of $L'$ the elements $\bigvee A$ and $\bigwedge A$, as defined in $L$, are actually in $L'$.

In the 1930's Birkhoff introduced the lattice of equivalence relations on a set, which is especially important in the study of quotient structures.

**Definition 4.4.** Let $A$ be a set. Recall that a binary relation $r$ on $A$ is a subset of $A^2$. If $\langle a,b\rangle \in r$ we also write $arb$. If $r_1$ and $r_2$ are binary relations on $A$ then the *relational product* $r_1 \circ r_2$ is the binary relation on $A$ defined by $\langle a,b\rangle \in r_1 \circ r_2$ iff there is a $c \in A$ such that $\langle a,c\rangle \in r_1$ and $\langle c,b\rangle \in r_2$. Inductively one defines $r_1 \circ r_2 \circ \cdots \circ r_n = (r_1 \circ r_2 \circ \cdots \circ r_{n-1}) \circ r_n$. The *inverse* of a binary relation $r$ is given by $r^{\smallsmile} = \{\langle a,b\rangle \in A^2 : \langle b,a\rangle \in r\}$. The *diagonal relation* $\Delta_A$ on $A$ is the set $\{\langle a,a\rangle : a \in A\}$ and the *all relation* $A^2$ is denoted by $\nabla_A$. (We write simply $\Delta$ (read: delta) and $\nabla$ (read: nabla) when there is no confusion.) A binary relation $r$ on $A$ is an *equivalence relation* on $A$ if, for any $a,b,c$ from $A$, it satisfies:

| | |
|---|---|
| E1: $ara$ | (reflexivity) |
| E2: $arb$ implies $bra$ | (symmetry) |
| E3: $arb$ and $brc$ imply $arc$ | (transitivity). |

$\mathrm{Eq}(A)$ is the *set of all equivalence relations on $A$*.

**Theorem 4.5.** *The poset* $\mathrm{Eq}(A)$, *with* $\subseteq$ *as the partial ordering, is a complete lattice.*

PROOF. Note that $\mathrm{Eq}(A)$ is closed under arbitrary intersections. □

For $\theta_1$ and $\theta_2$ in $\mathrm{Eq}(A)$ it is clear that $\theta_1 \wedge \theta_2 = \theta_1 \cap \theta_2$. Next we look at a (constructive) description of $\theta_1 \vee \theta_2$.

**Theorem 4.6.** *If* $\theta_1$ *and* $\theta_2$ *are two equivalence relations on* $A$ *then*

$$\theta_1 \vee \theta_2 = \theta_1 \cup (\theta_1 \circ \theta_2) \cup (\theta_1 \circ \theta_2 \circ \theta_1) \cup (\theta_1 \circ \theta_2 \circ \theta_1 \circ \theta_2) \cup \cdots,$$

*or equivalently,* $\langle a,b\rangle \in \theta_1 \vee \theta_2$ *iff there is a sequence of elements* $c_1,c_2,\ldots,c_n$ *from* $A$ *such that*

$$\langle c_i,c_{i+1}\rangle \in \theta_1 \quad \text{or} \quad \langle c_i,c_{i+1}\rangle \in \theta_2$$

*for* $i = 1,\ldots,n-1$, *and* $a = c_1$, $b = c_n$.

PROOF. It is not difficult to see that the right-hand side of the above equation is indeed an equivalence relation, and also that each of the relational products in parentheses is contained in $\theta_1 \vee \theta_2$. □

If $\{\theta_i\}_{i\in I}$ is a subset of Eq($A$) then it is also easy to see that $\bigwedge_{i\in I}\theta_i$ is just $\bigcap_{i\in I}\theta_i$. The following straightforward generalization of the previous theorem describes arbitrary sups in Eq($A$).

**Theorem 4.7.** *If* $\theta_i \in \text{Eq}(A)$ *for* $i \in I$, *then*

$$\bigvee_{i\in I}\theta_i = \bigcup\{\theta_{i_0}\circ\theta_{i_1}\circ\cdots\circ\theta_{i_k}: i_0,\ldots,i_k \in I,\ k<\omega\}.$$

**Definition 4.8.** Let $\theta$ be a member of Eq($A$). For $a \in A$, the *equivalence class* (or *coset*) *of a modulo* $\theta$ is the set $a/\theta = \{b \in A: \langle b,a\rangle \in \theta\}$. The set $\{a/\theta: a \in A\}$ is denoted by $A/\theta$.

**Theorem 4.9.** *For* $\theta \in \text{Eq}(A)$ *and* $a,b \in A$ *we have*

(a) $A = \bigcup_{a\in A} a/\theta$.
(b) $a/\theta \neq b/\theta$ *implies* $a/\theta \cap b/\theta = \varnothing$.

PROOF. (Exercise.) □

An alternative approach to equivalence relations is given by partitions, in view of 4.9.

**Definition 4.10.** A *partition* $\pi$ of a set $A$ is a family of nonempty pairwise disjoint subsets of $A$ such that $A = \bigcup\pi$. The sets in $\pi$ are called the *blocks* of $\pi$. The set of all partitions of $A$ is denoted by $\Pi(A)$.

For $\pi$ in $\Pi(A)$, let us define an equivalence relation $\theta(\pi)$ by $\theta(\pi) = \{\langle a,b\rangle \in A^2: \{a,b\} \subseteq B \text{ for some } B \text{ in } \pi\}$. Note that the mapping $\pi \mapsto \theta(\pi)$ is a bijection between $\Pi(A)$ and Eq($A$). Define a relation $\leq$ on $\Pi(A)$ by $\pi_1 \leq \pi_2$ iff each block of $\pi_1$ is contained in some block of $\pi_2$.

**Theorem 4.11.** *With the above ordering* $\Pi(A)$ *is a complete lattice, and it is isomorphic to the lattice* Eq($A$) *under the mapping* $\pi \mapsto \theta(\pi)$.

The verification of this result is left to the reader.

**Definition 4.12.** The lattice $\Pi(A)$ is called the *lattice of partitions* of $A$.

The last class of lattices which we introduce is that of algebraic lattices.

**Definition 4.13.** Let $L$ be a lattice. An element $a$ in $L$ is *compact* iff whenever $\bigvee A$ exists and $a \le \bigvee A$ for $A \subseteq L$, then $a \le \bigvee B$ for some finite $B \subseteq A$. $L$ is *compactly generated* iff every element in $L$ is a sup of compact elements. A lattice $L$ is *algebraic* if it is complete and compactly generated.

The reader will readily see the similarity between the definition of a compact element in a lattice and that of a compact subset of a topological space. Algebraic lattices originated with Komatu and Nachbin in the 1940's and Büchi in the early 1950's; the original definition was somewhat different, however.

EXAMPLES. (1) The lattice of subsets of a set is an algebraic lattice (where the compact elements are finite sets).

(2) The lattice of subgroups of a group is an algebraic lattice (in which "compact" = "finitely generated").

(3) Finite lattices are algebraic lattices.

(4) The subset $[0,1]$ of the real line is a complete lattice, but is not algebraic.

In the next chapter we will encounter two situations where algebraic lattices arise, namely as lattices of subuniverses of algebras and as lattices of congruences on algebras.

## EXERCISES §4

1. Show that the binary relations on a set $A$ form a lattice under $\subseteq$.
2. Show that the right-hand side of the equation in Theorem 4.6 is indeed an equivalence relation on $A$.
3. If $I$ is a closed and bounded interval of the real line with the usual ordering, and $P$ a nonempty subset of $I$ with the same ordering, show that $P$ is a complete sublattice iff $P$ is a closed subset of $I$.
4. If $L$ is a complete chain show that $L$ is algebraic iff for every $a_1,a_2 \in L$ with $a_1 < a_2$ there are $b_1,b_2 \in L$ with $a_1 \le b_1 \prec b_2 \le a_2$.
5. Draw the Hasse diagram of the lattice of partitions of a set with $n$ elements for $1 \le n \le 4$. For $|A| \ge 4$ show that $\Pi(A)$ is not a modular lattice.
6. If $L$ is an algebraic lattice and $D$ is a subset of $L$ such that for $d_1,d_2 \in D$ there is a $d_3 \in D$ with $d_1 \le d_3$, $d_2 \le d_3$ (i.e., $D$ is upward directed) then, for $a \in L$, $a \wedge \bigvee D = \bigvee_{d \in D}(a \wedge d)$.
7. If $L$ is a distributive algebraic lattice then, for any $A \subseteq L$, we have $a \wedge \bigvee A = \bigvee_{d \in A} (a \wedge d)$.

8. If $a$ and $b$ are compact elements of a lattice $L$, show that $a \vee b$ is also compact. Is $a \wedge b$ always compact?

9. If $L$ is a lattice with at least one compact element, let $C(L)$ be the poset of compact elements of $L$ with the partial order on $C(L)$ agreeing with the partial order on $L$. An *ideal* of $C(L)$ is a nonempty subset $I$ of $C(L)$ such that $a,b \in I$ implies $a \vee b \in I$, and $a \in I$, $b \in C(L)$ with $b \leq a$ implies $b \in I$. Show that the ideals of $C(L)$ form a lattice under $\subseteq$ if $L$ has a least element and that the lattice of ideals of $C(L)$ is isomorphic to $L$ if $L$ is an algebraic lattice.

## §5. Closure Operators

One way of producing, and recognizing, complete [algebraic] lattices is through [algebraic] closure operators. Tarski developed one of the most fascinating applications of closure operators during the 1930's in his study of "consequences" in logic.

**Definition 5.1.** If we are given a set $A$, a mapping $C:\mathrm{Su}(A) \to \mathrm{Su}(A)$ is called a *closure operator* on $A$ if, for $X,Y \subseteq A$, it satisfies:

C1: $X \subseteq C(X)$ (extensive)
C2: $C^2(X) = C(X)$ (idempotent)
C3: $X \subseteq Y$ implies $C(X) \subseteq C(Y)$ (isotone).

A subset $X$ of $A$ is called a *closed subset* if $C(X) = X$. The poset of closed subsets of $A$ with set inclusion as the partial ordering is denoted by $L_C$.

The definition of a closure operator is more general than that of a topological closure operator since we do not require that the union of two closed subsets be closed.

**Theorem 5.2.** *Let $C$ be a closure operator on a set $A$. Then $L_C$ is a complete lattice with*

$$\bigwedge_{i \in I} C(A_i) = \bigcap_{i \in I} C(A_i)$$

and

$$\bigvee_{i \in I} C(A_i) = C\left(\bigcup_{i \in I} A_i\right).$$

Proof. Let $(A_i)_{i \in I}$ be an indexed family of closed subsets of $A$. From

$$\bigcap_{i \in I} A_i \subseteq A_i,$$

for each $i$, we have

$$C\left(\bigcap_{i \in I} A_i\right) \subseteq C(A_i) = A_i,$$

so

$$C\left(\bigcap_{i\in I} A_i\right) \subseteq \bigcap_{i\in I} A_i;$$

hence

$$C\left(\bigcap_{i\in I} A_i\right) = \bigcap_{i\in I} A_i;$$

so $\bigcap_{i\in I} A_i$ is in $L_C$. Then, if one notes that $A$ itself is in $L_C$, it follows that $L_C$ is a complete lattice. The verification of the formulas for the $\bigwedge$'s and $\bigvee$'s of families of closed sets is straightforward. □

Interestingly enough, the converse of this theorem is also true, which shows that the lattices $L_C$ arising from closure operators provide typical examples of complete lattices.

**Theorem 5.3.** *Every complete lattice is isomorphic to the lattice of closed subsets of some set $A$ with a closure operator $C$.*

PROOF. Let $L$ be a complete lattice. For $X \subseteq L$ define

$$C(X) = \{a \in L : a \leq \sup X\}.$$

Then $C$ is a closure operator on $L$ and the mapping $a \mapsto \{b \in L : b \leq a\}$ gives the desired isomorphism between $L$ and $L_C$. □

The closure operators which give rise to algebraic lattices of closed subsets are called algebraic closure operators; actually the consequence operator of Tarski is an algebraic closure operator.

**Definition 5.4.** A closure operator $C$ on the set $A$ is an *algebraic closure operator* if for every $X \subseteq A$

C4: $C(X) = \bigcup\{C(Y) : Y \subseteq X \text{ and } Y \text{ is finite}\}$.

(Note that C4 implies C3.)

**Theorem 5.5.** *If $C$ is an algebraic closure operator on a set $A$ then $L_C$ is an algebraic lattice, and the compact elements of $L_C$ are precisely the closed sets $C(X)$, where $X$ is a finite subset of $A$.*

PROOF. First we will show that $C(X)$ is compact if $X$ is finite. Then by (C4), and in view of 5.2, $L_C$ is indeed an algebraic lattice. So suppose $X = \{a_1, \ldots, a_k\}$ and

$$C(X) \subseteq \bigvee_{i\in I} C(A_i) = C\left(\bigcup_{i\in I} A_i\right).$$

For each $a_j \in X$ we have by (C4) a finite $X_j \subseteq \bigcup_{i\in I} A_i$ with $a_j \in C(X_j)$.

Since there are finitely many $A_i$'s, say $A_{j1}, \ldots, A_{jn_j}$, such that

$$X_j \subseteq A_{j1} \cup \cdots \cup A_{jn_j},$$

then

$$a_j \in C(A_{j1} \cup \cdots \cup A_{jn_j}).$$

But then

$$X \subseteq \bigcup_{1 \le j \le k} C(A_{j1} \cup \cdots \cup A_{jn_j}),$$

so

$$X \subseteq C\left( \bigcup_{\substack{1 \le j \le k \\ 1 \le i \le n_j}} A_{ji} \right),$$

and hence

$$C(X) \subseteq C\left( \bigcup_{\substack{1 \le j \le k \\ 1 \le i \le n_j}} A_{ji} \right) = \bigvee_{\substack{1 \le j \le k \\ 1 \le i \le n_j}} C(A_{ji}),$$

so $C(X)$ is compact.

Now suppose $C(Y)$ is not equal to $C(X)$ for any finite $X$. From

$$C(Y) \subseteq \bigcup \{C(X) : X \subseteq Y \text{ and } X \text{ is finite}\}$$

it is easy to see that $C(Y)$ cannot be contained in any finite union of the $C(X)$'s; hence $C(Y)$ is not compact. □

**Definition 5.6.** If $C$ is a closure operator on $A$ and $Y$ is a closed subset of $A$, then we say a set $X$ is a *generating set* for $Y$ if $C(X) = Y$. The set $Y$ is *finitely generated* if there is a finite generating set for $Y$. The set $X$ is a *minimal* generating set for $Y$ if $X$ generates $Y$ and no proper subset of $X$ generates $Y$.

**Corollary 5.7.** *Let $C$ be an algebraic closure operator on $A$. Then the finitely generated subsets of $A$ are precisely the compact elements of $L_C$.*

**Theorem 5.8.** *Every algebraic lattice is isomorphic to the lattice of closed subsets of some set $A$ with an algebraic closure operator $C$.*

Proof. Let $L$ be an algebraic lattice, and let $A$ be the subset of compact elements. For $X \subseteq A$ define

$$C(X) = \{a \in A : a \le \bigvee X\}.$$

$C$ is a closure operator, and from the definition of compact elements it follows that $C$ is algebraic. The map $a \mapsto \{b \in A : b \le a\}$ gives the desired isomorphism as $L$ is compactly generated. □

References

1. P. M. Cohn [9]
2. A. Tarski [1930]

## Exercises §5

1. If $G$ is a group and $X \subseteq G$, let $C(X)$ be the subgroup of $G$ generated by $X$. Show that $C$ is an algebraic closure operator on $G$.

2. If $G$ is a group and $X \subseteq G$, let $C(X)$ be the normal subgroup generated by $X$. Show that $C$ is an algebraic closure operator on $G$.

3. If $R$ is a ring and $X \subseteq R$ let $C(X)$ be the ideal generated by $X$. Show that $C$ is an algebraic closure operator on $R$.

4. If $L$ is a lattice and $A \subseteq L$, let $u(A) = \{b \in L : a \leq b \text{ for } a \in A\}$, the set of upper bounds of $A$, and let $l(A) = \{b \in L : b \leq a \text{ for } a \in A\}$, the set of lower bounds of $A$. Show that $C(A) = l(u(A))$ is a closure operator on $A$, and that the map $\alpha : a \mapsto C(\{a\})$ gives an embedding of $L$ into the complete lattice $L_C$ (called the *Dedekind–MacNeille completion*). What is the Dedekind–MacNeille completion of the rational numbers?

5. If we are given a set $A$, a family $K$ of subsets of $A$ is called a *closed set system* for $A$ if there is a closure operator on $A$ such that the closed subsets of $A$ are precisely the members of $K$. If $K \subseteq \operatorname{Su}(A)$ show that $K$ is a closed set system for $A$ iff $K$ is closed under arbitrary intersections.

Given a set $A$ and a family $K$ of subsets of $A$, $K$ is said to be *closed under unions of chains* if whenever $C \subseteq K$ and $C$ is a chain (under $\subseteq$) then $\bigcup C \in K$; and $K$ is said to be *closed under unions of upward directed families of sets* if whenever $D \subseteq K$ is such that $A_1, A_2 \in D$ implies $A_1 \cup A_2 \subseteq A_3$ for some $A_3 \in D$, then $\bigcup D \in K$. A result of set theory says that $K$ is closed under unions of chains iff $K$ is closed under unions of upward directed families of sets.

6. (Schmidt). A closed set system $K$ for a set $A$ is called an *algebraic closed set system* for $A$ if there is an algebraic closure operator on $A$ such that the closed subsets of $A$ are precisely the members of $K$. If $K \subseteq \operatorname{Su}(A)$ show that $K$ is an algebraic closed set system iff $K$ is closed under (i) arbitrary intersections and (ii) unions of chains.

7. If $C$ is an algebraic closure operator on $S$ and $X$ is a finitely generated closed subset, then for any $Y$ which generates $X$ show there is a finite $Y_0 \subseteq Y$ such that $Y_0$ generates $X$.

8. Let $C$ be a closure operator on $S$. A closed subset $X \neq S$ is *maximal* if for any closed subset $Y$ with $X \subseteq Y \subseteq S$, either $X = Y$ or $Y = S$. Show that if $C$ is algebraic and $X \subseteq S$ with $C(X) \neq S$ then $X$ is contained in a maximal closed subset if $S$ is finitely generated. (In logic one applies this to show every consistent theory is contained in a complete theory.)

CHAPTER II

# The Elements of Universal Algebra

One of the aims of universal algebra is to extract, whenever possible, the common elements of several seemingly different types of algebraic structures. In achieving this one discovers general concepts, constructions, and results which not only generalize and unify the known special situations, thus leading to an economy of presentation, but, being at a higher level of abstraction, can also be applied to entirely new situations, yielding significant information and giving rise to new directions.

In this chapter we describe some of these concepts and their interrelationships. Of primary importance is the concept of an algebra; centered around this we discuss the notions of isomorphism, subalgebra, congruence, quotient algebra, homomorphism, direct product, subdirect product, term, identity, and free algebra.

## §1. Definition and Examples of Algebras

The definition of an algebra given below encompasses most of the well known algebraic structures, as we shall point out, as well as numerous lesser known algebras which are of current research interest. Although the need for such a definition was noted by several mathematicians such as Whitehead in 1898, and later by Noether, the credit for realizing this goal goes to Birkhoff in 1933. Perhaps it should be noted here that recent research in logic, recursive function theory, theory of automata, and computer science has revealed that Birkhoff's original notion could be fruitfully extended, for example to partial algebras and heterogeneous algebras, topics which lie outside the scope of this text. (Birkhoff's definition allowed infinitary operations; however, his main results were concerned with finitary operations.)

**Definition 1.1.** For $A$ a nonempty set and $n$ a nonnegative integer we define $A^0 = \{\emptyset\}$, and, for $n > 0$, $A^n$ is the set of $n$-tuples of elements from $A$. An *$n$-ary operation* (or *function*) *on* $A$ is any function $f$ from $A^n$ to $A$; $n$ is the *arity* (or *rank*) of $f$. A *finitary* operation is an $n$-ary operation, for some $n$. The image of $\langle a_1, \ldots, a_n \rangle$ under an $n$-ary operation $f$ is denoted by $f(a_1, \ldots, a_n)$. An operation $f$ on $A$ is called a *nullary* operation (or *constant*) if its arity is zero; it is completely determined by the image $f(\emptyset)$ in $A$ of the only element $\emptyset$ in $A^0$, and as such it is convenient to identify it with the element $f(\emptyset)$. Thus a nullary operation is thought of as an element of $A$. An operation $f$ on $A$ is *unary*, *binary*, or *ternary* if its arity is 1,2, or 3, respectively.

**Definition 1.2.** A *language* (or *type*) of algebras is a set $\mathscr{F}$ of *function symbols* such that a nonnegative integer $n$ is assigned to each member $f$ of $\mathscr{F}$. This integer is called the *arity* (or *rank*) of $f$, and $f$ is said to be an *$n$-ary function symbol*. The subset of $n$-ary function symbols in $\mathscr{F}$ is denoted by $\mathscr{F}_n$.

**Definition 1.3.** If $\mathscr{F}$ is a language of algebras then an *algebra* $\mathbf{A}$ of *type* $\mathscr{F}$ is an ordered pair $\langle A,F \rangle$ where $A$ is a nonempty set and $F$ is a family of finitary operations on $A$ indexed by the language $\mathscr{F}$ such that corresponding to each $n$-ary function symbol $f$ in $\mathscr{F}$ there is an $n$-ary operation $f^{\mathbf{A}}$ on $A$. The set $A$ is called the *universe* (or *underlying set*) of $\mathbf{A} = \langle A,F \rangle$, and the $f^{\mathbf{A}}$'s are called the *fundamental operations of* $\mathbf{A}$. (In practice we prefer to write just $f$ for $f^{\mathbf{A}}$—this convention creates an ambiguity which seldom causes a problem. However in this chapter we will be unusually careful.) If $\mathscr{F}$ is finite, say $\mathscr{F} = \{f_1, \ldots, f_k\}$, we often write $\langle A, f_1, \ldots, f_k \rangle$ for $\langle A,F \rangle$, usually adopting the convention:

$$\text{arity } f_1 \geq \text{arity } f_2 \geq \cdots \geq \text{arity } f_k.$$

An algebra $\mathbf{A}$ is *unary* if all of its operations are unary, and it is *mono-unary* if it has just one unary operation. $\mathbf{A}$ is a *groupoid* if it has just one binary operation; this operation is usually denoted by $+$ or $\cdot$, and we write $a + b$ or $a \cdot b$ (or just $ab$) for the image of $\langle a,b \rangle$ under this operation, and call it the *sum* or *product* of $a$ and $b$, respectively. An algebra $\mathbf{A}$ is *finite* if $|A|$ is finite, and *trivial* if $|A| = 1$.

It is a curious fact that the algebras that have been most extensively studied in conventional (albeit modern!) algebra do not have fundamental operations of arity greater than two. (However see IV§7 Ex. 8.)

Not all of the following examples of algebras are well-known, but they are of considerable importance in current research. In particular we would like to point out the role of recent directions in logic aimed at providing algebraic models for certain logical systems. The reader will notice that all of the different kinds of algebras listed below are distinguished from each other by their fundamental operations and the fact that they satisfy certain identities. One of the early achievements of Birkhoff was to clarify the role of identities (see §11).

EXAMPLES. (1) *Groups*. A *group* **G** is an algebra $\langle G, \cdot, ^{-1}, 1\rangle$ with a binary, a unary, and a nullary operation in which the following identities are true:

G1: $x \cdot (y \cdot z) \approx (x \cdot y) \cdot z$
G2: $x \cdot 1 \approx 1 \cdot x \approx x$
G3: $x \cdot x^{-1} \approx x^{-1} \cdot x \approx 1.$

A group **G** is *Abelian* (or *commutative*) if the following identity is true:

G4: $x \cdot y \approx y \cdot x.$

Groups were one of the earliest concepts studied in algebra (groups of substitutions appeared about two hundred years ago). The definition given above is not the one which appears in standard texts on groups, for they use only one binary operation and axioms involving existential quantifiers. The reason for the above choice, and for the descriptions given below, will become clear in §2.

Groups are generalized to semigroups and monoids in one direction, and to quasigroups and loops in another direction.

(2) SEMIGROUPS AND MONOIDS. A *semigroup* is a groupoid $\langle G, \cdot\rangle$ in which (G1) is true. It is *commutative* (or *Abelian*) if (G4) holds. A *monoid* is an algebra $\langle M, \cdot, 1\rangle$ with a binary and a nullary operation satisfying (G1) and (G2).

(3) QUASIGROUPS AND LOOPS. A *quasigroup* is an algebra $\langle Q, /, \cdot, \backslash\rangle$ with three binary operations satisfying the following identities:

Q1: $x\backslash(x \cdot y) \approx y$; $(x \cdot y)/y \approx x$
Q2: $x \cdot (x\backslash y) \approx y$; $(x/y) \cdot y \approx x.$

A *loop* is a quasigroup with identity, i.e., an algebra $\langle Q, /, \cdot, \backslash, 1\rangle$ which satisfies (Q1), (Q2) and (G2). Quasigroups and loops will play a major role in Chapter III.

(4) RINGS. A *ring* is an algebra $\langle R, +, \cdot, -, 0\rangle$, where $+$ and $\cdot$ are binary, $-$ is unary and 0 is nullary, satisfying the following conditions:

R1: $\langle R, +, -, 0\rangle$ is an Abelian group
R2: $\langle R, \cdot\rangle$ is a semigroup
R3: $x \cdot (y + z) \approx (x \cdot y) + (x \cdot z)$
$\quad (x + y) \cdot z \approx (x \cdot z) + (y \cdot z).$

A *ring with identity* is an algebra $\langle R, +, \cdot, -, 0, 1\rangle$ such that (R1)–(R3) and (G2) hold.

(5) MODULES OVER A (FIXED) RING. Let **R** be a given ring. A (*left*) **R**-*module* is an algebra $\langle M,+,-,0,(f_r)_{r\in R}\rangle$ where $+$ is binary, $-$ is unary, 0 is nullary, and each $f_r$ is unary, such that the following hold:

M1: $\langle M,+,-,0\rangle$ is an abelian group
M2: $f_r(x+y)\approx f_r(x)+f_r(y)$, for $r\in R$
M3: $f_{r+s}(x)\approx f_r(x)+f_s(x)$, for $r,s\in R$
M4: $f_r(f_s(x))\approx f_{rs}(x)$ for $r,s\in R$.

Let **R** be a ring with identity. A *unitary* **R**-*module* is an algebra as above satisfying (M1)–(M4) and

M5: $f_1(x)\approx x$.

(6) ALGEBRAS OVER A RING. Let **R** be a ring with identity. An *algebra over* **R** is an algebra $\langle A,+,\cdot,-,0,(f_r)_{r\in R}\rangle$ such that the following hold:

A1: $\langle A,+,-,0,(f_r)_{r\in R}\rangle$ is a unitary $R$-module
A2: $\langle A,+,\cdot,-,0\rangle$ is a ring
A3: $f_r(x\cdot y)\approx (f_r(x))\cdot y\approx x\cdot f_r(y)$ for $r\in R$.

(7) SEMILATTICES. A *semilattice* is a semigroup $\langle S,\cdot\rangle$ which satisfies the commutative law (G4) and the idempotent law

S1: $x\cdot x\approx x$.

Two definitions of a lattice were given in the last chapter. We reformulate the first definition given there in order that it be a special case of algebras as defined in this chapter.

(8) LATTICES. A *lattice* is an algebra $\langle L,\vee,\wedge\rangle$ with two binary operations which satisfies (L1)–(L4) of I§1.

(9) BOUNDED LATTICES. An algebra $\langle L,\vee,\wedge,0,1\rangle$ with two binary and two nullary operations is a *bounded lattice* if it satisfies:

BL1: $\langle L,\vee,\wedge\rangle$ is a lattice
BL2: $x\wedge 0\approx 0$; $x\vee 1\approx 1$.

(10) BOOLEAN ALGEBRAS. A *Boolean algebra* is an algebra $\langle B,\vee,\wedge,',0,1\rangle$ with two binary, one unary, and two nullary operations which satisfies:

B1: $\langle B,\vee,\wedge\rangle$ is a distributive lattice
B2: $x\wedge 0\approx 0$; $x\vee 1\approx 1$
B3: $x\wedge x'\approx 0$; $x\vee x'\approx 1$.

Boolean algebras were of course discovered as a result of Boole's investigations into the underlying laws of correct reasoning. Since then they have become vital to electrical engineering, computer science, axiomatic set theory, model theory, and other areas of science and mathematics. We will return to them in Chapter IV.

(11) HEYTING ALGEBRAS. An algebra $\langle H, \vee, \wedge, \rightarrow, 0, 1\rangle$ with three binary and two nullary operations is a *Heyting algebra* if it satisfies:

H1: $\langle H, \vee, \wedge\rangle$ is a distributive lattice
H2: $x \wedge 0 \approx 0$; $x \vee 1 \approx 1$
H3: $x \rightarrow x \approx 1$
H4: $(x \rightarrow y) \wedge y \approx y$; $x \wedge (x \rightarrow y) \approx x \wedge y$
H5: $x \rightarrow (y \wedge z) \approx (x \rightarrow y) \wedge (x \rightarrow z)$; $(x \vee y) \rightarrow z \approx (x \rightarrow z) \wedge (y \rightarrow z)$.

These were introduced by Birkhoff under a different name, *Brouwerian algebras*, and with a different notation ($v{:}u$ for $u \rightarrow v$).

(12) $n$-VALUED POST ALGEBRAS. An algebra $\langle A, \vee, \wedge, ', 0, 1\rangle$ with two binary, one unary, and two nullary operations is an *n-valued Post algebra* if it satisfies every identity satisfied by the algebra $\mathbf{P}_n = \langle\{0,1,\ldots,n-1\}, \vee, \wedge, ', 0, 1\rangle$ where $\langle\{0,1,\ldots,n-1\}, \vee, \wedge, 0, 1\rangle$ is a bounded chain with $0 < n-1 < n-2 < \cdots < 2 < 1$, and $1' = 2, 2' = 3, \ldots, (n-2)' = n-1$, $(n-1)' = 0$, and $0' = 1$. See Figure 8, where the unary operation $'$ is depicted by arrows. In IV§7 we will give a structure theorem for all $n$-valued Post algebras, and in V§4 show that they can be defined by a finite set of equations.

Figure 8 The Post algebra $\mathbf{P}_n$

(13) CYLINDRIC ALGEBRAS OF DIMENSION $n$. If we are given $n \in \omega$, then an algebra $\langle A, \vee, \wedge, ', c_0, \ldots, c_{n-1}, 0, 1, d_{00}, d_{01}, \ldots, d_{n-1,n-1}\rangle$ with two binary operations, $n+1$ unary operations, and $n^2+2$ nullary operations is a *cylindric algebra of dimension n* if it satisfies the following, where $0 \leq i,j,k < n$:

C1: $\langle A, \vee, \wedge, ', 0, 1\rangle$ is a Boolean algebra
C2: $c_i 0 \approx 0$
C3: $x \leq c_i x$

C4: $c_i(x \wedge c_i y) \approx (c_i x) \wedge (c_i y)$
C5: $c_i c_j x \approx c_j c_i x$
C6: $d_{ii} \approx 1$
C7: $d_{ik} \approx c_j(d_{ij} \wedge d_{jk})$ if $i \neq j \neq k$
C8: $c_i(d_{ij} \wedge x) \wedge c_i(d_{ij} \wedge x') \approx 0$ if $i \neq j$.

Cylindric algebras were introduced by Tarski and Thompson to provide an algebraic version of the predicate logic.

(14) Ortholattices. An algebra $\langle L, \vee, \wedge, ', 0, 1\rangle$ with two binary, one unary and two nullary operations is an ortholattice if it satisfies:

Q1: $\langle L, \vee, \wedge, 0, 1\rangle$ is a bounded lattice
Q2: $x \wedge x' \approx 0$; $x \vee x' \approx 1$
Q3: $(x \wedge y)' \approx x' \vee y'$; $(x \vee y)' \approx x' \wedge y'$
Q4: $(x')' \approx x$.

An *orthomodular lattice* is an ortholattice which satisfies

Q5: $x \leq y \rightarrow x \vee (x' \wedge y) \approx y$.

References

1. P. M. Cohn [9]
2. G. Grätzer [16]
3. A. G. Kurosh [22]
4. A. I. Mal'cev [25]
5. B. H. Neumann [27]
6. R. S. Pierce [28]
7. W. Taylor [35]

Exercises §1

1. An algebra $\langle A, F\rangle$ is the *reduct* of an algebra $\langle A, F^*\rangle$ to $\mathscr{F}$ if $\mathscr{F} \subseteq \mathscr{F}^*$, and $F$ is the restriction of $F^*$ to $\mathscr{F}$. Given $n \geq 1$, find equations $\Sigma$ for semigroups such that $\Sigma$ will hold in a semigroup $\langle S, \cdot\rangle$ iff $\langle S, \cdot\rangle$ is a reduct of a group $\langle S, \cdot, ^{-1}, 1\rangle$ of exponent $n$ (i.e., every element of $S$ is such that its order divides $n$).

2. Two elements $a,b$ of a bounded lattice $\langle L, \vee, \wedge, 0, 1\rangle$ are *complements* if $a \vee b = 1$, $a \wedge b = 0$. In this case each of $a,b$ is the complement of the other. A *complemented lattice* is a bounded lattice in which every element has a complement.

   (a) Show that in a bounded distributive lattice an element can have at most one complement.

   (b) Show that the class of complemented distributive lattices is precisely the class of reducts of Boolean algebras (to $\{\vee, \wedge, 0, 1\}$).

3. If $\langle B, \vee, \wedge, ', 0, 1\rangle$ is a Boolean algebra and $a,b \in B$ define $a \rightarrow b$ to be $a' \vee b$. Show that $\langle B, \vee, \wedge, \rightarrow, 0, 1\rangle$ is a Heyting algebra.

4. Show that every Boolean algebra is an ortholattice, but not conversely.

5. (a) If $\langle H, \vee, \wedge, \rightarrow, 0, 1\rangle$ is a Heyting algebra and $a,b \in H$ show that $a \rightarrow b$ is the largest element $c$ of $H$ (in the lattice sense) such that $a \wedge c \leq b$.

(b) Show that the class of bounded distributive lattices $\langle L, \vee, \wedge, 0, 1\rangle$ such that for each $a,b \in L$ there is a largest $c \in L$ with $a \wedge c \leq b$ is precisely the class of reducts of Heyting algebras (to $\{\vee, \wedge, 0, 1\}$).

(c) Show how one can construct a Heyting algebra from the open subsets of a topological space.

(d) Show that every finite distributive lattice is a reduct of a Heyting algebra.

6. Let $\langle M, \cdot, 1\rangle$ be a monoid and suppose $A \subseteq M$. For $a \in A$ define $f_a : M \to M$ by $f_a(s) = a \cdot s$. Show that the unary algebra $\langle M, (f_a)_{a \in A}\rangle$ satisfies $f_{a_1} \cdots f_{a_n}(x) \approx f_{b_1} \cdots f_{b_k}(x)$ iff $a_1 \cdots a_n = b_1 \cdots b_k$. (This observation of Mal'cev [24] allows one to translate undecidability results about word problems for monoids into undecidability results about equations of unary algebras. This idea has been refined and developed by McNulty [1976] and by Murskiĭ [1971]).

## §2. Isomorphic Algebras, and Subalgebras

The concepts of isomorphism in group theory, ring theory, and lattice theory are special cases of the notion of isomorphism between algebras.

**Definition 2.1.** Let **A** and **B** be two algebras of the same type $\mathscr{F}$. Then a function $\alpha : A \to B$ is an *isomorphism* from **A** to **B** if $\alpha$ is one-to-one and onto, and for every $n$-ary $f \in \mathscr{F}$, for $a_1, \ldots, a_n \in A$, we have

$$\alpha f^{\mathbf{A}}(a_1, \ldots, a_n) = f^{\mathbf{B}}(\alpha a_1, \ldots, \alpha a_n). \qquad (*)$$

We say **A** is *isomorphic* to **B**, written $\mathbf{A} \cong \mathbf{B}$, if there is an isomorphism from **A** to **B**. If $\alpha$ is an isomorphism from **A** to **B** we may simply say "$\alpha : \mathbf{A} \to \mathbf{B}$ is an isomorphism".

As is well-known, following Felix Klein's Erlanger Programm, algebra is often considered as the study of those properties of algebras which are invariant under isomorphism, and such properties are called algebraic properties. Thus from an algebraic point of view, isomorphic algebras can be regarded as equal or the same, as they would have the same algebraic structure, and would differ only in the nature of the elements; the phrase "they are equal up to isomorphism" is often used.

There are several important methods of constructing new algebras from given ones. Three of the most fundamental are the formation of subalgebras, homomorphic images, and direct products. These will occupy us for the next few sections.

**Definition 2.2.** Let **A** and **B** be two algebras of the same type. Then **B** is a *subalgebra* of **A** if $B \subseteq A$ and every fundamental operation of **B** is the restriction of the corresponding operation of **A**, i.e., for each function symbol $f$, $f^{\mathbf{B}}$ is $f^{\mathbf{A}}$ restricted to $B$; we write simply $\mathbf{B} \leq \mathbf{A}$. A *subuniverse* of **A** is a subset $B$ of $A$ which is closed under the fundamental operations of **A**, i.e., if $f$ is a fundamental $n$-ary operation of **A** and $a_1, \ldots, a_n \in B$ we would require $f(a_1, \ldots, a_n) \in B$.

Thus if **B** is a subalgebra of **A**, then $B$ is a subuniverse of **A**. Note that the empty set may be a subuniverse, but it is not the underlying set of any subalgebra. If **A** has nullary operations then every subuniverse contains them as well.

It is the above definition of subalgebra which motivated the choice of fundamental operations for the several examples given in §1. For example, we would like a subalgebra of a group to again be a group. If we were to consider a group as an algebra with only the usual binary operation then, unfortunately, subalgebra would only mean subsemigroup (for example the positive integers are a subsemigroup, but not a subgroup, of the group of all integers). Similar remarks apply to rings, modules, etc. By considering a suitable modification (enlargement) of the set of fundamental operations the concept of subalgebra as defined above coincides with the usual notion for the several examples in §1.

A slight generalization of the notion of isomorphism leads to the following definition.

**Definition 2.3.** Let **A** and **B** be of the same type. A function $\alpha: A \to B$ is an *embedding* of **A** into **B** if $\alpha$ is one-to-one and satisfies (*) of 2.1 (such an $\alpha$ is also called a *monomorphism*). For brevity we simply say "$\alpha$: $\mathbf{A} \to \mathbf{B}$ is an embedding". We say **A** can be *embedded* in **B** if there is an embedding of **A** into **B**.

**Theorem 2.4.** *If* $\alpha: \mathbf{A} \to \mathbf{B}$ *is an embedding, then* $\alpha(A)$ *is a subuniverse of* **B**.

PROOF. Let $\alpha: \mathbf{A} \to \mathbf{B}$ be an embedding. Then for an $n$-ary function symbol $f$ and $a_1, \ldots, a_n \in A$,

$$f^{\mathbf{B}}(\alpha a_1, \ldots, \alpha a_n) = \alpha f^{\mathbf{A}}(a_1, \ldots, a_n) \in \alpha(A),$$

hence $\alpha(A)$ is a subuniverse of **B**. □

**Definition 2.5.** If $\alpha: \mathbf{A} \to \mathbf{B}$ is an embedding, $\alpha(\mathbf{A})$ denotes the subalgebra of **B** with universe $\alpha(A)$.

A problem of general interest to algebraists may be formulated as follows. Let $K$ be a class of algebras and let $K_1$ be a proper subclass of $K$. (In practice, $K$ may have been obtained from the process of abstraction of certain properties of $K_1$, or $K_1$ may be obtained from $K$ by certain additional, more desirable, properties.) Two basic questions arise in the quest for structure theorems.

(1) Is every member of $K$ isomorphic to some member of $K_1$?
(2) Is every member of $K$ embeddable in some member of $K_1$?

For example, every Boolean algebra is isomorphic to a field of sets (see IV§1), every group is isomorphic to a group of permutations, a finite Abelian group is isomorphic to a direct product of cyclic groups, and a finite distributive lattice can be embedded in a power of the two-element distributive lattice. Structure theorems are certainly a major theme in Chapter IV.

## §3. Algebraic Lattices and Subuniverses

We shall now describe one of the natural ways that algebraic lattices arise in universal algebra.

**Definition 3.1.** Given an algebra $\mathbf{A}$ define, for every $X \subseteq A$,

$$\mathrm{Sg}(X) = \bigcap\{B : X \subseteq B \text{ and } B \text{ is a subuniverse of } \mathbf{A}\}.$$

We read Sg($X$) as "the subuniverse generated by $X$".

**Theorem 3.2.** *If we are given an algebra* $\mathbf{A}$, *then* Sg *is an algebraic closure operator on* $A$.

Proof. Observe that an arbitrary intersection of subuniverses of $\mathbf{A}$ is again a subuniverse, hence Sg is a closure operator on $A$ whose closed sets are precisely the subuniverses of $A$. Now, for any $X \subseteq A$ define

$$E(X) = X \cup \{f(a_1, \ldots, a_n) : f \text{ is a fundamental } n\text{-ary operation on } A \text{ and } a_1, \ldots, a_n \in X\}.$$

Then define $E^n(X)$ for $n \geq 0$ by

$$E^0(X) = X$$
$$E^{n+1}(X) = E(E^n(X)).$$

As all the fundamental operations on $A$ are finitary and

$$X \subseteq E(X) \subseteq E^2(X) \subseteq \cdots$$

one can show that (Exercise 1)

$$\mathrm{Sg}(X) = X \cup E(X) \cup E^2(X) \cup \cdots,$$

and from this it follows that if $a \in \mathrm{Sg}(X)$ then $a \in E^n(X)$ for some $n < \omega$; hence for some finite $Y \subseteq X$, $a \in E^n(Y)$. Thus $a \in \mathrm{Sg}(Y)$. But this says Sg is an algebraic closure operator. □

**Corollary 3.3.** *If* **A** *is an algebra then* $\mathbf{L}_{\mathrm{Sg}}$, *the lattice of subuniverses of* **A**, *is an algebraic lattice.*

The corollary says that the subuniverses of **A**, with $\subseteq$ as the partial order, form an algebraic lattice.

**Definition 3.4.** Given an algebra **A**, Sub(**A**) denotes the set of subuniverses of **A**, and **Sub(A)** is the corresponding algebraic lattice, the *lattice of subuniverses of* **A**. For $X \subseteq A$ we say $X$ *generates* **A** (or **A** *is generated by* $X$, or $X$ *is a set of generators of* **A**) if $\mathrm{Sg}(X) = A$. The algebra **A** is *finitely generated* if it has a finite set of generators.

One cannot hope to find any further essentially new lattice properties which hold for the class of lattices of subuniverses since every algebraic lattice is isomorphic to the lattice of subuniverses of some algebra.

**Theorem 3.5** (Birkhoff and Frink). *If* **L** *is an algebraic lattice, then* $\mathbf{L} \cong$ **Sub(A)**, *for some algebra* **A**.

Proof. Let $C$ be an algebraic closure operator on a set $A$ such that $\mathbf{L} \cong \mathbf{L}_C$ (such exists by I§5.8). For each finite subset $B$ of $A$ and each $b \in C(B)$ define an $n$-ary function $f_{B,b}$ on $A$, where $n = |B|$, by

$$f_{B,b}(a_1, \ldots, a_n) = \begin{cases} b & \text{if } B = \{a_1, \ldots, a_n\} \\ a_1 & \text{otherwise,} \end{cases}$$

and call the resulting algebra **A**. Then clearly

$$f_{B,b}(a_1, \ldots, a_n) \in C(\{a_1, \ldots, a_n\}),$$

hence for $X \subseteq A$,

$$\mathrm{Sg}(X) \subseteq C(X).$$

On the other hand

$$C(X) = \bigcup\{C(B) : B \subseteq X \text{ and } B \text{ is finite}\}$$

and, for $B$ finite,

$$\begin{aligned} C(B) &= \{f_{B,b}(a_1, \ldots, a_n) : B = \{a_1, \ldots, a_n\}, b \in C(B)\} \\ &\subseteq \mathrm{Sg}(B) \\ &\subseteq \mathrm{Sg}(X) \end{aligned}$$

imply

$$C(X) \subseteq \mathrm{Sg}(X);$$

hence

$$C(X) = \mathrm{Sg}(X).$$

Thus $\mathbf{L}_C =$ **Sub(A)**, so **Sub(A)** $\cong$ **L**. □

The following set-theoretic result is used to justify the possibility of certain constructions in universal algebra—in particular it shows that for a given type there cannot be "too many" algebras (up to isomorphism) generated by sets no larger than a given cardinality. Recall that $\omega$ is the smallest infinite cardinal.

**Corollary 3.6.** *If* $\mathbf{A}$ *is an algebra and* $X \subseteq A$ *then* $|\mathrm{Sg}(X)| \leq |X| + |\mathscr{F}| + \omega$.

PROOF. Using induction on $n$ one has $|E^n(X)| \leq |X| + |\mathscr{F}| + \omega$, so the result follows from the proof of 3.2. □

REFERENCE

1. G. Birkhoff and O. Frink [1948]

EXERCISE §3

1. Show $\mathrm{Sg}(X) = X \cup E(X) \cup E^2(X) \cup \cdots$.

## §4. The Irredundant Basis Theorem

Recall that finitely generated vector spaces have the property that all minimal generating sets have the same cardinality. It is a rather rare phenomenon, though, to have a "dimension." For example, consider the Abelian group $\mathbf{Z}_6$—it has both $\{1\}$ and $\{2,3\}$ as minimal generating sets.

**Definition 4.1.** Let $C$ be a closure operator on $A$. For $n < \omega$, let $C_n$ be the function defined on $\mathrm{Su}(A)$ by

$$C_n(X) = \bigcup\{C(Y): Y \subseteq X, |Y| \leq n\}.$$

We say that $C$ is $n$-ary if

$$C(X) = C_n(X) \cup C_n^2(X) \cup \cdots,$$

where

$$C_n^1(X) = C_n(X),$$
$$C_n^{k+1}(X) = C_n(C_n^k(X)).$$

**Lemma 4.2.** *Let* $\mathbf{A}$ *be an algebra all of whose fundamental operations have arity at most n. Then* Sg *is an n-ary closure operator on A.*

PROOF. Note that (using the $E$ of the proof of 3.2)

$$E(X) \subseteq (\mathrm{Sg})_n(X) \subseteq \mathrm{Sg}(X);$$

hence

$$\begin{aligned}\mathrm{Sg}(X) &= X \cup E(X) \cup E^2(X) \cup \cdots \\ &\subseteq (\mathrm{Sg})_n(X) \cup (\mathrm{Sg})_n^2(X) \cup \cdots \\ &\subseteq \mathrm{Sg}(X),\end{aligned}$$

so

$$\mathrm{Sg}(X) = (\mathrm{Sg})_n(X) \cup (\mathrm{Sg})_n^2(X) \cup \cdots. \qquad \square$$

**Definition 4.3.** Suppose $C$ is a closure operator on $S$. A minimal generating set of $S$ is called an *irredundant basis*. Let $\mathrm{IrB}(C) = \{n < \omega : S$ has an irredundant basis of $n$ elements$\}$.

The next result shows that the length of the finite gaps in $\mathrm{IrB}(C)$ is bounded by $n - 2$ if $C$ is an $n$-ary closure operator.

**Theorem 4.4** (Tarski). *If $C$ is an $n$-ary closure operator on $S$ with $n \geq 2$, and if $i < j$ with $i,j \in \mathrm{IrB}(C)$ such that*

$$\{i + 1, \ldots, j - 1\} \cap \mathrm{IrB}(C) = \varnothing, \qquad (*)$$

*then $j - i \leq n - 1$. In particular, if $n = 2$ then $\mathrm{IrB}(C)$ is a convex subset of $\omega$, i.e., a sequence of consecutive numbers.*

Proof. Let $B$ be an irredundant basis with $|B| = j$. Let $K$ be the set of irredundant bases $A$ with $|A| \leq i$.

The idea of the proof is simple. We will think of $B$ as the center of $S$, and measure the distance from $B$ using the "rings" $C_n^{k+1}(B) - C_n^k(B)$. We want to choose a basis $A_0$ in $K$ such that $A_0$ is as close as possible to $B$, and such that the last ring which contains elements of $A_0$ contains as few elements of $A_0$ as possible. We choose one of the latter elements $a_0$ and replace it by $n$ or fewer closer elements $b_1, \ldots, b_m$ to obtain a new generating set $A_1$, with $|A_1| < i + n$. Then $A_1$ contains an irredundant basis $A_2$. By the 'minimal distance' condition on $A_0$ we see that $A_2 \notin K$, hence $|A_2| > i$, so $|A_2| \geq j$ by (*). Thus $j < i + n$.

Now for the details of this proof, choose $A_0 \in K$ such that

$$A_0 \nsubseteq C_n^k(B) \quad \text{implies} \quad A \nsubseteq C_n^k(B)$$

for $A \in K$ (see Figure 9). Let $t$ be such that

$$A_0 \subseteq C_n^{t+1}(B), \qquad A_0 \nsubseteq C_n^t(B).$$

We can assume that

$$|A_0 \cap (C_n^{t+1}(B) - C_n^t(B))| \leq |A \cap (C_n^{t+1}(B) - C_n^t(B))|$$

for all $A \in K$ with $A \subseteq C_n^{t+1}(B)$. Choose

$$a_0 \in [C_n^{t+1}(B) - C_n^t(B)] \cap A_0.$$

Then there must exist $b_1, \ldots, b_m \in C_n^t(B)$, for some $m \leq n$, with

$$a_0 \in C_n(\{b_1, \ldots, b_m\}),$$

so

$$A_0 \subseteq C_n(A_1),$$

where

$$A_1 = (A_0 - \{a_0\}) \cup \{b_1, \ldots, b_m\};$$

hence

$$C(A_0) \subseteq C(A_1),$$

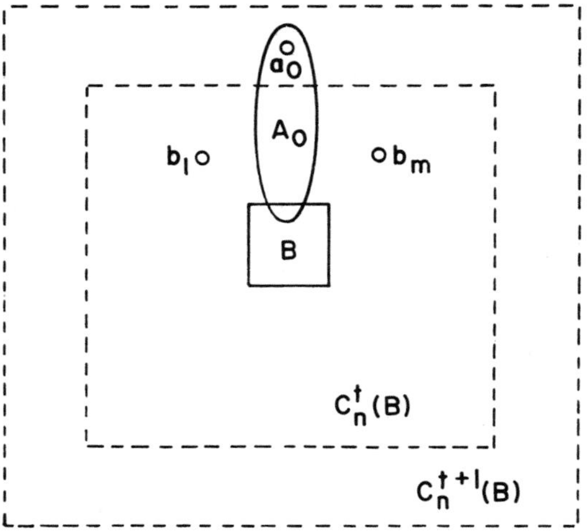


Figure 9

which says $A_1$ is a set of generators of $S$. Consequently, there is an irredundant basis $A_2 \subseteq A_1$. Now $|A_2| < |A_0| + n$. If $|A_0| + n \le j$, we see that the existence of $A_2$ contradicts the choice of $A_0$ as then we would have

$$A_2 \in K, \qquad A_2 \subseteq C_n^{t+1}(B)$$

and

$$|A_2 \cap (C_n^{t+1}(B) - C_n^t(B))| < |A_0 \cap (C_n^{t+1}(B) - C_n^t(B))|.$$

Thus $|A_0| + n > j$. As $|A_0| \le i$, we have $j - i < n$. □

EXAMPLE. If **A** is an algebra all of whose fundamental operations have arity not exceeding 2 then IrB(Sg) is a convex set. This applies to all the examples given in §1.

## REFERENCES

1. G. F. McNulty and W. Taylor [1975]
2. A. Tarski [1975]

## EXERCISES §4

1. Find IrB(Sg), where Sg is the subuniverse closure operator on the group of integers **Z**.
2. If $C$ is a closure operator on a set $S$ and $X$ is a closed subset of $S$, show that 4.4 applies to the irredundant bases of $X$.
3. If **A** is a unary algebra show that $|\text{IrB(Sg)}| \le 1$.
4. Give an example of an algebra **A** such that IrB(Sg) is not convex.

# §5. Congruences and Quotient Algebras

The concepts of congruence, quotient algebra, and homomorphism are all closely related. These will be the subjects of this and the next section.

Normal subgroups, which were introduced by Galois at the beginning of the last century, play a fundamental role in defining quotient groups and in the so-called homomorphism and isomorphism theorems which are so basic to the general development of group theory. Ideals, introduced in the second half of the last century by Dedekind, play an analogous role in defining quotient rings, and in the corresponding homomorphism and isomorphism theorems in ring theory. Given such a parallel situation, it was inevitable that mathematicians should seek a general common formulation. In these two sections the reader will see that congruences do indeed form the unifying concept, and furthermore they provide another meeting place for lattice theory and universal algebra.

**Definition 5.1.** Let **A** be an algebra of type $\mathscr{F}$ and let $\theta \in \mathrm{Eq}(A)$. Then $\theta$ is a *congruence* on **A** if $\theta$ satisfies the following *compatibility property*:

CP: For each $n$-ary function symbol $f \in \mathscr{F}$ and elements $a_i,b_i \in A$, if $a_i\theta b_i$ holds for $1 \le i \le n$ then

$$f^{\mathbf{A}}(a_1,\ldots,a_n)\theta f^{\mathbf{A}}(b_1,\ldots,b_n)$$

holds.

The compatibility property is an obvious condition for introducing an algebraic structure on the set of equivalence classes $A/\theta$, an algebraic structure which is inherited from the algebra **A**. For if $a_1,\ldots,a_n$ are elements of $A$ and $f$ is an $n$-ary symbol in $\mathscr{F}$, then the easiest choice of an equivalence class to be the value of $f$ applied to $\langle a_1/\theta,\ldots,a_n/\theta\rangle$ would be simply $f^{\mathbf{A}}(a_1,\ldots,a_n)/\theta$. This will indeed define a function on $A/\theta$ iff (CP) holds. We illustrate (CP) for a binary operation in Figure 10 by subdividing $A$ into

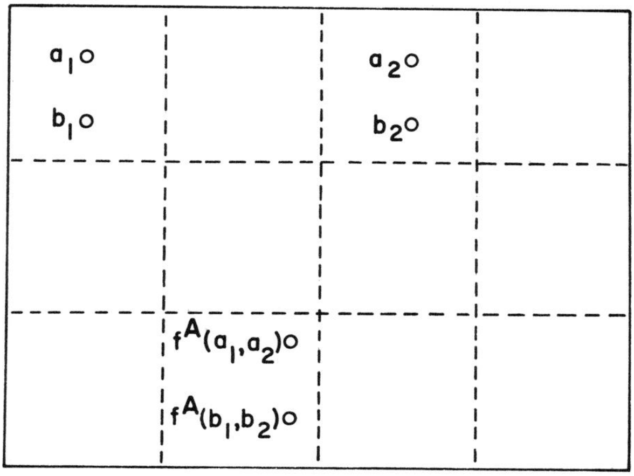


Figure 10

the equivalence classes of $\theta$; then selecting $a_1,b_1$ in the same equivalence class and $a_2,b_2$ in the same equivalence class we want $f^{\mathbf{A}}(a_1,b_1)$, $f^{\mathbf{A}}(a_2,b_2)$ to be in the same equivalence class.

**Definition 5.2.** The *set of all congruences on an algebra* **A** is denoted by Con **A**. Let $\theta$ be a congruence on an algebra **A**. Then the *quotient algebra of* **A** *by* $\theta$, written $\mathbf{A}/\theta$, is the algebra whose universe is $A/\theta$ and whose fundamental operations satisfy

$$f^{\mathbf{A}/\theta}(a_1/\theta, \ldots, a_n/\theta) = f^{\mathbf{A}}(a_1, \ldots, a_n)/\theta$$

where $a_1, \ldots, a_n \in A$ and $f$ is an $n$-ary function symbol in $\mathscr{F}$.

Note that quotient algebras of **A** are of the same type as **A**.

Examples. (1) Let **G** be a group. Then one can establish the following connection between congruences on **G** and normal subgroups of **G**:

(a) If $\theta \in \mathrm{Con}\, \mathbf{G}$ then $1/\theta$ is the universe of a normal subgroup of **G**, and for $a,b \in G$ we have $\langle a,b\rangle \in \theta$ iff $a \cdot b^{-1} \in 1/\theta$;

(b) If **N** is a normal subgroup of **G**, then the binary relation defined on $G$ by

$$\langle a,b\rangle \in \theta \quad \text{iff} \quad a \cdot b^{-1} \in N$$

is a congruence on **G** with $1/\theta = N$.

Thus the mapping $\theta \longmapsto 1/\theta$ is an order-preserving bijection between congruences on **G** and normal subgroups of **G**.

(2) Let **R** be a ring. The following establishes a similar connection between the congruences on **R** and ideals of **R**:

(a) If $\theta \in \mathrm{Con}\, \mathbf{R}$ then $0/\theta$ is an ideal of **R**, and for $a,b \in R$ we have $\langle a,b\rangle \in \theta$ iff $a - b \in 0/\theta$;

(b) If $I$ is an ideal of $R$ then the binary relation $\theta$ defined on $R$ by

$$\langle a,b\rangle \in \theta \quad \text{iff} \quad a - b \in I$$

is a congruence on **R** with $0/\theta = I$.

Thus the mapping $\theta \longmapsto 0/\theta$ is an order-preserving bijection between congruences on **R** and ideals of **R**.

These two examples are a bit misleading in that they suggest any congruence on an algebra might be determined by a single equivalence class of the congruence. The next example shows this need not be the case.

(3) Let **L** be a lattice which is a chain, and let $\theta$ be an equivalence relation on $L$ such that the equivalence classes of $\theta$ are convex subsets of $L$ (i.e., if $a\theta b$ and $a \leq c \leq b$ then $a\theta c$). Then $\theta$ is a congruence on **L**.

We will delay further discussion of quotient algebras until the next section and instead concentrate now on the lattice structure of Con **A**.

**Theorem 5.3.** $\langle \text{Con } \mathbf{A}, \subseteq \rangle$ *is a complete sublattice of* $\langle \text{Eq}(A), \subseteq \rangle$, *the lattice of equivalence relations on* $A$.

PROOF. To verify that Con **A** is closed under arbitrary intersection is straightforward. For arbitrary joins in Con **A** suppose $\theta_i \in \text{Con } \mathbf{A}$ for $i \in I$. Then, if $f$ is a fundamental $n$-ary operation of **A** and

$$\langle a_1,b_1 \rangle, \ldots, \langle a_n,b_n \rangle \in \bigvee_{i \in I} \theta_i,$$

where $\bigvee$ is the join of Eq($A$), then from I§4.7 it follows that one can find $i_0, \ldots, i_k \in I$ such that

$$\langle a_i,b_i \rangle \in \theta_{i_0} \circ \theta_{i_1} \circ \cdots \circ \theta_{i_k}, \qquad 0 \leq i \leq n.$$

An easy argument then suffices to show that

$$\langle f(a_1, \ldots, a_n), f(b_1, \ldots, b_n) \rangle \in \theta_{i_0} \circ \theta_{i_1} \circ \cdots \circ \theta_{i_k};$$

hence $\bigvee_{i \in I} \theta_i$ is a congruence relation on **A**. □

**Definition 5.4.** The *congruence lattice of* **A**, denoted by **Con A**, is the lattice whose universe is Con **A**, and meets and joins are calculated the same as when working with equivalence relations (see I§4).

The following theorem suggests the abstract characterization of congruence lattices of algebras.

**Theorem 5.5.** *For* **A** *an algebra, there is an algebraic closure operator* $\Theta$ *on* $A \times A$ *such that the closed subsets of* $A \times A$ *are precisely the congruences on* **A**. *Hence* **Con A** *is an algebraic lattice.*

PROOF. Let us start by setting up an appropriate algebraic structure on $A \times A$. First, for each $n$-ary function symbol $f$ in the type of **A** let us define a corresponding $n$-ary function $f$ on $A \times A$ by

$$f(\langle a_1,b_1 \rangle, \ldots, \langle a_n,b_n \rangle) = \langle f^{\mathbf{A}}(a_1, \ldots, a_n), f^{\mathbf{A}}(b_1, \ldots, b_n) \rangle.$$

Then we add the nullary operations $\langle a,a \rangle$ for each $a \in A$, a unary operation $s$ defined by

$$s(\langle a,b \rangle) = \langle b,a \rangle,$$

and a binary operation $t$ defined by

$$t(\langle a,b \rangle, \langle c,d \rangle) = \begin{cases} \langle a,d \rangle & \text{if } b = c \\ \langle a,b \rangle & \text{otherwise.} \end{cases}$$

Now it is an interesting exercise to verify that $B$ is a subuniverse of this new algebra iff $B$ is a congruence on **A**. Let $\Theta$ be the Sg closure operator on

$A \times A$ for the algebra we have just described. Thus, by 3.3, **Con A** is an algebraic lattice. □

The compact members of **Con A** are, by I§5.7, the finitely generated members $\Theta(\langle a_1,b_1\rangle, \ldots, \langle a_n,b_n\rangle)$ of **Con A**.

**Definition 5.6.** For **A** an algebra and $a_1, \ldots, a_n \in A$ let $\Theta(a_1, \ldots, a_n)$ denote the congruence generated by $\{\langle a_i,a_j\rangle : 1 \leq i, j \leq n\}$, i.e., the smallest congruence such that $a_1, \ldots, a_n$ are in the same equivalence class. The congruence $\Theta(a_1,a_2)$ is called a *principal congruence*. For arbitrary $X \subseteq A$, let $\Theta(X)$ be defined to mean the congruence generated by $X \times X$.

Finitely generated congruences will play a key role in II§12, in Chapter IV, and Chapter V. In certain cases we already know a good description of principal congruences.

EXAMPLES. (1) If **G** is a group and $a,b,c,d \in G$ then $\langle a,b\rangle \in \Theta(c,d)$ iff $ab^{-1}$ is a product of conjugates of $cd^{-1}$ and conjugates of $dc^{-1}$. This follows from the fact that the smallest normal subgroup of **G** containing a given element $e$ has as its universe the set of all products of conjugates of $e$ and conjugates of $e^{-1}$.

(2) If **R** is a ring with unity and $a,b,c,d \in R$ then $\langle a,b\rangle \in \Theta(c,d)$ iff $a - b$ is of the form $\sum_{1 \leq i \leq n} r_i(c - d)s_i$ where $r_i,s_i \in R$. This follows from the fact that the smallest ideal of **R** containing a given element $e$ of $R$ is precisely the set $\{\sum_{1 \leq i \leq n} r_i e s_i : r_i,s_i \in R, n \geq 1\}$.

Some useful facts about congruences which depend primarily on the fact that $\Theta$ is an algebraic closure operator are given in the following.

**Theorem 5.7.** *Let* **A** *be an algebra, and suppose* $a_1,b_1, \ldots, a_n,b_n \in A$ *and* $\theta \in$ Con **A**. *Then*

(a) $\Theta(a_1,b_1) = \Theta(b_1,a_1)$
(b) $\Theta(\langle a_1,b_1\rangle, \ldots, \langle a_n,b_n\rangle) = \Theta(a_1,b_1) \vee \cdots \vee \Theta(a_n,b_n)$
(c) $\Theta(a_1, \ldots, a_n) = \Theta(a_1,a_2) \vee \Theta(a_2,a_3) \vee \cdots \vee \Theta(a_{n-1},a_n)$
(d) $\theta = \bigcup\{\Theta(a,b) : \langle a,b\rangle \in \theta\} = \bigvee\{\Theta(a,b) : \langle a,b\rangle \in \theta\}$
(e) $\theta = \bigcup\{\Theta(\langle a_1,b_1\rangle, \ldots, \langle a_n,b_n\rangle) : \langle a_i,b_i\rangle \in \theta, n \geq 1\}$.

PROOF. (a) As

$$\langle b_1,a_1\rangle \in \Theta(a_1,b_1)$$

we have

$$\Theta(b_1,a_1) \subseteq \Theta(a_1,b_1);$$

hence, by symmetry,

$$\Theta(a_1,b_1) = \Theta(b_1,a_1).$$

(b) For $1 \leq i \leq n$,

$$\langle a_i,b_i\rangle \in \Theta(\langle a_1,b_1\rangle, \ldots, \langle a_n,b_n\rangle);$$

hence

$$\Theta(a_i,b_i) \subseteq \Theta(\langle a_1,b_1\rangle, \ldots, \langle a_n,b_n\rangle),$$

so

$$\Theta(a_1,b_1) \vee \cdots \vee \Theta(a_n,b_n) \subseteq \Theta(\langle a_1,b_1\rangle, \ldots, \langle a_n,b_n\rangle).$$

On the other hand, for $1 \le i \le n$,

$$\langle a_i,b_i\rangle \in \Theta(a_i,b_i) \subseteq \Theta(a_1,b_1) \vee \cdots \vee \Theta(a_n,b_n),$$

so

$$\{\langle a_1,b_1\rangle, \ldots, \langle a_n,b_n\rangle\} \subseteq \Theta(a_1,b_1) \vee \cdots \vee \Theta(a_n,b_n);$$

hence

$$\Theta(\langle a_1,b_1\rangle, \ldots, \langle a_n,b_n\rangle) \subseteq \Theta(a_1,b_1) \vee \cdots \vee \Theta(a_n,b_n),$$

so

$$\Theta(\langle a_1,b_1\rangle, \ldots, \langle a_n,b_n\rangle) = \Theta(a_1,b_1) \vee \cdots \vee \Theta(a_n,b_n).$$

(c) For $1 \le i \le n-1$,

$$\langle a_i,a_{i+1}\rangle \in \Theta(a_1, \ldots, a_n),$$

so

$$\Theta(a_i,a_{i+1}) \subseteq \Theta(a_1, \ldots, a_n);$$

hence

$$\Theta(a_1,a_2) \vee \cdots \vee \Theta(a_{n-1},a_n) \subseteq \Theta(a_1, \ldots, a_n).$$

Conversely, for $1 \le i < j \le n$,

$$\langle a_i,a_j\rangle \in \Theta(a_i,a_{i+1}) \circ \cdots \circ \Theta(a_{j-1},a_j)$$

so, by I§4.7

$$\langle a_i,a_j\rangle \in \Theta(a_i,a_{i+1}) \vee \cdots \vee \Theta(a_{j-1},a_j);$$

hence

$$\langle a_i,a_j\rangle \in \Theta(a_1,a_2) \vee \cdots \vee \Theta(a_{n-1},a_n).$$

In view of (a) this leads to

$$\Theta(a_1, \ldots, a_n) \subseteq \Theta(a_1,a_2) \vee \cdots \vee \Theta(a_{n-1},a_n),$$

so

$$\Theta(a_1, \ldots, a_n) = \Theta(a_1,a_2) \vee \cdots \vee \Theta(a_{n-1},a_n).$$

(d) For $\langle a,b\rangle \in \theta$ clearly

$$\langle a,b\rangle \in \Theta(a,b) \subseteq \theta$$

so

$$\theta \subseteq \bigcup\{\Theta(a,b) : \langle a,b\rangle \in \theta\} \subseteq \bigvee\{\Theta(a,b) : \langle a,b\rangle \in \theta\} \subseteq \theta;$$

hence

$$\theta = \bigcup\{\Theta(a,b) : \langle a,b\rangle \in \theta\} = \bigvee\{\Theta(a,b) : \langle a,b\rangle \in \theta\}.$$

(e) (Similar to (d)). □

One cannot hope for a further sharpening of the abstract characterization of congruence lattices of algebras in 5.5 because in 1963 Grätzer and Schmidt proved that for every algebraic lattice **L** there is an algebra **A** such that **L** ≅ **Con A**. Of course, for particular classes of algebras one might find that some additional properties hold for the corresponding classes of congruence lattices. For example, the congruence lattices of lattices satisfy the distributive law, and the congruence lattices of groups (or rings) satisfy the modular law. One of the major themes of universal algebra has been to study the consequences of special assumptions about the congruence lattices (or congruences) of algebras (see §12 as well as Chapters IV and V). For this purpose we introduce the following terminology.

**Definition 5.8.** An algebra **A** is *congruence-distributive* (*congruence-modular*) if **Con A** is a distributive (modular) lattice. If $\theta_1,\theta_2 \in \operatorname{Con} \mathbf{A}$ and

$$\theta_1 \circ \theta_2 = \theta_2 \circ \theta_1$$

then we say $\theta_1$ and $\theta_2$ are *permutable*, or $\theta_1$ and $\theta_2$ *permute*. **A** is *congruence-permutable* if every pair of congruences on **A** permutes. A class $K$ of algebras is congruence-distributive, congruence-modular, respectively congruence-permutable iff every algebra in $K$ has the desired property.

We have already looked at distributivity and modularity, so we will finish this section with two results on permutable congruences.

**Theorem 5.9.** *Let* **A** *be an algebra and suppose* $\theta_1,\theta_2 \in \operatorname{Con} \mathbf{A}$. *Then the following are equivalent*:

(a) $\theta_1 \circ \theta_2 = \theta_2 \circ \theta_1$
(b) $\theta_1 \vee \theta_2 = \theta_1 \circ \theta_2$
(c) $\theta_1 \circ \theta_2 \subseteq \theta_2 \circ \theta_1$.

PROOF. (a) $\Rightarrow$ (b): For any equivalence relation $\theta$ we have $\theta \circ \theta = \theta$, so from (a) it follows that the expression for $\theta_1 \vee \theta_2$ given in I§4.6 reduces to $\theta_1 \cup (\theta_1 \circ \theta_2)$, and hence to $\theta_1 \circ \theta_2$.

(c) $\Rightarrow$ (a): Given (c) we have to show that

$$\theta_2 \circ \theta_1 \subseteq \theta_1 \circ \theta_2.$$

This, however, follows easily from applying the relational inverse operation to (c), namely we have

$$(\theta_1 \circ \theta_2)^{\smile} \subseteq (\theta_2 \circ \theta_1)^{\smile},$$

and hence (as the reader can easily verify)

$$\theta_2^{\smile} \circ \theta_1^{\smile} \subseteq \theta_1^{\smile} \circ \theta_2^{\smile}.$$

Since the inverse of an equivalence relation is just that equivalence relation, we have established (a).

(b) $\Rightarrow$ (c): Since

$$\theta_2 \circ \theta_1 \subseteq \theta_1 \vee \theta_2,$$

from (b) we could deduce

$$\theta_2 \circ \theta_1 \subseteq \theta_1 \circ \theta_2,$$

and then from the previous paragraph it would follow that

$$\theta_2 \circ \theta_1 = \theta_1 \circ \theta_2;$$

hence (c) holds. ☐

**Theorem 5.10** (Birkhoff). *If* **A** *is congruence-permutable, then* **A** *is congruence-modular.*

PROOF. Let $\theta_1, \theta_2, \theta_3 \in \operatorname{Con} \mathbf{A}$ with $\theta_1 \subseteq \theta_2$. We want to show that

$$\theta_2 \cap (\theta_1 \vee \theta_3) \subseteq \theta_1 \vee (\theta_2 \cap \theta_3),$$

so suppose $\langle a,b \rangle$ is in $\theta_2 \cap (\theta_1 \vee \theta_3)$. By 5.9 there is an element $c$ such that

$$a\theta_1 c\theta_3 b$$

holds as

$$\theta_1 \vee \theta_3 = \theta_1 \circ \theta_3.$$

By symmetry

$$\langle c,a \rangle \in \theta_1;$$

hence

$$\langle c,a \rangle \in \theta_2,$$

and then by transitivity

$$\langle c,b \rangle \in \theta_2.$$

Thus

$$\langle c,b \rangle \in \theta_2 \cap \theta_3,$$

so from

$$a\theta_1 c(\theta_2 \cap \theta_3)b$$

follows

$$\langle a,b \rangle \in \theta_1 \circ (\theta_2 \cap \theta_3);$$

hence

$$\langle a,b \rangle \in \theta_1 \vee (\theta_2 \cap \theta_3).$$ ☐

We would like to note that in 1953 Jónsson improved on Birkhoff's result above by showing that one could derive the so-called *Arguesian identity* for lattices from congruence-permutability. In §12 we will concern ourselves again with congruence-distributivity and -permutability.

REFERENCES

1. G. Birkhoff [3]
2. G. Grätzer and E. T. Schmidt [1963]
3. B. Jónsson [1953]
4. P. Pudlák [1976]

EXERCISES §5

1. Verify the connection between normal subgroups and congruences on a group stated in Example 1 (after 5.2).
2. Verify the connection between ideals and congruences on rings stated in Example 2 (after 5.2).

3. Show that the normal subgroups of a group form an algebraic lattice which is modular.

4. Show that every group and ring is congruence-permutable, but not necessarily congruence-distributive.

5. Show that every lattice is congruence-distributive, but not necessarily congruence-permutable.

6. In the proof of 5.5, verify that subuniverses of the new algebra are precisely the congruences on **A**.

7. Show that $\Theta$ is a 2-ary closure operator. [Hint: replace each $n$-ary $f$ of **A** by unary operations

$$f(a_1, \ldots, a_{i-1}, x, a_{i+1}, \ldots, a_n),\ a_1, \ldots, a_{i-1}, a_{i+1}, \ldots, a_n \in A$$

and show this gives a unary algebra with the same congruences.]

8. If **A** is a unary algebra and $B$ is a subuniverse define $\theta$ by $\langle a,b \rangle \in \theta$ iff $a = b$ or $\{a,b\} \subseteq B$. Show that $\theta$ is a congruence on **A**.

9. Let **S** be a semilattice. Define $a \leq b$ for $a,b \in S$ if $a \cdot b = a$. Show that $\leq$ is a partial order on $S$. Next, given $a \in S$ define

$$\theta_a = \{\langle b,c \rangle \in S \times S\text{: both or neither of } a \leq b,\ a \leq c \text{ hold}\}.$$

Show $\theta_a$ is a congruence on **S**.

An algebra **A** has the *congruence extension property* (CEP) if for every $\mathbf{B} \leq \mathbf{A}$ and $\theta \in \operatorname{Con} \mathbf{B}$ there is a $\phi \in \operatorname{Con} \mathbf{A}$ such that $\theta = \phi \cap B^2$. A class $K$ of algebras has the CEP if every algebra in the class has the CEP.

10. Show that the class of Abelian groups has the CEP. Does the class of lattices have the CEP?

11. If **L** is a distributive lattice and $a,b,c,d \in L$ show that $\langle a,b \rangle \in \Theta(c,d)$ iff $c \wedge d \wedge a = c \wedge d \wedge b$ and $c \vee d \vee a = c \vee d \vee b$.

An algebra **A** has 3-*permutable congruences* if for all $\theta,\phi \in \operatorname{Con} \mathbf{A}$ we have $\theta \circ \phi \circ \theta \subseteq \phi \circ \theta \circ \phi$.

12. (Jónsson) Show that if **A** has 3-permutable congruences then **A** is congruence-modular.

## §6. Homomorphisms and the Homomorphism and Isomorphism Theorems

Homomorphisms are a natural generalization of the concept of isomorphism, and, as we shall see, go hand in hand with congruences.

**Definition 6.1.** Suppose **A** and **B** are two algebras of the same type $\mathscr{F}$. A mapping $\alpha: A \to B$ is called a *homomorphism* from **A** to **B** if

$$\alpha f^{\mathbf{A}}(a_1, \ldots, a_n) = f^{\mathbf{B}}(\alpha a_1, \ldots, \alpha a_n)$$

for each $n$-ary $f$ in $\mathscr{F}$ and each sequence $a_1, \ldots, a_n$ from $A$. If, in addition, the mapping $\alpha$ is onto then $\mathbf{B}$ is said to be a *homomorphic image* of $\mathbf{A}$, and $\alpha$ is called an *epimorphism*. (In this terminology an isomorphism is a homomorphism which is one-to-one and onto.) In case $\mathbf{A} = \mathbf{B}$ a homomorphism is also called an *endomorphism* and an isomorphism is referred to as an *automorphism*. The phrase "$\alpha : \mathbf{A} \to \mathbf{B}$ is a homomorphism" is often used to express the fact that $\alpha$ is a homomorphism from $\mathbf{A}$ to $\mathbf{B}$.

EXAMPLES. Lattice, group, ring, module, and monoid homomorphisms are all special cases of homomorphisms as defined above.

**Theorem 6.2.** *Let* $\mathbf{A}$ *be an algebra generated by a set* $X$. *If* $\alpha : \mathbf{A} \to \mathbf{B}$ *and* $\beta : \mathbf{A} \to \mathbf{B}$ *are two homomorphisms which agree on* $X$ *(i.e.,* $\alpha(a) = \beta(a)$ *for* $a \in X$*), then* $\alpha = \beta$.

PROOF. Recall the definition of $E$ in §3. Note that if $\alpha$ and $\beta$ agree on $X$ then $\alpha$ and $\beta$ agree on $E(X)$, for if $f$ is an $n$-ary function symbol and $a_1, \ldots, a_n \in X$ then

$$\begin{aligned} \alpha f^{\mathbf{A}}(a_1, \ldots, a_n) &= f^{\mathbf{B}}(\alpha a_1, \ldots, \alpha a_n) \\ &= f^{\mathbf{B}}(\beta a_1, \ldots, \beta a_n) \\ &= \beta f^{\mathbf{A}}(a_1, \ldots, a_n). \end{aligned}$$

Thus by induction, if $\alpha$ and $\beta$ agree on $X$ then they agree on $E^n(X)$ for $n < \omega$, and hence they agree on $\operatorname{Sg}(X)$. □

**Theorem 6.3.** *Let* $\alpha : \mathbf{A} \to \mathbf{B}$ *be a homomorphism. Then the image of a subuniverse* of $\mathbf{A}$ *under* $\alpha$ *is a subuniverse of* $\mathbf{B}$, *and the inverse image of a subuniverse of* $\mathbf{B}$ *is a subuniverse of* $\mathbf{A}$.

PROOF. Let $S$ be a subuniverse of $\mathbf{A}$, let $f$ be an $n$-ary member of $\mathscr{F}$, and let $a_1, \ldots, a_n \in S$. Then

$$f^{\mathbf{B}}(\alpha a_1, \ldots, \alpha a_n) = \alpha f^{\mathbf{A}}(a_1, \ldots, a_n) \in \alpha(S),$$

so $\alpha(S)$ is a subuniverse of $\mathbf{B}$. If we now assume that $S$ is a subuniverse of $\mathbf{B}$ (instead of $\mathbf{A}$) and $\alpha(a_1), \ldots, \alpha(a_n) \in S$ then $\alpha f^{\mathbf{A}}(a_1, \ldots, a_n) \in S$ follows from the above equation, so $f^{\mathbf{A}}(a_1, \ldots, a_n)$ is in $\alpha^{-1}(S)$. Thus $\alpha^{-1}(S)$ is a subuniverse of $\mathbf{A}$. □

**Definition 6.4.** If $\alpha : \mathbf{A} \to \mathbf{B}$ is a homomorphism and $\mathbf{C} \leq \mathbf{A}$, $\mathbf{D} \leq \mathbf{B}$, let $\alpha(\mathbf{C})$ be the subalgebra of $\mathbf{B}$ with universe $\alpha(C)$, and let $\alpha^{-1}(\mathbf{D})$ be the subalgebra of $\mathbf{A}$ with universe $\alpha^{-1}(D)$, provided $\alpha^{-1}(D) \neq \varnothing$.

**Theorem 6.5.** *Suppose* $\alpha : \mathbf{A} \to \mathbf{B}$ *and* $\beta : \mathbf{B} \to \mathbf{C}$ *are homomorphisms. Then the composition* $\beta \circ \alpha$ *is a homomorphism from* $\mathbf{A}$ *to* $\mathbf{C}$.

Proof. For $f$ an $n$-ary function symbol and $a_1,\ldots,a_n \in A$, we have

$$\begin{aligned}(\beta \circ \alpha)f^{\mathbf{A}}(a_1,\ldots,a_n) &= \beta(\alpha f^{\mathbf{A}}(a_1,\ldots,a_n))\\ &= \beta f^{\mathbf{B}}(\alpha a_1,\ldots,\alpha a_n)\\ &= f^{\mathbf{C}}(\beta(\alpha a_1),\ldots,\beta(\alpha a_n))\\ &= f^{\mathbf{C}}((\beta \circ \alpha)a_1,\ldots,(\beta \circ \alpha)a_n).\end{aligned}$$

□

The next result says that homomorphisms commute with subuniverse closure operators.

**Theorem 6.6.** *If $\alpha: \mathbf{A} \to \mathbf{B}$ is a homomorphism and $X$ is a subset of $\mathbf{A}$ then*

$$\alpha\,\mathrm{Sg}(X) = \mathrm{Sg}(\alpha X).$$

Proof. From the definition of $E$ (see §3) and the fact that $\alpha$ is a homomorphism we have

$$\alpha E(Y) = E(\alpha Y)$$

for all $Y \subseteq A$. Thus, by induction on $n$,

$$\alpha E^n(X) = E^n(\alpha X)$$

for $n \geq 1$; hence

$$\begin{aligned}\alpha\,\mathrm{Sg}(X) &= \alpha(X \cup E(X) \cup E^2(X) \cup \ldots)\\ &= \alpha X \cup \alpha E(X) \cup \alpha E^2(X) \cup \ldots\\ &= \alpha X \cup E(\alpha X) \cup E^2(\alpha X) \cup \ldots\\ &= \mathrm{Sg}(\alpha X).\end{aligned}$$

□

**Definition 6.7.** Let $\alpha: \mathbf{A} \to \mathbf{B}$ be a homomorphism. Then the *kernel of* $\alpha$, written $\ker(\alpha)$, is defined by

$$\ker(\alpha) = \{\langle a,b\rangle \in A^2 : \alpha(a) = \alpha(b)\}.$$

**Theorem 6.8.** *Let $\alpha: \mathbf{A} \to \mathbf{B}$ be a homomorphism. Then* $\ker(\alpha)$ *is a congruence on* $\mathbf{A}$.

Proof. If $\langle a_i,b_i\rangle \in \ker(\alpha)$ for $1 \leq i \leq n$ and $f$ is $n$-ary in $\mathscr{F}$, then

$$\begin{aligned}\alpha f^{\mathbf{A}}(a_1,\ldots,a_n) &= f^{\mathbf{B}}(\alpha a_1,\ldots,\alpha a_n)\\ &= f^{\mathbf{B}}(\alpha b_1,\ldots,\alpha b_n)\\ &= \alpha f^{\mathbf{A}}(b_1,\ldots,b_n);\end{aligned}$$

hence

$$\langle f^{\mathbf{A}}(a_1,\ldots,a_n), f^{\mathbf{A}}(b_1,\ldots,b_n)\rangle \in \ker(\alpha).$$

Clearly $\ker(\alpha)$ is an equivalence relation, so it follows that $\ker(\alpha)$ is actually a congruence on $\mathbf{A}$. □

When studying groups it is usual to refer to the kernel of a homomorphism as a normal subgroup, namely the inverse image of the identity element under the homomorphism. This does not cause any real problems since we have already pointed out in §5 that a congruence on a group is determined by the equivalence class of the identity element, which is a normal subgroup. Similarly, in the study of rings one refers to the kernel of a homomorphism as a certain ideal.

We are now ready to look at the straightforward generalizations to abstract algebras of the homomorphism and isomorphism theorems usually encountered in a first course on group theory.

**Definition 6.9.** Let **A** be an algebra and let $\theta \in \operatorname{Con} \mathbf{A}$. The *natural map* $\nu_\theta : A \to A/\theta$ is defined by $\nu_\theta(a) = a/\theta$. (When there is no ambiguity we write simply $\nu$ instead of $\nu_\theta$.) Figure 11 shows how one might visualize the natural map.

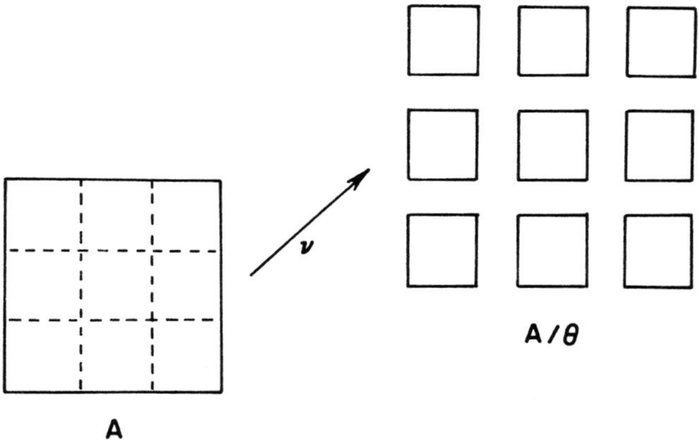


Figure 11

**Theorem 6.10.** *The natural map from an algebra to a quotient of the algebra is an onto homomorphism.*

Proof. Let $\theta \in \operatorname{Con} \mathbf{A}$ and let $\nu : A \to A/\theta$ be the natural map. Then for $f$ an $n$-ary function symbol and $a_1, \ldots, a_n \in A$ we have

$$\begin{aligned} \nu f^{\mathbf{A}}(a_1, \ldots, a_n) &= f^{\mathbf{A}}(a_1, \ldots, a_n)/\theta \\ &= f^{\mathbf{A}/\theta}(a_1/\theta, \ldots, a_n/\theta) \\ &= f^{\mathbf{A}/\theta}(\nu a_1, \ldots, \nu a_n), \end{aligned}$$

so $\nu$ is a homomorphism. Clearly $\nu$ is onto. □

**Definition 6.11.** The *natural homomorphism* from an algebra to a quotient of the algebra is given by the natural map.

**Theorem 6.12** (Homomorphism Theorem). *Suppose* $\alpha$: $\mathbf{A} \to \mathbf{B}$ *is a homomorphism onto B. Then there is an isomorphism* $\beta$ *from* $\mathbf{A}/\ker(\alpha)$ *to* $\mathbf{B}$ *defined by* $\alpha = \beta \circ v$, *where* $v$ *is the natural homomorphism from* $\mathbf{A}$ *to* $\mathbf{A}/\ker(\alpha)$. (*See Figure 12*).

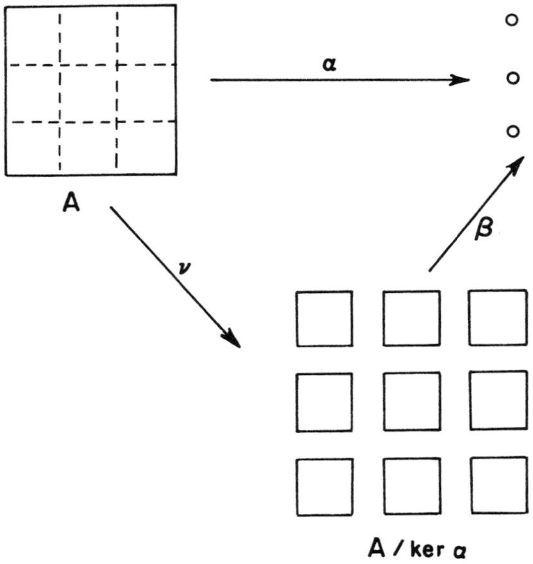


Figure 12

PROOF. First note that if $\alpha = \beta \circ v$ then we must have $\beta(a/\theta) = \alpha(a)$. The second of these equalities does indeed define a function $\beta$, and $\beta$ satisfies $\alpha = \beta \circ v$. It is not difficult to verify that $\beta$ is a bijection. To show that $\beta$ is actually an isomorphism, suppose $f$ is an $n$-ary function symbol and $a_1, \ldots, a_n \in A$. Then

$$\begin{aligned}\beta(f^{\mathbf{A}/\theta}(a_1/\theta, \ldots, a_n/\theta)) &= \beta(f^{\mathbf{A}}(a_1, \ldots, a_n)/\theta)\\ &= \alpha f^{\mathbf{A}}(a_1, \ldots, a_n)\\ &= f^{\mathbf{B}}(\alpha a_1, \ldots, \alpha a_n)\\ &= f^{\mathbf{B}}(\beta(a_1/\theta), \ldots, \beta(a_n/\theta)).\end{aligned}$$ □

Combining Theorems 6.5 and 6.12 we see that an algebra is a homomorphic image of an algebra **A** iff it is isomorphic to a quotient of the algebra **A**. Thus the "external" problem of finding all homomorphic images of **A** reduces to the "internal" problem of finding all congruences on **A**. The homomorphism theorem is also called "the first isomorphism theorem".

**Definition 6.13.** Suppose $\mathbf{A}$ is an algebra and $\phi,\theta \in \operatorname{Con} \mathbf{A}$ with $\theta \subseteq \phi$. Then let

$$\phi/\theta = \{\langle a/\theta,b/\theta\rangle \in (A/\theta)^2 : \langle a,b\rangle \in \phi\}.$$

**Lemma 6.14.** *If $\phi,\theta \in \operatorname{Con} \mathbf{A}$ and $\theta \subseteq \phi$, then $\phi/\theta$ is a congruence on $\mathbf{A}/\theta$.*

PROOF. Let $f$ be an $n$-ary function symbol and suppose $\langle a_i/\theta,b_i/\theta\rangle \in \phi/\theta$, $1 \le i \le n$. Then $\langle a_i,b_i\rangle \in \phi$ (why?), so

$$\langle f^{\mathbf{A}}(a_1,\ldots,a_n), f^{\mathbf{A}}(b_1,\ldots,b_n)\rangle \in \phi,$$

and thus

$$\langle f^{\mathbf{A}}(a_1,\ldots,a_n)/\theta, f^{\mathbf{A}}(b_1,\ldots,b_n)/\theta\rangle \in \phi/\theta.$$

From this it follows that

$$\langle f^{\mathbf{A}/\theta}(a_1/\theta,\ldots,a_n/\theta), f^{\mathbf{A}/\theta}(b_1/\theta,\ldots,b_n/\theta)\rangle \in \phi/\theta. \qquad \square$$

**Theorem 6.15** (Second Isomorphism Theorem). *If $\phi,\theta \in \operatorname{Con} \mathbf{A}$ and $\theta \subseteq \phi$, then the map*

$$\alpha : (A/\theta)/(\phi/\theta) \to A/\phi$$

*defined by*

$$\alpha((a/\theta)/(\phi/\theta)) = a/\phi$$

*is an isomorphism from $(\mathbf{A}/\theta)/(\phi/\theta)$ to $\mathbf{A}/\phi$. (See Figure 13.)*

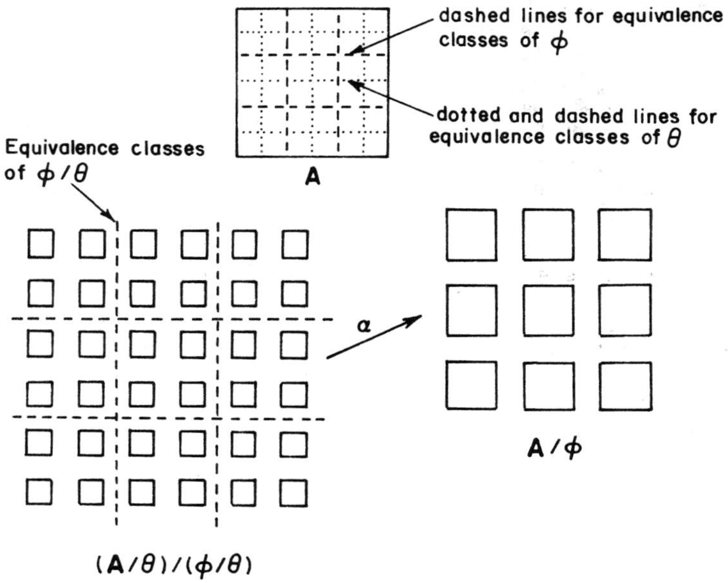


Figure 13

PROOF. Let $a,b \in A$. Then from

$$(a/\theta)/(\phi/\theta) = (b/\theta)/(\phi/\theta) \quad \text{iff} \quad a/\phi = b/\phi$$

it follows that $\alpha$ is a well-defined bijection. Now, for $f$ an $n$-ary function symbol and $a_1, \ldots, a_n \in A$ we have

$$\begin{aligned}\alpha f^{(\mathbf{A}/\theta)/(\phi/\theta)}((a_1/\theta)/(\phi/\theta), \ldots, (a_n/\theta)/(\phi/\theta)) \\ &= \alpha(f^{\mathbf{A}/\theta}(a_1/\theta, \ldots, a_n/\theta)/(\phi/\theta)) \\ &= \alpha((f^{\mathbf{A}}(a_1, \ldots, a_n)/\theta)/(\phi/\theta)) \\ &= f^{\mathbf{A}}(a_1, \ldots, a_n)/\phi \\ &= f^{\mathbf{A}/\phi}(a_1/\phi, \ldots, a_n/\phi) \\ &= f^{\mathbf{A}/\phi}(\alpha((a_1/\theta)/(\phi/\theta)), \ldots, \alpha((a_n/\theta)/(\phi/\theta))),\end{aligned}$$

so $\alpha$ is an isomorphism. □

**Definition 6.16.** Suppose $B$ is a subset of $A$ and $\theta$ is a congruence on $\mathbf{A}$. Let $B^\theta = \{a \in A : B \cap a/\theta \neq \emptyset\}$. Let $\mathbf{B}^\theta$ be the subalgebra of $\mathbf{A}$ generated by $B^\theta$. Also define $\theta\restriction_B$ to be $\theta \cap B^2$, *the restriction of* $\theta$ *to* $B$. (See Figure 14, where the dashed-line subdivisions of $A$ are the equivalence classes of $\theta$.)

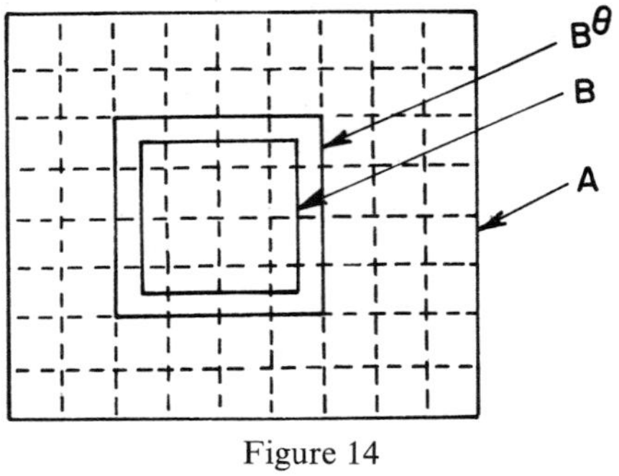


Figure 14

**Lemma 6.17.** *If* $\mathbf{B}$ *is a subalgebra of* $\mathbf{A}$ *and* $\theta \in \operatorname{Con} \mathbf{A}$, *then*

(a) *The universe of* $\mathbf{B}^\theta$ *is* $B^\theta$.
(b) $\theta\restriction_B$ *is a congruence on* $\mathbf{B}$.

PROOF. Suppose $f$ is an $n$-ary function symbol. For (a) let $a_1, \ldots, a_n \in B^\theta$. Then one can find $b_1, \ldots, b_n \in B$ such that

$$\langle a_i, b_i \rangle \in \theta, \qquad 1 \leq i \leq n,$$

hence

$$\langle f^{\mathbf{A}}(a_1, \ldots, a_n), f^{\mathbf{A}}(b_1, \ldots, b_n) \rangle \in \theta,$$

so

$$f^{\mathbf{A}}(a_1, \ldots, a_n) \in B^\theta.$$

Thus $B^\theta$ is a subuniverse of $\mathbf{A}$. Next, to verify that $\theta \restriction_B$ is a congruence on $\mathbf{B}$ is straightforward. □

**Theorem 6.18** (Third Isomorphism Theorem). *If* $\mathbf{B}$ *is a subalgebra of* $\mathbf{A}$ *and* $\theta \in \text{Con } \mathbf{A}$, *then (see Figure 15)*

$$\mathbf{B}/\theta\restriction_B \cong \mathbf{B}^\theta/\theta\restriction_{B^\theta}.$$

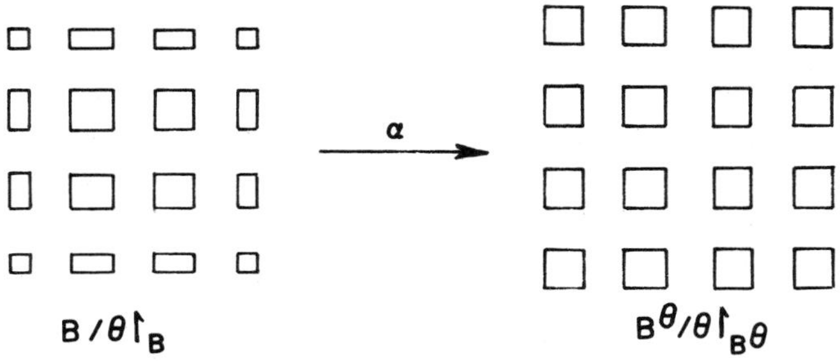


Figure 15

PROOF. We leave it to the reader to verify that the map $\alpha$ defined by $\alpha(b/\theta\restriction_B) = b/\theta\restriction_{B^\theta}$ is the desired isomorphism. □

The last theorem in this section will be quite important in the subsequent study of subdirectly irreducible algebras. Before looking at this theorem let us note that if $\mathbf{L}$ is a lattice and $a,b \in L$ with $a \leq b$ then the interval $[a,b]$ is a subuniverse of $\mathbf{L}$.

**Definition 6.19.** For $[a,b]$ a closed interval of a lattice $\mathbf{L}$, where $a \leq b$, let $[\mathbf{a},\mathbf{b}]$ denote the corresponding sublattice of $\mathbf{L}$.

**Theorem 6.20** (Correspondence Theorem). *Let* $\mathbf{A}$ *be an algebra and let* $\theta \in \text{Con } \mathbf{A}$. *Then the mapping* $\alpha$ *defined on* $[\theta,\nabla_A]$ *by*

$$\alpha(\phi) = \phi/\theta$$

*is a lattice isomorphism from* $[\theta, \nabla_A]$ *to* $\mathbf{Con}\ \mathbf{A}/\theta$, *where* $[\theta, \nabla_A]$ *is a sublattice of* $\mathbf{Con}\ \mathbf{A}$. *(See Figure 16.)*

PROOF. To see that $\alpha$ is one-to-one, let $\phi,\psi \in [\theta,\nabla_A]$ with $\phi \neq \psi$. Then, without loss of generality, we can assume that there are elements $a,b \in A$ with $\langle a,b\rangle \in \phi - \psi$. Thus

$$\langle a/\theta,b/\theta\rangle \in (\phi/\theta) - (\psi/\theta),$$

so

$$\alpha(\phi) \neq \alpha(\psi).$$

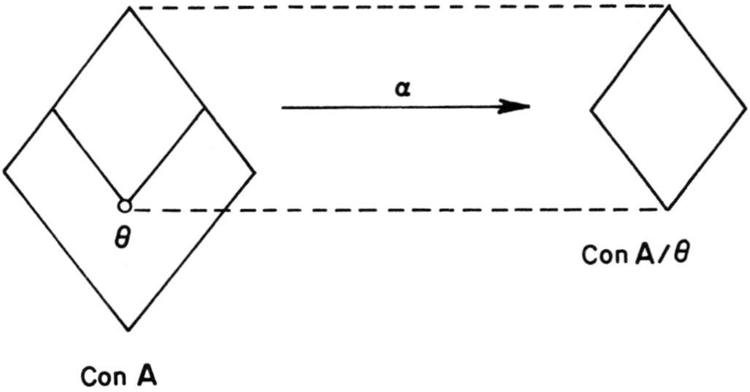


Figure 16

To show that $\alpha$ is onto, let $\psi \in \text{Con } \mathbf{A}/\theta$ and define $\phi$ to be $\ker(v_\psi v_\theta)$. Then for $a,b \in A$,

$$\begin{aligned} &\langle a/\theta, b/\theta \rangle \in \phi/\theta \\ \text{iff} \quad &\langle a,b \rangle \in \phi \\ \text{iff} \quad &\langle a/\theta, b/\theta \rangle \in \psi, \end{aligned}$$

so

$$\phi/\theta = \psi.$$

Finally, we will show that $\alpha$ is an isomorphism. If $\phi,\psi \in [\theta, \nabla_A]$ then it is clear that

$$\begin{aligned} &\phi \subseteq \psi \\ \text{iff} \quad &\phi/\theta \subseteq \psi/\theta \\ \text{iff} \quad &\alpha\phi \subseteq \alpha\psi. \end{aligned}$$ □

One can readily translate 6.12, 6.15, 6.18, and 6.20 into the (usual) theorems used in group theory and in ring theory.

## Exercises §6

1. Show that, under composition, the endomorphisms of an algebra form a monoid, and the automorphisms form a group.
2. Translate the isomorphism theorems and the correspondence theorem into results about groups [rings], replacing congruences by normal subgroups [ideals].
3. Show that a homomorphism $\alpha$ is an embedding iff $\ker \alpha = \Delta$.
4. If $\theta \in \text{Con } \mathbf{A}$ and $\mathbf{Con\ A}$ is a modular [distributive] lattice then show $\mathbf{Con\ A}/\theta$ is also a modular [distributive] lattice.
5. Let $\alpha: \mathbf{A} \to \mathbf{B}$ be a homomorphism, and $X \subseteq A$. Show that $\langle a,b \rangle \in \Theta(X) \Rightarrow \langle \alpha a, \alpha b \rangle \in \Theta(\alpha X)$.
6. Given two homomorphisms $\alpha: \mathbf{A} \to \mathbf{B}$ and $\beta: \mathbf{A} \to \mathbf{C}$, if $\ker \beta \subseteq \ker \alpha$ and $\beta$ is onto, show that there is a homomorphism $\gamma: \mathbf{C} \to \mathbf{B}$ such that $\alpha = \gamma \circ \beta$.

## §7. Direct Products, Factor Congruences, and Directly Indecomposable Algebras

The constructions we have looked at so far, namely subalgebras and quotient algebras, do not give a means of creating algebras of larger cardinality than what we start with, or of combining several algebras into one.

**Definition 7.1.** Let $\mathbf{A}_1$ and $\mathbf{A}_2$ be two algebras of the same type $\mathscr{F}$. Define the *(direct) product* $\mathbf{A}_1 \times \mathbf{A}_2$ to be the algebra whose universe is the set $A_1 \times A_2$, and such that for $f \in \mathscr{F}_n$ and $a_i \in A_1$, $a_i' \in A_2$, $1 \le i \le n$,

$$f^{\mathbf{A}_1 \times \mathbf{A}_2}(\langle a_1,a_1'\rangle, \ldots, \langle a_n,a_n'\rangle) = \langle f^{\mathbf{A}_1}(a_1, \ldots, a_n), f^{\mathbf{A}_2}(a_1', \ldots, a_n')\rangle.$$

In general neither $\mathbf{A}_1$ nor $\mathbf{A}_2$ is embeddable in $\mathbf{A}_1 \times \mathbf{A}_2$, although in special cases like groups this is possible because there is always a trivial subalgebra. However both $\mathbf{A}_1$ and $\mathbf{A}_2$ are homomorphic images of $\mathbf{A}_1 \times \mathbf{A}_2$.

**Definition 7.2.** The mapping

$$\pi_i : A_1 \times A_2 \to A_i, \qquad i \in \{1,2\},$$

defined by

$$\pi_i(\langle a_1,a_2\rangle) = a_i,$$

is called the *projection map on the* $i$th *coordinate* of $A_1 \times A_2$.

**Theorem 7.3.** *For $i = 1$ or $2$ the mapping $\pi_i : A_1 \times A_2 \to A_i$ is a surjective homomorphism from $\mathbf{A} = \mathbf{A}_1 \times \mathbf{A}_2$ to $\mathbf{A}_i$. Furthermore, in* **Con** $\mathbf{A}_1 \times \mathbf{A}_2$ *we have*

$$\ker \pi_1 \cap \ker \pi_2 = \Delta,$$

$$\ker \pi_1 \textit{ and } \ker \pi_2 \textit{ permute},$$

*and*

$$\ker \pi_1 \vee \ker \pi_2 = \nabla.$$

Proof. Clearly $\pi_i$ is surjective. If $f \in \mathscr{F}_n$ and $a_i \in A_1$, $a_i' \in A_2$, $1 \le i \le n$, then

$$\begin{aligned}\pi_1(f^{\mathbf{A}}(\langle a_1,a_1'\rangle, \ldots, \langle a_n,a_n'\rangle)) &= \pi_1(\langle f^{\mathbf{A}_1}(a_1, \ldots, a_n), f^{\mathbf{A}_2}(a_1', \ldots, a_n')\rangle) \\ &= f^{\mathbf{A}_1}(a_1, \ldots, a_n) \\ &= f^{\mathbf{A}_1}(\pi_1(\langle a_1,a_1'\rangle), \ldots, \pi_1(\langle a_n,a_n'\rangle)),\end{aligned}$$

so $\pi_1$ is a homomorphism; and similarly $\pi_2$ is a homomorphism.

Now

$$\begin{aligned} & \langle\langle a_1,a_2\rangle, \langle b_1,b_2\rangle\rangle \in \ker \pi_i \\ \text{iff} \quad & \pi_i(\langle a_1,a_2\rangle) = \pi_i(\langle b_1,b_2\rangle) \\ \text{iff} \quad & a_i = b_i. \end{aligned}$$

Thus

$$\ker \pi_1 \cap \ker \pi_2 = \Delta.$$

Also if $\langle a_1,a_2\rangle$, $\langle b_1,b_2\rangle$ are any two elements of $A_1 \times A_2$ then

$$\langle a_1,a_2\rangle \ker \pi_1 \langle a_1,b_2\rangle \ker \pi_2 \langle b_1,b_2\rangle,$$

so

$$\nabla = \ker \pi_1 \circ \ker \pi_2.$$

But then $\ker \pi_1$ and $\ker \pi_2$ permute, and their join is $\nabla$. □

The last half of Theorem 7.3 motivates the following definition.

**Definition 7.4.** A congruence $\theta$ on $\mathbf{A}$ is a *factor congruence* if there is a congruence $\theta^*$ on $\mathbf{A}$ such that

$$\theta \cap \theta^* = \Delta,$$
$$\theta \vee \theta^* = \nabla,$$

and

$$\theta \text{ permutes with } \theta^*.$$

The pair $\theta,\theta^*$ is called a *pair of factor congruences* on $\mathbf{A}$.

**Theorem 7.5.** *If $\theta,\theta^*$ is a pair of factor congruences on $\mathbf{A}$, then*

$$\mathbf{A} \cong \mathbf{A}/\theta \times \mathbf{A}/\theta^*$$

*under the map*

$$\alpha(a) = \langle a/\theta, a/\theta^*\rangle.$$

Proof. If $a,b \in A$ and

$$\alpha(a) = \alpha(b)$$

then

$$a/\theta = b/\theta \quad \text{and} \quad a/\theta^* = b/\theta^*,$$

so

$$\langle a,b\rangle \in \theta \quad \text{and} \quad \langle a,b\rangle \in \theta^*;$$

hence

$$a = b.$$

This means that $\alpha$ is injective. Next, given $a,b \in A$ there is a $c \in A$ with

$$a\theta c\theta^* b;$$

hence

$$\begin{aligned}\alpha(c) &= \langle c/\theta, c/\theta^*\rangle \\ &= \langle a/\theta, b/\theta^*\rangle,\end{aligned}$$

so $\alpha$ is onto. Finally, for $f \in \mathscr{F}_n$ and $a_1, \ldots, a_n \in A$,

$$\begin{aligned}\alpha f^{\mathbf{A}}(a_1,\ldots,a_n) &= \langle f^{\mathbf{A}}(a_1,\ldots,a_n)/\theta, f^{\mathbf{A}}(a_1,\ldots,a_n)/\theta^*\rangle \\ &= \langle f^{\mathbf{A}/\theta}(a_1/\theta,\ldots,a_n/\theta), f^{\mathbf{A}/\theta^*}(a_1/\theta^*,\ldots,a_n/\theta^*)\rangle \\ &= f^{\mathbf{A}/\theta \times \mathbf{A}/\theta^*}(\langle a_1/\theta,a_1/\theta^*\rangle,\ldots,\langle a_n/\theta,a_n/\theta^*\rangle) \\ &= f^{\mathbf{A}/\theta \times \mathbf{A}/\theta^*}(\alpha a_1,\ldots,\alpha a_n);\end{aligned}$$

hence $\alpha$ is indeed an isomorphism. □

Thus we see that factor congruences come from and give rise to direct products.

**Definition 7.6.** An algebra $\mathbf{A}$ is *(directly) indecomposable* if $\mathbf{A}$ is not isomorphic to a direct product of two nontrivial algebras.

EXAMPLE. Any finite algebra $\mathbf{A}$ with $|A|$ a prime number must be directly indecomposable.

From Theorems 7.3 and 7.5 we have the following.

**Corollary 7.7.** $\mathbf{A}$ *is directly indecomposable iff the only factor congruences on* $\mathbf{A}$ *are* $\Delta$ *and* $\nabla$.

We can easily generalize the definition of $\mathbf{A}_1 \times \mathbf{A}_2$ as follows.

**Definition 7.8.** Let $(\mathbf{A}_i)_{i \in I}$ be an indexed family of algebras of type $\mathscr{F}$. The *(direct) product* $\mathbf{A} = \prod_{i \in I} \mathbf{A}_i$ is an algebra with universe $\prod_{i \in I} A_i$ and such that for $f \in \mathscr{F}_n$ and $a_1, \ldots, a_n \in \prod_{i \in I} A_i$,

$$f^{\mathbf{A}}(a_1, \ldots, a_n)(i) = f^{\mathbf{A}_i}(a_1(i), \ldots, a_n(i))$$

for $i \in I$, i.e., $f^{\mathbf{A}}$ is defined coordinate-wise. The empty product $\prod \varnothing$ is the trivial algebra with universe $\{\varnothing\}$. As before we have *projection maps*

$$\pi_j : \prod_{i \in I} A_i \to A_j$$

for $j \in I$ defined by

$$\pi_j(a) = a(j)$$

which give surjective homomorphisms

$$\pi_j : \prod_{i \in I} \mathbf{A}_i \to \mathbf{A}_j.$$

If $I = \{1, 2, \ldots, n\}$ we also write $\mathbf{A}_1 \times \cdots \times \mathbf{A}_n$. If $I$ is arbitrary but $\mathbf{A}_i = \mathbf{A}$ for all $i \in I$, then we usually write $\mathbf{A}^I$ for the direct product, and call it a *(direct) power* of $\mathbf{A}$. $\mathbf{A}^{\varnothing}$ is a trivial algebra.

A direct product $\prod_{i \in I} A_i$ of sets is often visualized as a rectangle with base $I$ and vertical cross sections $A_i$. An element $a$ of $\prod_{i \in I} A_i$ is then a curve as indicated in Figure 17. Two elementary facts about direct products are stated next.

**Theorem 7.9.** *If* $\mathbf{A}_1, \mathbf{A}_2$, *and* $\mathbf{A}_3$ *are of type* $\mathscr{F}$ *then*

(a) $\mathbf{A}_1 \times \mathbf{A}_2 \cong \mathbf{A}_2 \times \mathbf{A}_1$ *under* $\alpha(\langle a_1, a_2 \rangle) = \langle a_2, a_1 \rangle$.
(b) $\mathbf{A}_1 \times (\mathbf{A}_2 \times \mathbf{A}_3) \cong \mathbf{A}_1 \times \mathbf{A}_2 \times \mathbf{A}_3$ *under* $\alpha(\langle a_1, \langle a_2, a_3 \rangle \rangle) = \langle a_1, a_2, a_3 \rangle$.

PROOF. (Exercise.) □

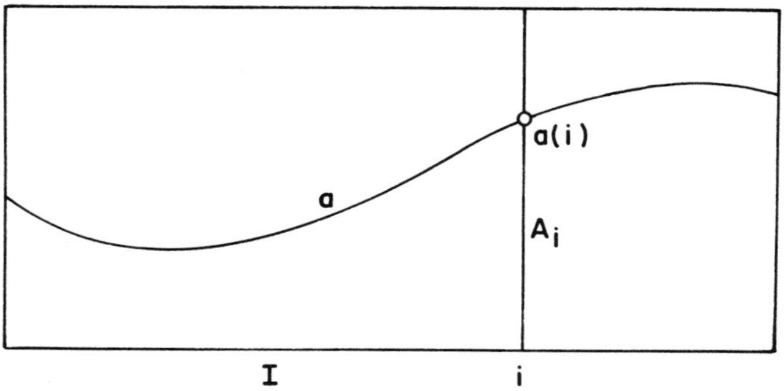


Figure 17

In Chapter IV we will see that there is up to isomorphism only one nontrivial directly indecomposable Boolean algebra, namely a two-element Boolean algebra, hence by cardinality considerations it follows that a countably infinite Boolean algebra cannot be isomorphic to a direct product of directly indecomposable algebras. On the other hand for finite algebras we have the following.

**Theorem 7.10.** *Every finite algebra is isomorphic to a direct product of directly indecomposable algebras.*

Proof. Let $\mathbf{A}$ be a finite algebra. If $\mathbf{A}$ is trivial then $\mathbf{A}$ is indecomposable. We proceed by induction on the cardinality of $A$. Suppose $\mathbf{A}$ is a nontrivial finite algebra such that for every $\mathbf{B}$ with $|B| < |A|$ we know that $\mathbf{B}$ is isomorphic to a product of indecomposable algebras. If $\mathbf{A}$ is indecomposable we are finished. If not, then $\mathbf{A} \cong \mathbf{A}_1 \times \mathbf{A}_2$ with $1 < |A_1|, |A_2|$. Then, $|A_1|, |A_2| < |A|$, so by the induction hypothesis,

$$\mathbf{A}_1 \cong \mathbf{B}_1 \times \cdots \times \mathbf{B}_m,$$
$$\mathbf{A}_2 \cong \mathbf{C}_1 \times \cdots \times \mathbf{C}_n,$$

where the $\mathbf{B}_i$ and $\mathbf{C}_j$ are indecomposable. Consequently,

$$\mathbf{A} \cong \mathbf{B}_1 \times \cdots \times \mathbf{B}_m \times \mathbf{C}_1 \times \cdots \times \mathbf{C}_n. \qquad \square$$

Using direct products there are two obvious ways (which occur a number of times in practice) of combining families of homomorphisms into single homomorphisms.

**Definition 7.11.** (i) If we are given maps $\alpha_i : A \to A_i$, $i \in I$, then the *natural map*

$$\alpha : A \to \prod_{i \in I} A_i$$

is defined by

$$(\alpha a)(i) = \alpha_i a.$$

(ii) If we are given maps $\alpha_i:A_i \to B_i$, $i \in I$, then the *natural map*

$$\alpha:\prod_{i\in I} A_i \to \prod_{i\in I} B_i$$

is defined by

$$(\alpha a)(i) = \alpha_i(a(i)).$$

**Theorem 7.12.** (a) *If $\alpha_i:\mathbf{A} \to \mathbf{A}_i$, $i \in I$, is an indexed family of homomorphisms, then the natural map $\alpha$ is a homomorphism from $\mathbf{A}$ to $\mathbf{A}^* = \prod_{i\in I} \mathbf{A}_i$.*

(b) *If $\alpha_i:\mathbf{A}_i \to \mathbf{B}_i$, $i \in I$, is an indexed family of homomorphisms, then the natural map $\alpha$ is a homomorphism from $\mathbf{A}^* = \prod_{i\in I} \mathbf{A}_i$ to $\mathbf{B}^* = \prod_{i\in I} \mathbf{B}_i$.*

Proof. Suppose $\alpha_i:\mathbf{A} \to \mathbf{A}_i$ is a homomorphism for $i \in I$. Then for $a_1, \dots, a_n \in A$ and $f \in \mathscr{F}_n$ we have, for $i \in I$,

$$\begin{aligned}(\alpha f^{\mathbf{A}}(a_1, \dots ,a_n))(i) &= \alpha_i f^{\mathbf{A}}(a_1, \dots ,a_n)\\ &= f^{\mathbf{A}_i}(\alpha_i a_1, \dots ,\alpha_i a_n)\\ &= f^{\mathbf{A}_i}((\alpha a_1)(i), \dots ,(\alpha a_n)(i))\\ &= f^{\mathbf{A}^*}(\alpha a_1, \dots ,\alpha a_n)(i);\end{aligned}$$

hence

$$\alpha f^{\mathbf{A}}(a_1, \dots, a_n) = f^{\mathbf{A}^*}(\alpha a_1, \dots, \alpha a_n),$$

so $\alpha$ is indeed a homomorphism in (a) above. Case (b) is a consequence of (a) using the homomorphisms $\alpha_i \circ \pi_i$. □

**Definition 7.13.** If $a_1,a_2 \in A$ and $\alpha:A \to B$ is a map we say $\alpha$ *separates* $a_1$ and $a_2$ if

$$\alpha a_1 \neq \alpha a_2.$$

The maps $\alpha_i:A \to A_i$, $i \in I$, *separate points* if for each $a_1,a_2 \in A$ with $a_1 \neq a_2$ there is an $\alpha_i$ such that

$$\alpha_i(a_1) \neq \alpha_i(a_2).$$

**Lemma 7.14.** *For an indexed family of maps $\alpha_i:A \to A_i$, $i \in I$, the following are equivalent*:

(a) *The maps $\alpha_i$ separate points.*
(b) *$\alpha$ is injective ($\alpha$ is the natural map of* 7.11(a)).
(c) $\bigcap_{i\in I} \ker \alpha_i = \Delta$.

Proof. (a) $\Rightarrow$ (b): Suppose $a_1,a_2 \in A$ and $a_1 \neq a_2$. Then for some $i$,

$$\alpha_i(a_1) \neq \alpha_i(a_2);$$

hence

$$(\alpha a_1)(i) \neq (\alpha a_2)(i)$$

so

$$\alpha a_1 \neq \alpha a_2.$$

(b) ⇒ (c): For $a_1,a_2 \in A$ with $a_1 \neq a_2$, we have

$$\alpha a_1 \neq \alpha a_2;$$

hence

$$(\alpha a_1)(i) \neq (\alpha a_2)(i)$$

for some $i$, so

$$\alpha_i a_1 \neq \alpha_i a_2$$

for some $i$, and this implies

$$\langle a_1,a_2\rangle \notin \ker \alpha_i,$$

so

$$\bigcap_{i\in I} \ker \alpha_i = \Delta.$$

(c) ⇒ (a): For $a_1,a_2 \in A$ with $a_1 \neq a_2$,

$$\langle a_1,a_2\rangle \notin \bigcap_{i\in I} \ker \alpha_i$$

so, for some $i$,

$$\langle a_1,a_2\rangle \notin \ker \alpha_i;$$

hence

$$\alpha_i a_1 \neq \alpha_i a_2. \qquad \square$$

**Theorem 7.15.** *If we are given an indexed family of homomorphisms* $\alpha_i$: $\mathbf{A} \to \mathbf{A}_i$, $i \in I$, *then the natural homomorphism* $\alpha: \mathbf{A} \to \prod_{i\in I} \mathbf{A}_i$ *is an embedding iff* $\bigcap_{i\in I} \ker \alpha_i = \Delta$ *iff the maps* $\alpha_i$ *separate points.*

PROOF. This is immediate from 7.14. □

EXERCISES §7

1. If $\theta,\theta^* \in \operatorname{Con} \mathbf{A}$ show that they form a pair of factor congruences on $\mathbf{A}$ iff $\theta \cap \theta^* = \Delta$ and $\theta \circ \theta^* = \nabla$.
2. Show that $(\mathbf{Con}\ \mathbf{A}_1) \times (\mathbf{Con}\ \mathbf{A}_2)$ can be embedded in $\mathbf{Con}\ \mathbf{A}_1 \times \mathbf{A}_2$.
3. Give examples of arbitrarily large directly indecomposable finite distributive lattices.
4. If $\mathbf{Con}\ \mathbf{A}$ is a distributive lattice show that the factor congruences on $\mathbf{A}$ form a complemented sublattice of $\mathbf{Con}\ \mathbf{A}$.
5. Find two algebras $\mathbf{A}_1,\mathbf{A}_2$ such that neither can be embedded in $\mathbf{A}_1 \times \mathbf{A}_2$.

## §8. Subdirect Products, Subdirectly Irreducible Algebras, and Simple Algebras

Although every finite algebra is isomorphic to a direct product of directly indecomposable algebras, the same does not hold for infinite algebras in general. For example, we see that a denumerable vector space over a finite field cannot be isomorphic to a direct product of one-dimensional spaces by merely considering cardinalities. The quest for general building blocks in the study of universal algebra led Birkhoff to consider subdirectly irreducible algebras.

**Definition 8.1.** An algebra $\mathbf{A}$ is a *subdirect product* of an indexed family $(\mathbf{A}_i)_{i \in I}$ of algebras if

(i) $\mathbf{A} \leq \prod_{i \in I} \mathbf{A}_i$

and

(ii) $\pi_i(\mathbf{A}) = \mathbf{A}_i$ for each $i \in I$.

An embedding $\alpha : \mathbf{A} \to \prod_{i \in I} \mathbf{A}_i$ is *subdirect* if $\alpha(\mathbf{A})$ is a subdirect product of the $\mathbf{A}_i$.

Note that if $I = \varnothing$ then $\mathbf{A}$ is a subdirect product of $\varnothing$ iff $\mathbf{A} = \prod \varnothing$, a trivial algebra.

**Lemma 8.2.** *If* $\theta_i \in \operatorname{Con} \mathbf{A}$ *for* $i \in I$ *and* $\bigcap_{i \in I} \theta_i = \Delta$, *then the natural homomorphism*

$$\nu : \mathbf{A} \to \prod_{i \in I} \mathbf{A}/\theta_i$$

*defined by*

$$\nu(a)(i) = a/\theta_i$$

*is a subdirect embedding.*

PROOF. Let $\nu_i$ be the natural homomorphism from $\mathbf{A}$ to $\mathbf{A}/\theta_i$, for $i \in I$. As $\ker \nu_i = \theta_i$, it follows from 7.15 that $\nu$ is an embedding. Since each $\nu_i$ is surjective, $\nu$ is a subdirect embedding. □

**Definition 8.3.** An algebra $\mathbf{A}$ is *subdirectly irreducible* if for every subdirect embedding

$$\alpha : \mathbf{A} \to \prod_{i \in I} \mathbf{A}_i$$

there is an $i \in I$ such that

$$\pi_i \circ \alpha : \mathbf{A} \to \mathbf{A}_i$$

is an isomorphism.

The following characterization of subdirectly irreducible algebras is most useful in practice.

**Theorem 8.4.** *An algebra* $\mathbf{A}$ *is subdirectly irreducible iff* $\mathbf{A}$ *is trivial or there is a minimum congruence in* $\operatorname{Con} \mathbf{A} - \{\Delta\}$. *In the latter case the minimum element is* $\bigcap (\operatorname{Con} \mathbf{A} - \{\Delta\})$, *a principal congruence, and the congruence lattice of* $\mathbf{A}$ *looks like the diagram in Figure 18.*

PROOF. ($\Rightarrow$) If $\mathbf{A}$ is not trivial and $\operatorname{Con} \mathbf{A} - \{\Delta\}$ has no minimum element then $\bigcap(\operatorname{Con} \mathbf{A} - \{\Delta\}) = \Delta$. Let $I = \operatorname{Con} \mathbf{A} - \{\Delta\}$. Then the natural map

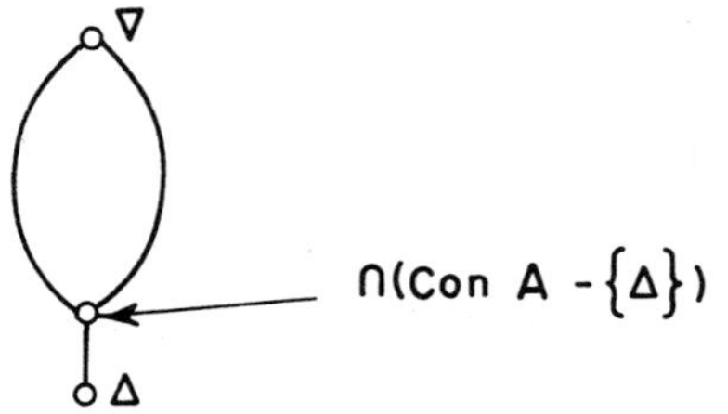


Figure 18

$\alpha:\mathbf{A} \to \prod_{\theta \in I} \mathbf{A}/\theta$ is a subdirect embedding by Lemma 8.2, and as the natural map $\mathbf{A} \to \mathbf{A}/\theta$ is not injective for $\theta \in I$, it follows that $\mathbf{A}$ is not subdirectly irreducible.

($\Leftarrow$) If $\mathbf{A}$ is trivial and $\alpha:\mathbf{A} \to \prod_{i \in I} \mathbf{A}_i$ is a subdirect embedding then each $\mathbf{A}_i$ is trivial; hence each $\pi_i \circ \alpha$ is an isomorphism. So suppose $\mathbf{A}$ is not trivial, and let $\theta = \bigcap(\mathrm{Con}\,\mathbf{A} - \{\Delta\}) \neq \Delta$. Choose $\langle a,b \rangle \in \theta$, $a \neq b$. If $\alpha:\mathbf{A} \to \prod_{i \in I} \mathbf{A}_i$ is a subdirect embedding then for some $i$, $(\alpha a)(i) \neq (\alpha b)(i)$; hence $(\pi_i \circ \alpha)(a) \neq (\pi_i \circ \alpha)(b)$. Thus $\langle a,b \rangle \notin \ker(\pi_i \circ \alpha)$ so $\theta \nsubseteq \ker(\pi_i \circ \alpha)$. But this implies $\ker(\pi_i \circ \alpha) = \Delta$, so $\pi_i \circ \alpha:\mathbf{A} \to \mathbf{A}_i$ is an isomorphism. Consequently $\mathbf{A}$ is subdirectly irreducible.

If Con $\mathbf{A} - \{\Delta\}$ has a minimum element $\theta$ then for $a \neq b$ and $\langle a,b \rangle \in \theta$ we have $\Theta(a,b) \subseteq \theta$, hence $\theta = \Theta(a,b)$. □

Using 8.4, we can readily list some subdirectly irreducible algebras.

Examples. (1) A finite Abelian group $\mathbf{G}$ is subdirectly irreducible iff it is cyclic and $|G| = \mathrm{p}^n$ for some prime $p$.

(2) The group $\mathbf{Z}_{p^\infty}$ is subdirectly irreducible.

(3) Every simple group is subdirectly irreducible.

(4) A vector space over a field $F$ is subdirectly irreducible iff it is trivial or one-dimensional.

(5) Any two-element algebra is subdirectly irreducible.

A directly indecomposable algebra need not be subdirectly irreducible. For example consider a three-element chain as a lattice. But the converse does indeed hold.

**Theorem 8.5.** *A subdirectly irreducible algebra is directly indecomposable.*

Proof. Clearly the only factor congruences on a subdirectly irreducible algebra are $\Delta$ and $\nabla$, so by 7.7 such an algebra is directly indecomposable. □

**Theorem 8.6** (Birkhoff). *Every algebra* $\mathbf{A}$ *is isomorphic to a subdirect product of subdirectly irreducible algebras (which are homomorphic images of* $\mathbf{A}$*).*

PROOF. As trivial algebras are subdirectly irreducible we only need to consider the case of nontrivial **A**. For $a,b \in A$ with $a \neq b$ we can find, using Zorn's lemma, a congruence $\theta_{a,b}$ on **A** which is maximal with respect to the property $\langle a,b \rangle \notin \theta_{a,b}$. Then clearly $\Theta(a,b) \vee \theta_{a,b}$ is the smallest congruence in $[\theta_{a,b},\nabla] - \{\theta_{a,b}\}$, so by 6.20 and 8.4 we see that $\mathbf{A}/\theta_{a,b}$ is subdirectly irreducible. As $\bigcap\{\theta_{a,b}: a \neq b\} = \Delta$ we can apply 8.2 to show that **A** is subdirectly embeddable in the product of the indexed family of subdirectly irreducible algebras $(\mathbf{A}/\theta_{a,b})_{a \neq b}$.

An immediate consequence of 8.6 is the following.

**Corollary 8.7.** *Every finite algebra is isomorphic to a subdirect product of a finite number of subdirectly irreducible finite algebras.*

Although subdirectly irreducible algebras do form the building blocks of algebra, the subdirect product construction is so flexible that one is often unable to draw significant conclusions for a class of algebras by studying its subdirectly irreducible members. In some special yet interesting cases we can derive an improved version of Birkhoff's theorem which permits a much deeper insight—this will be the theme of Chapter IV.

Next we look at a special kind of subdirectly irreducible algebra. This definition extends the usual notion of a simple group or a simple ring to arbitrary algebras.

**Definition 8.8.** An algebra **A** is *simple* if Con $\mathbf{A} = \{\Delta,\nabla\}$. A congruence $\theta$ on an algebra **A** is *maximal* if the interval $[\theta,\nabla]$ of Con **A** has exactly two elements.

Many algebraists prefer to require that a simple algebra be nontrivial. For our development, particularly for the material in Chapter IV, we find the discussion smoother by admitting trivial algebras.

Just as the quotient of a group by a normal subgroup is simple and nontrivial iff the normal subgroup is maximal, we have a similar result for arbitrary algebras.

**Theorem 8.9.** *Let $\theta \in$ Con* **A**. *Then $\mathbf{A}/\theta$ is a simple algebra iff $\theta$ is a maximal congruence on* **A** *or $\theta = \nabla$.*

PROOF. We know that

$$\text{Con } \mathbf{A}/\theta \cong [\theta,\nabla_A]$$

by 6.20, so the theorem is an immediate consequence of 8.8. □

REFERENCE

1. G. Birkhoff [1944]

EXERCISES §8

1. Represent the three-element chain as a subdirect product of subdirectly irreducible lattices.

2. Verify that the examples following 8.4 are indeed subdirectly irreducible algebras.

3. (Wenzel). Describe all subdirectly irreducible mono-unary algebras. [In particular show that they are countable.]

4. (Taylor). Let A be the set of functions from $\omega$ to $\{0,1\}$. Define the bi-unary algebra $\langle A,f,g\rangle$ by letting
$$f(a)(i) = a(i+1)$$
$$g(a)(i) = a(0).$$
Show that **A** is subdirectly irreducible.

5. (Taylor). Given an infinite cardinal $\lambda$ show that one can construct a unary algebra **A** of size $2^\lambda$ with $\lambda$ unary operations such that **A** is subdirectly irreducible.

6. Describe all subdirectly irreducible Abelian groups.

7. If **S** is a subdirectly irreducible semilattice show that $|S| \leq 2$. (Use §5 Exercise 9.) Hence show that every semilattice is isomorphic to a semilattice of the form $\langle A,\cap\rangle$, where $A$ is a family of sets closed under finite intersection.

8. A congruence $\theta$ on **A** is *completely meet irreducible* if whenever $\theta = \bigcap_{i\in I}\theta_i$, $\theta_i \in$ Con **A**, we have $\theta = \theta_i$, for some $i \in I$. Show that $\mathbf{A}/\theta$ is subdirectly irreducible iff $\theta$ is completely meet irreducible. (Hence, in particular, **A** is subdirectly irreducible iff $\Delta$ is completely meet irreducible.)

9. If $\mathbf{H} = \langle H,\vee,\wedge,\rightarrow,0,1\rangle$ is a Heyting algebra and $a \in H$ define $\theta_a = \{\langle b,c\rangle \in H^2: (b \rightarrow c) \wedge (c \rightarrow b) \geq a\}$. Show that $\theta_a$ is a congruence on **H**. From this show that **H** is subdirectly irreducible iff $|H| = 1$ or there is an element $e \neq 1$ such that $b \neq 1 \Rightarrow b \leq e$ for $b \in H$.

10. Show that the lattice of partitions $\langle \Pi(A),\subseteq\rangle$ of a set $A$ is a simple lattice.

11. If **A** is an algebra and $\theta_i \in$ Con **A**, $i \in I$, let $\theta = \bigcap_{i\in I}\theta_i$. Show that $\mathbf{A}/\theta$ can be subdirectly embedded in $\prod_{i\in I}\mathbf{A}/\theta_i$.

## §9. Class Operators and Varieties

A major theme in universal algebra is the study of classes of algebras of the same type closed under one or more constructions.

**Definition 9.1.** We introduce the following operators mapping classes of algebras to classes of algebras (all of the same type):

$\mathbf{A} \in I(K)$ iff **A** is isomorphic to some member of $K$
$\mathbf{A} \in S(K)$ iff **A** is a subalgebra of some member of $K$
$\mathbf{A} \in H(K)$ iff **A** is a homomorphic image of some member of $K$

$\mathbf{A} \in P(K)$ iff $\mathbf{A}$ is a direct product of a nonempty family of algebras in $K$

$\mathbf{A} \in P_s(K)$ iff $\mathbf{A}$ is a subdirect product of a nonempty family of algebras in $K$.

If $O_1$ and $O_2$ are two operators on classes of algebras we write $O_1 O_2$ for the composition of the two operators, and $\leq$ denotes the usual partial ordering, i.e., $O_1 \leq O_2$ if $O_1(K) \subseteq O_2(K)$ for all classes of algebras $K$. An operator $O$ is *idempotent* if $O^2 = O$. A class $K$ of algebras is *closed* under an operator $O$ if $O(K) \subseteq K$.

Our convention that $P$ and $P_s$ apply only to non-empty indexed families of algebras is the convention followed by model theorists. Thus for any operator $O$ above, $O(\varnothing) = \varnothing$. Many algebraists prefer to include $\prod \varnothing$, guaranteeing that $P(K)$ and $P_s(K)$ always contain a trivial algebra. However this leads to problems formulating certain preservation theorems–see V§2. For us $\prod \varnothing$ is really used only in IV§1, §5 and §7.

**Lemma 9.2.** *The following inequalities hold*: $SH \leq HS$, $PS \leq SP$, *and* $PH \leq HP$. *Also the operators* $H$, $S$, *and* $IP$ *are idempotent.*

Proof. Suppose $\mathbf{A} \in SH(K)$. Then for some $\mathbf{B} \in K$ and onto homomorphism $\alpha: \mathbf{B} \to \mathbf{C}$, we have $\mathbf{A} \leq \mathbf{C}$. Thus $\alpha^{-1}(\mathbf{A}) \leq \mathbf{B}$, and as $\alpha(\alpha^{-1}(\mathbf{A})) = \mathbf{A}$, we have $\mathbf{A} \in HS(K)$.

If $\mathbf{A} \in PS(K)$ then $\mathbf{A} = \prod_{i \in I} \mathbf{A}_i$ for suitable $\mathbf{A}_i \leq \mathbf{B}_i \in K$, $i \in I$. As $\prod_{i \in I} \mathbf{A}_i \leq \prod_{i \in I} \mathbf{B}_i$, we have $\mathbf{A} \in SP(K)$.

Next if $\mathbf{A} \in PH(K)$, then there are algebras $\mathbf{B}_i \in K$ and epimorphisms $\alpha_i: \mathbf{B}_i \to \mathbf{A}_i$ such that $\mathbf{A} = \prod_{i \in I} \mathbf{A}_i$. It is easy to check that the mapping $\alpha: \prod_{i \in I} \mathbf{B}_i \to \prod_{i \in I} \mathbf{A}_i$ defined by $\alpha(b)(i) = \alpha_i(b(i))$ is an epimorphism; hence $\mathbf{A} \in HP(K)$.

Finally it is a routine exercise to verify that $H^2 = H$, etc. □

**Definition 9.3.** A nonempty class $K$ of algebras of type $\mathscr{F}$ is called a *variety* if it is closed under subalgebras, homomorphic images, and direct products.

As the intersection of a class of varieties of type $\mathscr{F}$ is again a variety, and as all algebras of type $\mathscr{F}$ form a variety, we can conclude that for every class $K$ of algebras of the same type there is a smallest variety containing $K$.

**Definition 9.4.** If $K$ is a class of algebras of the same type let $V(K)$ denote the smallest variety containing $K$. We say that $V(K)$ is *the variety generated by* $K$. If $K$ has a single member $\mathbf{A}$ we write simply $V(\mathbf{A})$. A variety $V$ is *finitely generated* if $V = V(K)$ for some finite set $K$ of finite algebras.

**Theorem 9.5** (Tarski). $V = HSP$.

Proof. Since $HV = SV = IPV = V$ and $I \leq V$ it follows that $HSP \leq HSPV = V$. From Lemma 9.2 we see that $H(HSP) = HSP$, $S(HSP) \leq HSSP = HSP$, and $P(HSP) \leq HPSP \leq HSPP \leq HSIPIP = HSIP \leq HSHP \leq HHSP = HSP$; hence for any $K$, $HSP(K)$ is closed under $H$, $S$, and $P$. As $V(K)$ is the smallest class containing $K$ and closed under $H$, $S$, and $P$, we must have $V = HSP$. □

Another description of the operator $V$ will be given at the end of §11. The following version of Birkhoff's Theorem 8.6 is useful in studying varieties.

**Theorem 9.6.** *If K is a variety, then every member of K is isomorphic to a subdirect product of subdirectly irreducible members of K.*

**Corollary 9.7.** *A variety is determined by its subdirectly irreducible members.*

References

1. E. Nelson [1967]
2. D. Pigozzi [1972]
3. A. Tarski [1946]

Exercises §9

1. Show that $ISP(K)$ is the smallest class containing $K$ and closed under $I$, $S$, and $P$.
2. Show $HS \neq SH$, $HP \neq IPH$, $ISP \neq IPS$.
3. Show $ISPHS \neq ISHPS \neq IHSP$.
4. (Pigozzi). Show that there are 18 distinct class operators of the form $IO_1 \cdots O_n$ where $O_i \in \{H,S,P\}$ for $1 \leq i \leq n$.
5. Show that if $V$ has the CEP (see §5 Exercise 10) then for $K \subseteq V$, $HS(K) = SH(K)$.

## §10. Terms, Term Algebras, and Free Algebras

Given an algebra **A** there are usually many functions besides the fundamental operations which are compatible with the congruences on **A** and which "preserve" subalgebras of **A**. The most obvious functions of this type are those obtained by compositions of the fundamental operations. This leads us to the study of terms.

**Definition 10.1.** Let $X$ be a set of (distinct) objects called *variables*. Let $\mathscr{F}$ be a type of algebras. The set $T(X)$ of *terms of type* $\mathscr{F}$ *over* $X$ is the smallest set such that

(i) $X \cup \mathscr{F}_0 \subseteq T(X)$.

(ii) If $p_1, \ldots, p_n \in T(X)$ and $f \in \mathscr{F}_n$ then the "string" $f(p_1, \ldots, p_n) \in T(X)$.

For a binary function symbol $\cdot$ we usually prefer $p_1 \cdot p_2$ to $\cdot(p_1,p_2)$. For $p \in T(X)$ we often write $p$ as $p(x_1, \ldots ,x_n)$ to indicate that the variables occurring in $p$ are *among* $x_1, \ldots ,x_n$. A term $p$ is *n-ary* if the number of variables appearing explicitly in $p$ is $\leq n$.

EXAMPLES. (1) Let $\mathscr{F}$ consist of a single binary function symbol $\cdot$, and let $X = \{x,y,z\}$. Then

$$x,\ y,\ z,\ x \cdot y,\ y \cdot z,\ x \cdot (y \cdot z),\ \text{and}\ (x \cdot y) \cdot z$$

are some of the terms over $X$.

(2) Let $\mathscr{F}$ consist of two binary operation symbols $+$ and $\cdot$, and let $X$ be as before. Then

$$x,\ y,\ z,\ x \cdot (y + z),\ \text{and}\ (x \cdot y) + (x \cdot z)$$

are some of the terms over $X$.

(3) The classical polynomials over the field of real numbers **R** are really the terms as defined above of the type $\mathscr{F}$ consisting of $+$, $\cdot$, and $-$ together with a nullary function symbol $r$ for each $r \in R$.

In elementary algebra one often thinks of an $n$-ary polynomial over **R** as a function from $R^n$ to $R$ for some $n$. This can be applied to terms as well.

**Definition 10.2.** Given a term $p(x_1, \ldots ,x_n)$ of type $\mathscr{F}$ over some set $X$ and given an algebra **A** of type $\mathscr{F}$ we define a mapping $p^{\mathbf{A}}: A^n \to A$ as follows:

(1) if $p$ is a variable $x_i$, then

$$p^{\mathbf{A}}(a_1, \ldots ,a_n) = a_i$$

for $a_1, \ldots ,a_n \in A$, i.e., $p^{\mathbf{A}}$ is the $i$th projection map;

(2) if $p$ is of the form $f(p_1(x_1, \ldots ,x_n), \ldots , p_k(x_1, \ldots ,x_n))$, where $f \in \mathscr{F}_k$, then

$$p^{\mathbf{A}}(a_1, \ldots ,a_n) = f^{\mathbf{A}}(p_1^{\mathbf{A}}(a_1, \ldots ,a_n), \ldots , p_k^{\mathbf{A}}(a_1, \ldots ,a_n)).$$

In particular if $p = f \in \mathscr{F}$ then $p^{\mathbf{A}} = f^{\mathbf{A}}$. $p^{\mathbf{A}}$ is the *term function* on **A** corresponding to the term $p$. (Often we will drop the superscript **A**.)

The next theorem gives some useful properties of term functions, namely they behave like fundamental operations insofar as congruences and homomorphisms are concerned, and they can be used to describe the closure operator Sg of §3 in a most efficient manner.

**Theorem 10.3.** *For any type $\mathscr{F}$ and algebras* **A**, **B** *of type $\mathscr{F}$ we have the following.*

(a) *Let $p$ be an n-ary term of type $\mathscr{F}$, let $\theta \in$* Con **A**, *and suppose $\langle a_i,b_i \rangle \in \theta$ for $1 \leq i \leq n$. Then*

$$p^{\mathbf{A}}(a_1, \ldots ,a_n)\theta p^{\mathbf{A}}(b_1, \ldots ,b_n).$$

(b) *If p is an n-ary term of type $\mathscr{F}$ and $\alpha:\mathbf{A}\to\mathbf{B}$ is a homomorphism, then*

$$\alpha p^{\mathbf{A}}(a_1,\ldots,a_n) = p^{\mathbf{B}}(\alpha a_1,\ldots,\alpha a_n)$$

*for* $a_1,\ldots,a_n \in A$.

(c) *Let S be a subset of A. Then*

$$\mathrm{Sg}(S) = \{p^{\mathbf{A}}(a_1,\ldots,a_n)\colon p \text{ is an } n\text{-ary term of type } \mathscr{F}, n<\omega, \text{and } a_1,\ldots,a_n \in S\}.$$

Proof. Given a term $p$ define the length $l(p)$ of $p$ to be the number of occurrences of $n$-ary operation symbols in $p$ for $n \geq 1$. Note that $l(p) = 0$ iff $p \in X \cup \mathscr{F}_0$.

(a) We proceed by induction on $l(p)$. If $l(p) = 0$, then either $p = x_i$ for some $i$, whence

$$\langle p^{\mathbf{A}}(a_1,\ldots,a_n), p^{\mathbf{A}}(b_1,\ldots,b_n)\rangle = \langle a_i,b_i\rangle \in \theta$$

or $p = a$ for some $a \in \mathscr{F}_0$, whence

$$\langle p^{\mathbf{A}}(a_1,\ldots,a_n), p^{\mathbf{A}}(b_1,\ldots,b_n)\rangle = \langle a^{\mathbf{A}},a^{\mathbf{A}}\rangle \in \theta.$$

Now suppose $l(p) > 0$ and the assertion holds for every term $q$ with $l(q) < l(p)$. Then we know $p$ is of the form

$$f(p_1(x_1,\ldots,x_n),\ldots,p_k(x_1,\ldots,x_n)),$$

and as $l(p_i) < l(p)$ we must have, for $1 \leq i \leq k$,

$$\langle p_i^{\mathbf{A}}(a_1,\ldots,a_n), p_i^{\mathbf{A}}(b_1,\ldots,b_n)\rangle \in \theta;$$

hence

$$\langle f^{\mathbf{A}}(p_1^{\mathbf{A}}(a_1,\ldots,a_n),\ldots,p_k^{\mathbf{A}}(a_1,\ldots,a_n)), f^{\mathbf{A}}(p_1^{\mathbf{A}}(b_1,\ldots,b_n),\ldots,p_k^{\mathbf{A}}(b_1,\ldots,b_n))\rangle \in \theta,$$

and consequently

$$\langle p^{\mathbf{A}}(a_1,\ldots,a_n), p^{\mathbf{A}}(b_1,\ldots,b_n)\rangle \in \theta.$$

(b) The proof of this is an induction argument on $l(p)$.

(c) Referring to §3 one can give an induction proof, for $k \geq 1$, of

$$E^k(S) = \{p^{\mathbf{A}}(a_1,\ldots,a_n)\colon p \text{ is an } n\text{-ary term}, l(p) \leq k, n < \omega, a_1,\ldots,a_n \in S\},$$

and thus

$$\mathrm{Sg}(S) = \bigcup_{k<\omega} E^k(S) = \{p^{\mathbf{A}}(a_1,\ldots,a_n)\colon p \text{ is an } n\text{-ary term}, n < \omega, a_1,\ldots,a_n \in S\}. \quad \square$$

One can, in a natural way, transform the set $T(X)$ into an algebra.

**Definition 10.4.** Given $\mathscr{F}$ and $X$, if $T(X) \neq \varnothing$ then the *term algebra* of type $\mathscr{F}$ over $X$, written $\mathbf{T}(X)$, has as its universe the set $T(X)$, and the fundamental operations satisfy

$$f^{\mathbf{T}(X)}:\langle p_1,\ldots,p_n\rangle \mapsto f(p_1,\ldots,p_n)$$

for $f \in \mathscr{F}_n$ and $p_i \in T(X)$, $1 \leq i \leq n$. ($\mathbf{T}(\varnothing)$ exists iff $\mathscr{F}_0 \neq \varnothing$.)

Note that $\mathbf{T}(X)$ is indeed generated by $X$. Term algebras provide us with the simplest examples of algebras with the universal mapping property.

**Definition 10.5.** Let $K$ be a class of algebras of type $\mathscr{F}$ and let $\mathbf{U}(X)$ be an algebra of type $\mathscr{F}$ which is generated by $X$. If for every $\mathbf{A} \in K$ and for every map

$$\alpha: X \to A$$

there is a homomorphism

$$\beta: \mathbf{U}(X) \to \mathbf{A}$$

which extends $\alpha$ (i.e., $\beta(x) = \alpha(x)$ for $x \in X$), then we say $\mathbf{U}(X)$ has the *universal mapping property for $K$ over $X$*, $X$ is called a set of *free generators* of $\mathbf{U}(X)$, and $\mathbf{U}(X)$ is said to be *freely generated* by $X$.

**Lemma 10.6.** *Suppose $\mathbf{U}(X)$ has the universal mapping property for $K$ over $X$. Then if we are given $\mathbf{A} \in K$ and $\alpha: X \to A$, there is a unique extension $\beta$ of $\alpha$ such that $\beta$ is a homomorphism from $\mathbf{U}(X)$ to $\mathbf{A}$.*

PROOF. This follows simply from noting that a homomorphism is completely determined by how it maps a set of generators (see 6.2) from the domain. □

The next result says that for a given cardinal $m$ there is, up to isomorphism, at most one algebra *in* a class $K$ which has the universal mapping property for $K$ over a set of free generators of size $m$.

**Theorem 10.7.** *Suppose $\mathbf{U}_1(X_1)$ and $\mathbf{U}_2(X_2)$ are two algebras in a class $K$ with the universal mapping property for $K$ over the indicated sets. If $|X_1| = |X_2|$, then $\mathbf{U}_1(X_1) \cong \mathbf{U}_2(X_2)$.*

PROOF. First note that the identity map

$$\iota_j: X_j \to X_j, \qquad j = 1,2,$$

has as its unique extension to a homomorphism from $\mathbf{U}_j(X_j)$ to $\mathbf{U}_j(X_j)$ the identity map. Now let

$$\alpha: X_1 \to X_2$$

be a bijection. Then we have a homomorphism

$$\beta: \mathbf{U}_1(X_1) \to \mathbf{U}_2(X_2)$$

extending $\alpha$, and a homomorphism

$$\gamma: \mathbf{U}_2(X_2) \to \mathbf{U}_1(X_1)$$

extending $\alpha^{-1}$. As $\beta \circ \gamma$ is an endomorphism of $\mathbf{U}_2(X_2)$ extending $\iota_2$, it follows by 10.6 that $\beta \circ \gamma$ is the identity map on $\mathbf{U}_2(X_2)$. Likewise $\gamma \circ \beta$ is the identity map on $\mathbf{U}_1(X_1)$. Thus $\beta$ is a bijection, so $\mathbf{U}_1(X_1) \cong \mathbf{U}_2(X_2)$. □

**Theorem 10.8.** *For any type $\mathscr{F}$ and set $X$ of variables, where $X \neq \varnothing$ if $\mathscr{F}_0 = \varnothing$, the term algebra $\mathbf{T}(X)$ has the universal mapping property for the class of all algebras of type $\mathscr{F}$ over $X$.*

Proof. Let $\alpha: X \to A$ where $\mathbf{A}$ is of type $\mathscr{F}$. Define

$$\beta: T(X) \to A$$

recursively by

$$\beta x = \alpha x$$

for $x \in X$, and

$$\beta(f(p_1, \ldots, p_n)) = f^{\mathbf{A}}(\beta p_1, \ldots, \beta p_n)$$

for $p_1, \ldots, p_n \in T(X)$ and $f \in \mathscr{F}_n$. Then $\beta(p(x_1, \ldots, x_n)) = p^{\mathbf{A}}(\alpha x_1, \ldots, \alpha x_n)$, and $\beta$ is the desired homomorphism extending $\alpha$. □

Thus given any class $K$ of algebras the term algebras provide algebras which have the universal mapping property for $K$. To study properties of classes of algebras we often try to find special kinds of algebras in these classes which yield the desired information. Directly indecomposable and subdirectly irreducible algebras are two examples which we have already encountered. In order to find algebras with the universal mapping property for $K$ which give more insight into $K$ we will introduce $K$-free algebras. Unfortunately not every class $K$ contains algebras with the universal mapping property for $K$. Nonetheless we will be able to show that any class closed under $I$, $S$, and $P$ contains its $K$-free algebras. There is reasonable difficulty in providing transparent descriptions of $K$-free algebras for most $K$. However, most of the applications of $K$-free algebras come directly from the universal mapping property, the fact that they exist in varieties, and their relation to identities holding in $K$ (which we will examine in the next section). A proper understanding of free algebras is essential in our development of universal algebra—we use them to show varieties are the same as classes defined by equations (Birkhoff), to give useful characterizations (Mal'cev conditions) of important properties of varieties, and to show every nontrivial variety contains a nontrivial simple algebra (Magari).

**Definition 10.9.** Let $K$ be a family of algebras of type $\mathscr{F}$. Given a set $X$ of variables define the congruence $\theta_K(X)$ on $\mathbf{T}(X)$ by

$$\theta_K(X) = \bigcap \Phi_K(X),$$

where

$$\Phi_K(X) = \{\phi \in \operatorname{Con} \mathbf{T}(X) : \mathbf{T}(X)/\phi \in IS(K)\};$$

and then define $\mathbf{F}_K(\bar{X})$, *the K-free algebra over* $\bar{X}$, by

$$\mathbf{F}_K(\bar{X}) = \mathbf{T}(X)/\theta_K(X),$$

where

$$\bar{X} = X/\theta_K(X).$$

For $x \in X$ we write $\bar{x}$ for $x/\theta_K(X)$, and for $p = p(x_1, \ldots, x_n) \in T(X)$ we write $\bar{p}$ for $p^{\mathbf{F}_K(\bar{X})}(\bar{x}_1, \ldots, \bar{x}_n)$. If $X$ is finite, say $X = \{x_1, \ldots, x_n\}$, we often write $\mathbf{F}_K(\bar{x}_1, \ldots, \bar{x}_n)$ for $\mathbf{F}_K(\bar{X})$. $F_K(\bar{X})$ is the universe of $\mathbf{F}_K(\bar{X})$.

*Remarks.*

(1) $\mathbf{F}_K(\bar{X})$ exists iff $\mathbf{T}(X)$ exists iff $X \neq \varnothing$ or $\mathscr{F}_0 \neq \varnothing$. (2) If $\mathbf{F}_K(\bar{X})$ exists, then $\bar{X}$ is a set of generators of $\mathbf{F}_K(\bar{X})$ as $X$ generates $\mathbf{T}(X)$. (3) If $\mathscr{F}_0 \neq \varnothing$, then the algebra $\mathbf{F}_K(\bar{\varnothing})$ is often referred to as an *initial object* by category theorists and computer scientists. (4) If $K = \varnothing$ or $K$ consists solely of trivial algebras, then $\mathbf{F}_K(\bar{X})$ is a trivial algebra as $\theta_K(X) = \nabla$. (5) If $K$ has a nontrivial algebra $\mathbf{A}$ and $\mathbf{T}(X)$ exists, then $X \cap (x/\theta_K(X)) = \{x\}$ as distinct members $x,y$ of $X$ can be separated by some homomorphism $\alpha:\mathbf{T}(X) \to \mathbf{A}$. In this case $|\bar{X}| = |X|$. (6) If $|X| = |Y|$ and $\mathbf{T}(X)$ exists, then clearly $\mathbf{F}_K(\bar{X}) \cong \mathbf{F}_K(\bar{Y})$ under an isomorphism which maps $\bar{X}$ to $\bar{Y}$ as $\mathbf{T}(X) \cong \mathbf{T}(Y)$ under an isomorphism mapping $X$ to $Y$. Thus $\mathbf{F}_K(\bar{X})$ is determined, up to isomorphism, by $K$ and $|X|$.

**Theorem 10.10** (Birkhoff). *Suppose* $\mathbf{T}(X)$ *exists. Then* $\mathbf{F}_K(\bar{X})$ *has the universal mapping property for* $K$ *over* $\bar{X}$.

PROOF. Given $\mathbf{A} \in K$ let $\alpha$ be a map from $\bar{X}$ to $A$. Let $v:\mathbf{T}(X) \to \mathbf{F}_K(\bar{X})$ be the natural homomorphism. Then $\alpha \circ v$ maps $X$ into $A$, so by the universal mapping property of $\mathbf{T}(X)$ there is a homomorphism $\mu:\mathbf{T}(X) \to \mathbf{A}$ extending $\alpha \circ v\restriction_X$. From the definition of $\theta_K(X)$ it is clear that $\theta_K(X) \subseteq \ker \mu$ (as $\ker \mu \in \Phi_K(X)$). Thus there is a homomorphism $\beta:\mathbf{F}_K(\bar{X}) \to \mathbf{A}$ such that $\mu = \beta \circ v$ (see §6 Exercise 6) as $\ker v = \theta_K(X)$. But then, for $x \in X$,

$$\begin{aligned}\beta(\bar{x}) &= \beta \circ v(x) \\ &= \mu(x) \\ &= \alpha \circ v(x) \\ &= \alpha(\bar{x}),\end{aligned}$$

so $\beta$ extends $\alpha$. Thus $\mathbf{F}_K(\bar{X})$ has the universal mapping property for $K$ over $\bar{X}$. □

If $\mathbf{F}_K(\bar{X}) \in K$ then it is, up to isomorphism, the unique algebra in $K$ with the universal mapping property freely generated by a set of generators of size $|\bar{X}|$. Actually every algebra in $K$ with the universal mapping property for $K$ is isomorphic to a $K$-free algebra (see Exercise 6).

EXAMPLES. (1) It is clear that $\mathbf{T}(X)$ is isomorphic to the free algebra with respect to the class $K$ of all algebras of type $\mathscr{F}$ over $X$ since $\theta_K(X) = \Delta$. The corresponding free algebra is sometimes called the absolutely free algebra $\mathbf{F}(\bar{X})$ of type $\mathscr{F}$.

(2) Given $X$ let $X^*$ be the set of finite strings of elements of $X$, including the empty string. We can construct a monoid $\langle X^*, \cdot, 1\rangle$ by defining $\cdot$ to be concatenation, and 1 is the empty string. By checking the universal mapping property one sees that $\langle X^*, \cdot, 1\rangle$ is, up to isomorphism, the free monoid freely generated by $\bar{X}$.

**Corollary 10.11.** *If $K$ is a class of algebras of type $\mathscr{F}$ and $\mathbf{A} \in K$, then for sufficiently large $X$, $\mathbf{A} \in H(\mathbf{F}_K(\bar{X}))$.*

PROOF. Choose $|X| \geq |A|$ and let

$$\alpha : \bar{X} \to A$$

be a surjection. Then let

$$\beta : \mathbf{F}_K(\bar{X}) \to \mathbf{A}$$

be a homomorphism extending $\alpha$. □

In general $\mathbf{F}_K(\bar{X})$ is not isomorphic to a member of $K$ (for example, let $K = \{\mathbf{L}\}$ where $\mathbf{L}$ is a two-element lattice; then $\mathbf{F}_K(\bar{x}, \bar{y}) \notin I(K)$). However $\mathbf{F}_K(\bar{X})$ can be embedded in a product of members of $K$.

**Theorem 10.12** (Birkhoff). *Suppose $\mathbf{T}(X)$ exists. Then for $K \neq \varnothing$, $\mathbf{F}_K(\bar{X}) \in ISP(K)$. Thus if $K$ is closed under $I$, $S$, and $P$, in particular if $K$ is a variety, then $\mathbf{F}_K(\bar{X}) \in K$.*

PROOF. As

$$\theta_K(X) = \bigcap \Phi_K(X)$$

it follows (see §8 Exercise 11) that

$$\mathbf{F}_K(\bar{X}) = \mathbf{T}(X)/\theta_K(X) \in IP_S(\{\mathbf{T}(X)/\theta : \theta \in \Phi_K(X)\}),$$

so

$$\mathbf{F}_K(\bar{X}) \in IP_S IS(K),$$

and thus by 9.2 and the fact that $P_S \leq SP$,

$$\mathbf{F}_K(\bar{X}) \in ISP(K).$$ □

From an earlier theorem of Birkhoff we know that if a variety has a nontrivial algebra in it then it must have a nontrivial subdirectly irreducible algebra in it. The next result shows that such a variety must also contain a nontrivial simple algebra.

**Theorem 10.13** (Magari). *If we are given a variety $V$ with a nontrivial member, then $V$ contains a nontrivial simple algebra.*

**Proof.** Let $X = \{x,y\}$, and let

$$S = \{p(\bar{x}) : p \in T(\{x\})\}$$

a subset of $F_V(\bar{X})$. First suppose that $\Theta(S) \neq \nabla$ in Con $\mathbf{F}_V(\bar{X})$. Then by Zorn's lemma there is a maximal element in $[\Theta(S),\nabla] - \{\nabla\}$. (The key observation for this step is that for $\theta \in [\Theta(S),\nabla]$,

$$\theta = \nabla \quad \text{iff} \quad \langle \bar{x},\bar{y} \rangle \in \theta.$$

To see this note that if $\langle \bar{x},\bar{y} \rangle \in \theta$ and $\Theta(S) \subseteq \theta$, then for any term $p(x,y)$, with $\mathbf{F} = \mathbf{F}_V(\bar{X})$ we have

$$p^{\mathbf{F}}(\bar{x},\bar{y})\theta p^{\mathbf{F}}(\bar{x},\bar{x})\Theta(S)\bar{x};$$

hence $\theta = \nabla$.) Let $\theta_0$ be a maximal element in $[\Theta(S),\nabla] - \{\nabla\}$. Then $\mathbf{F}_V(\bar{X})/\theta_0$ is a simple algebra by 8.9, and it is in $V$.

If, however, $\Theta(S) = \nabla$, then since $\Theta$ is an algebraic closure operator by 5.5, it follows that for some finite subset $S_0$ of $S$ we must have $\langle \bar{x},\bar{y} \rangle \in \Theta(S_0)$. Let $\mathbf{S}$ be the subalgebra of $\mathbf{F}_V(\bar{X})$ with universe $S$ (note that $S = \mathrm{Sg}(\{\bar{x}\})$ by 10.3(c)). As $V$ is nontrivial we must have $\bar{x} \neq \bar{y}$ in $\mathbf{F}_V(\bar{X})$, and as $\langle \bar{x},\bar{y} \rangle \in \Theta(S)$ it follows that $S$ is nontrivial. Now we claim that $\nabla_S = \Theta(S_0)$, where $\Theta$ in this case is understood to be the appropriate closure operator on $S$. To see this let $p(\bar{x}) \in S$ and let

$$\alpha : \mathbf{F}_V(\bar{X}) \to \mathbf{S}$$

be the homomorphism defined by

$$\alpha(\bar{x}) = \bar{x}$$
$$\alpha(\bar{y}) = p(\bar{x}).$$

As

$$\langle \bar{x},\bar{y} \rangle \in \Theta(S_0) \quad \text{in } \mathbf{F}_V(\bar{X}),$$

it follows from 6.6 (see §6 Exercise 5) that

$$\langle \bar{x},p(\bar{x}) \rangle \in \Theta(S_0) \quad \text{in } \mathbf{S}$$

as

$$\alpha(S_0) = S_0.$$

This establishes our claim; hence using Zorn's lemma we can find a maximal congruence $\theta$ on $\mathbf{S}$ as $\nabla_S$ is finitely generated. Hence $\mathbf{S}/\theta$ is a simple algebra in $V$. □

Let us turn to another application of free algebras.

**Definition 10.14.** An algebra $\mathbf{A}$ is *locally finite* if every finitely generated subalgebra (see §3.4) is finite. A class $K$ of algebras is *locally finite* if every member of $K$ is locally finite.

**Theorem 10.15.** *A variety $V$ is locally finite iff*

$$|X| < \omega \Rightarrow |F_V(\bar{X})| < \omega.$$

Proof. The direction ($\Rightarrow$) is clear as $\bar{X}$ generates $\mathbf{F}_V(\bar{X})$. For ($\Leftarrow$) let $\mathbf{A}$ be a finitely generated member of $V$, and let $B \subseteq A$ be a finite set of generators. Choose $X$ such that we have a bijection

$$\alpha: \bar{X} \to B.$$

Extend this to a homomorphism

$$\beta: \mathbf{F}_V(\bar{X}) \to \mathbf{A}.$$

As $\beta(\mathbf{F}_V(\bar{X}))$ is a subalgebra of $\mathbf{A}$ containing $B$, it must equal $\mathbf{A}$. Thus $\beta$ is surjective, and as $\mathbf{F}_V(\bar{X})$ is finite so is $\mathbf{A}$. □

**Theorem 10.16.** *Let $K$ be a finite set of finite algebras. Then $V(K)$ is a locally finite variety.*

Proof. First verify that $P(K)$ is locally finite. To do this define an equivalence relation $\sim$ on $T(\{x_1, \ldots, x_n\})$ by $p \sim q$ if the term functions corresponding to $p$ and $q$ are the same for each member of $K$. Use the finiteness conditions to show that $\sim$ has finitely many equivalence classes. This, combined with 10.3(c), suffices. Then it easily follows that $V$ is locally finite since every finitely generated member of $HSP(K)$ is a homomorphic image of a finitely generated member of $SP(K)$. □

References

1. G. Birkhoff [1935]
2. R. Magari [1969]

Exercises §10

1. Let $\mathbf{L}$ be the four-element lattice $\langle\{0,a,b,1\}, \vee, \wedge\rangle$ where 0 is the least element, 1 is the largest element and $a \wedge b = 0$, $a \vee b = 1$ (the Hasse diagram is Figure 1(c)). Show that $\mathbf{L}$ has the universal mapping property for the class of lattices over the set $\{a,b\}$.
2. Let $\mathbf{A} = \langle \omega, f\rangle$ be the mono-unary algebra with $f(n) = n + 1$. Show $\mathbf{A}$ has the universal mapping property for the class of mono-unary algebras over the set $\{0\}$.
3. Let $p$ be a prime number, and let $Z_p$ be the set of integers modulo $p$. Let $\mathbf{Z}_p$ be the mono-unary algebra $\langle Z_p, f\rangle$ defined by $f(\bar{n}) = \overline{n+1}$. Show $\mathbf{Z}_p$ has the universal mapping property for $K$ over $\{\bar{1}\}$, where $K$ is the class of mono-unary algebras $\langle A, f\rangle$ satisfying $f^p(x) \approx x$.
4. Show that the group $\mathbf{Z} = \langle Z, +, -, 0\rangle$ of integers has the universal mapping property for the class of groups over $\{1\}$.
5. If $V$ is a variety and $|X| \le |Y|$ show $\mathbf{F}_V(\bar{X})$ can be embedded in $\mathbf{F}_V(\bar{Y})$ in a natural way.
6. If $\mathbf{U}(X) \in K$ and $\mathbf{U}(X)$ has the universal mapping property for $K$ over $X$ show that $\mathbf{U}(X) \cong \mathbf{F}_K(\bar{X})$ under a mapping $\alpha$ such that $\alpha(x) = \bar{x}$.

7. Show that for any algebra $\mathbf{A}$ and $a,b \in A$, $\Theta(\langle a,b\rangle) = t^*(s(\{\langle p(a,\vec{c}), p(b,\vec{c})\rangle : p(x,y_1,\ldots,y_n)$ is a term, $c_1,\ldots,c_n \in A\})) \cup \Delta_A$, where $t^*(\ )$ is the transitive closure operator, i.e., for $Y \subseteq A \times A$, $t^*(Y)$ is the smallest subset of $A \times A$ containing $Y$ and closed under $t$. (See the proof of 5.5.)

# §11. Identities, Free Algebras, and Birkhoff's Theorem

One of the most celebrated theorems of Birkhoff says that the classes of algebras defined by identities are precisely those which are closed under $H$, $S$, and $P$. In this section we study identities, their relation to free algebras, and then give several applications, including Birkhoff's theorem. We have already seen particular examples of identities, among which are the commutative law, the associative law, and the distributive laws. Now let us formalize the general notion of an identity, and the notion of an identity holding in an algebra $\mathbf{A}$, or in a class of algebras $K$.

**Definition 11.1.** An *identity* of type $\mathscr{F}$ over $X$ is an expression of the form

$$p \approx q$$

where $p,q \in T(X)$. Let $\mathrm{Id}(X)$ be the set of identities of type $\mathscr{F}$ over $X$. An algebra $\mathbf{A}$ of type $\mathscr{F}$ *satisfies* an identity

$$p(x_1,\ldots,x_n) \approx q(x_1,\ldots,x_n)$$

(or the identity *is true in* $\mathbf{A}$, or *holds in* $\mathbf{A}$), abbreviated by

$$\mathbf{A} \models p(x_1,\ldots,x_n) \approx q(x_1,\ldots,x_n),$$

or more briefly

$$\mathbf{A} \models p \approx q,$$

if for every choice of $a_1,\ldots,a_n \in A$ we have

$$p^{\mathbf{A}}(a_1,\ldots,a_n) = q^{\mathbf{A}}(a_1,\ldots,a_n).$$

A class $K$ of algebras satisfies $p \approx q$, written

$$K \models p \approx q,$$

if each member of $K$ satisfies $p \approx q$. If $\Sigma$ is a set of identities, we say $K$ satisfies $\Sigma$, written

$$K \models \Sigma,$$

if $K \models p \approx q$ for each $p \approx q \in \Sigma$. Given $K$ and $X$ let

$$\mathrm{Id}_K(X) = \{p \approx q \in \mathrm{Id}(X) : K \models p \approx q\}.$$

We use the symbol $\not\models$ for "does not satisfy."

We can reformulate the above definition of satisfaction using the notion of homomorphism.

**Lemma 11.2.** *If $K$ is a class of algebras of type $\mathscr{F}$ and $p \approx q$ is an identity of type $\mathscr{F}$ over $X$, then*

$$K \models p \approx q$$

*iff for every $\mathbf{A} \in K$ and for every homomorphism $\alpha: \mathbf{T}(X) \to \mathbf{A}$ we have*

$$\alpha p = \alpha q.$$

Proof. ($\Rightarrow$) Let $p = p(x_1, \ldots, x_n)$, $q = q(x_1, \ldots, x_n)$. Suppose $K \models p \approx q$, $\mathbf{A} \in K$, and $\alpha: \mathbf{T}(X) \to \mathbf{A}$ is a homomorphism. Then

$$\begin{aligned} &p^{\mathbf{A}}(\alpha x_1, \ldots, \alpha x_n) = q^{\mathbf{A}}(\alpha x_1, \ldots, \alpha x_n) \\ &\Rightarrow \alpha p^{\mathbf{T}(X)}(x_1, \ldots, x_n) = \alpha q^{\mathbf{T}(X)}(x_1, \ldots, x_n) \\ &\Rightarrow \alpha p = \alpha q. \end{aligned}$$

($\Leftarrow$) For the converse choose $\mathbf{A} \in K$ and $a_1, \ldots, a_n \in A$. By the universal mapping property of $\mathbf{T}(X)$ there is a homomorphism $\alpha: \mathbf{T}(X) \to \mathbf{A}$ such that

$$\alpha x_i = a_i, \qquad 1 \le i \le n.$$

But then

$$\begin{aligned} p^{\mathbf{A}}(a_1, \ldots, a_n) &= p^{\mathbf{A}}(\alpha x_1, \ldots, \alpha x_n) \\ &= \alpha p \\ &= \alpha q \\ &= q^{\mathbf{A}}(\alpha x_1, \ldots, \alpha x_n) \\ &= q^{\mathbf{A}}(a_1, \ldots, a_n), \end{aligned}$$

so $K \models p \approx q$. □

Next we see that the basic class operators preserve identities.

**Lemma 11.3.** *For any class $K$ of type $\mathscr{F}$ all of the classes $K$, $I(K)$, $S(K)$, $H(K)$, $P(K)$ and $V(K)$ satisfy the same identities over any set of variables $X$.*

Proof. Clearly $K$ and $I(K)$ satisfy the same identities. As

$$I \le IS, \qquad I \le H, \quad \text{and} \quad I \le IP,$$

we must have

$$\mathrm{Id}_K(X) \supseteq \mathrm{Id}_{S(K)}(X), \qquad \mathrm{Id}_{H(K)}(X), \quad \text{and} \quad \mathrm{Id}_{P(K)}(X).$$

For the remainder of the proof suppose

$$K \models p(x_1, \ldots, x_n) \approx q(x_1, \ldots, x_n).$$

Then if $\mathbf{B} \le \mathbf{A} \in K$ and $b_1, \ldots, b_n \in B$, then as $b_1, \ldots, b_n \in A$ we have

$$p^{\mathbf{A}}(b_1, \ldots, b_n) = q^{\mathbf{A}}(b_1, \ldots, b_n);$$

hence

$$p^{\mathbf{B}}(b_1, \ldots ,b_n) = q^{\mathbf{B}}(b_1, \ldots ,b_n),$$

so

$$\mathbf{B} \models p \approx q.$$

Thus

$$\mathrm{Id}_K(X) = \mathrm{Id}_{S(K)}(X).$$

Next suppose $\alpha:\mathbf{A} \to \mathbf{B}$ is a surjective homomorphism with $\mathbf{A} \in K$. If $b_1, \ldots ,b_n \in B$, choose $a_1, \ldots ,a_n \in A$ such that

$$\alpha(a_1) = b_1, \qquad \ldots, \qquad \alpha(a_n) = b_n.$$

Then

$$p^{\mathbf{A}}(a_1, \ldots ,a_n) = q^{\mathbf{A}}(a_1, \ldots ,a_n)$$

implies

$$\alpha p^{\mathbf{A}}(a_1, \ldots ,a_n) = \alpha q^{\mathbf{A}}(a_1, \ldots ,a_n);$$

hence

$$p^{\mathbf{B}}(b_1, \ldots ,b_n) = q^{\mathbf{B}}(b_1, \ldots ,b_n).$$

Thus

$$\mathbf{B} \models p \approx q,$$

so

$$\mathrm{Id}_K(X) = \mathrm{Id}_{H(K)}(X).$$

Lastly, suppose $\mathbf{A}_i \in K$ for $i \in I$. Then for $a_1, \ldots ,a_n \in A = \prod_{i \in I} A_i$ we have

$$p^{\mathbf{A}_i}(a_1(i), \ldots ,a_n(i)) = q^{\mathbf{A}_i}(a_1(i), \ldots ,a_n(i));$$

hence

$$p^{\mathbf{A}}(a_1, \ldots ,a_n)(i) = q^{\mathbf{A}}(a_1, \ldots ,a_n)(i)$$

for $i \in I$, so

$$p^{\mathbf{A}}(a_1, \ldots ,a_n) = q^{\mathbf{A}}(a_1, \ldots ,a_n).$$

Thus

$$\mathrm{Id}_K(X) = \mathrm{Id}_{P(K)}(X).$$

As $V = HSP$ by 9.5, the proof is complete. □

Now we will formulate the crucial connection between $K$-free algebras and identities.

**Theorem 11.4.** *Given a class $K$ of algebras of type $\mathscr{F}$ and terms $p,q \in T(X)$ of type $\mathscr{F}$ we have*

$$\begin{aligned} & K \models p \approx q \\ \Leftrightarrow\ & \mathbf{F}_K(\bar{X}) \models p \approx q \\ \Leftrightarrow\ & \bar{p} = \bar{q} \quad \text{in} \quad \mathbf{F}_K(\bar{X}) \\ \Leftrightarrow\ & \langle p,q \rangle \in \theta_K(X). \end{aligned}$$

Proof. Let $\mathbf{F} = \mathbf{F}_K(\bar{X})$, $p = p(x_1, \ldots ,x_n)$, $q = q(x_1, \ldots ,x_n)$, and let

$$v:\mathbf{T}(X) \to \mathbf{F}$$

be the natural homomorphism. Certainly $K \models p \approx q$ implies $\mathbf{F} \models p \approx q$ as $\mathbf{F} \in ISP(K)$. Suppose next that $\mathbf{F} \models p \approx q$. Then

$$p^{\mathbf{F}}(\bar{x}_1, \ldots, \bar{x}_n) = q^{\mathbf{F}}(\bar{x}_1, \ldots, \bar{x}_n),$$

hence $\bar{p} = \bar{q}$. Now suppose $\bar{p} = \bar{q}$ in $\mathbf{F}$. Then

$$v(p) = \bar{p} = \bar{q} = v(q),$$

so

$$\langle p,q \rangle \in \ker v = \theta_K(X).$$

Finally suppose $\langle p,q \rangle \in \theta_K(X)$. Given $\mathbf{A} \in K$ and $a_1, \ldots, a_n \in A$ choose $\alpha: \mathbf{T}(X) \to \mathbf{A}$ such that $\alpha x_i = a_i$, $1 \leq i \leq n$. As $\ker \alpha \in \Phi_K(X)$ we have

$$\ker \alpha \supseteq \ker v = \theta_K(X),$$

so it follows that there is a homomorphism $\beta: \mathbf{F} \to \mathbf{A}$ such that $\alpha = \beta \circ v$ (see §6 Exercise 6). Then

$$\alpha(p) = \beta \circ v(p) = \beta \circ v(q) = \alpha(q).$$

Consequently

$$K \models p \approx q$$

by 11.2. □

**Corollary 11.5.** *Let $K$ be a class of algebras of type $\mathscr{F}$, and suppose $p,q \in T(X)$. Then for any set of variables $Y$ with $|Y| \geq |X|$ we have*

$$K \models p \approx q \quad \text{iff} \quad \mathbf{F}_K(\bar{Y}) \models p \approx q.$$

Proof. The direction ($\Rightarrow$) is obvious as $\mathbf{F}_K(\bar{Y}) \in ISP(K)$. For the converse choose $X_0 \supseteq X$ such that $|X_0| = |Y|$. Then

$$\mathbf{F}_K(\bar{X}_0) \cong \mathbf{F}_K(\bar{Y}),$$

and as

$$K \models p \approx q \quad \text{iff} \quad \mathbf{F}_K(\bar{X}_0) \models p \approx q$$

by 11.4, it follows that

$$K \models p \approx q \quad \text{iff} \quad \mathbf{F}_K(\bar{Y}) \models p \approx q.$$ □

**Corollary 11.6.** *Suppose $K$ is a class of algebras of type $\mathscr{F}$ and $X$ is a set of variables. Then for any infinite set of variables $Y$,*

$$\mathrm{Id}_K(X) = \mathrm{Id}_{\mathbf{F}_K(\bar{Y})}(X).$$

Proof. For $p \approx q \in \mathrm{Id}_K(X)$, say $p = p(x_1, \ldots, x_n)$, $q = q(x_1, \ldots, x_n)$, we have $p,q \in T(\{x_1, \ldots, x_n\})$. As $|\{x_1, \ldots, x_n\}| < |Y|$, by 11.5

$$K \models p \approx q \quad \text{iff} \quad \mathbf{F}_K(\bar{Y}) \models p \approx q,$$

so the corollary is proved. □

As we have seen in §1, many of the most popular classes of algebras are defined by identities.

**Definition 11.7.** Let $\Sigma$ be a set of identities of type $\mathscr{F}$, and define $M(\Sigma)$ to be the class of algebras $\mathbf{A}$ satisfying $\Sigma$. A class $K$ of algebras is an *equational class* if there is a set of identities $\Sigma$ such that $K = M(\Sigma)$. In this case we say that $K$ is *defined*, or *axiomatized*, by $\Sigma$.

**Lemma 11.8.** *If $V$ is a variety and $X$ is an infinite set of variables, then $V = M(\mathrm{Id}_V(X))$.*

Proof. Let

$$V' = M(\mathrm{Id}_V(X)).$$

Clearly $V'$ is a variety by 11.3, $V' \supseteq V$, and

$$\mathrm{Id}_{V'}(X) = \mathrm{Id}_V(X).$$

So by 11.4,

$$\mathbf{F}_{V'}(\bar{X}) = \mathbf{F}_V(\bar{X}).$$

Now given any infinite set of variables $Y$, we have by 11.6

$$\mathrm{Id}_{V'}(Y) = \mathrm{Id}_{\mathbf{F}_{V'}(\bar{X})}(Y) = \mathrm{Id}_{\mathbf{F}_V(\bar{X})}(Y) = \mathrm{Id}_V(Y).$$

Thus again by 11.4,

$$\theta_{V'}(Y) = \theta_V(Y);$$

hence

$$\mathbf{F}_{V'}(\bar{Y}) = \mathbf{F}_V(\bar{Y}).$$

Now for $\mathbf{A} \in V'$ we have (by 10.11), for suitable infinite $Y$,

$$\mathbf{A} \in H(\mathbf{F}_{V'}(\bar{Y}));$$

hence

$$\mathbf{A} \in H(\mathbf{F}_V(\bar{Y})),$$

so $\mathbf{A} \in V$; hence $V' \subseteq V$, and thus $V' = V$. □

Now we have all the background needed to prove the famous theorem of Birkhoff.

**Theorem 11.9** (Birkhoff). *$K$ is an equational class iff $K$ is a variety.*

Proof. ($\Rightarrow$) Suppose

$$K = M(\Sigma).$$

Then

$$V(K) \models \Sigma$$

by 11.3; hence

$$V(K) \subseteq M(\Sigma),$$

so

$$V(K) = K,$$

i.e., $K$ is a variety.

($\Leftarrow$) This follows from 11.8. □

We can also use 11.4 to obtain a significant strengthening of 10.12.

**Corollary 11.10.** *Let $K$ be a class of algebras of type $\mathscr{F}$. If $\mathbf{T}(X)$ exists and $K'$ is any class of algebras such that $K \subseteq K' \subseteq V(K)$, then*

$$\mathbf{F}_{K'}(\bar{X}) = \mathbf{F}_K(\bar{X}).$$

*In particular it follows that*

$$\mathbf{F}_{K'}(\bar{X}) \in ISP(K).$$

PROOF. Since $\mathrm{Id}_K(X) = \mathrm{Id}_{V(K)}(X)$ by 11.3, it follows that $\mathrm{Id}_K(X) = \mathrm{Id}_{K'}(X)$. Thus $\theta_{K'}(X) = \theta_K(X)$, so $\mathbf{F}_{K'}(\bar{X}) = \mathbf{F}_K(\bar{X})$. The last statement of the corollary then follows from 10.12. □

So far we know that $K$-free algebras belong to $ISP(K)$. The next result partially sharpens this by showing that large $K$-free algebras are in $IP_S(K)$.

**Theorem 11.11.** *Let $K$ be a nonempty class of algebras of type $\mathscr{F}$. Then for some cardinal $m$, if $|X| \geq m$ we have*

$$\mathbf{F}_K(\bar{X}) \in IP_S(K).$$

PROOF. First choose a *subset* $K^*$ of $K$ such that for any $X$, $\mathrm{Id}_{K^*}(X) = \mathrm{Id}_K(X)$. (One can find such a $K^*$ by choosing an infinite set of variables $Y$ and then selecting, for each identity $p \approx q$ in $\mathrm{Id}(Y) - \mathrm{Id}_K(Y)$, an algebra $\mathbf{A} \in K$ such that $\mathbf{A} \nvDash p \approx q$.) Let $m$ be any infinite upper bound of $\{|A| : \mathbf{A} \in K^*\}$. (Since $K^*$ is a set such a cardinal $m$ must exist.)

Next let $\Psi_{K^*}(X)$, for any $X$, be $\{\phi \in \mathrm{Con}\,\mathbf{T}(X) : \mathbf{T}(X)/\phi \in I(K^*)\}$. Then $\Psi_{K^*}(X) \subseteq \Phi_{K^*}(X)$, hence $\bigcap \Psi_{K^*}(X) \supseteq \theta_{K^*}(X)$. To prove equality of these two congruences for $|X| \geq m$ suppose $\langle p,q\rangle \notin \theta_{K^*}(X)$. Then $K^* \nvDash p \approx q$ by 11.4; hence for some $\mathbf{A} \in K^*$, $\mathbf{A} \nvDash p \approx q$. If $p = p(x_1, \ldots, x_n)$, $q = q(x_1, \ldots, x_n)$, choose $a_1, \ldots, a_n \in A$ such that $p^{\mathbf{A}}(a_1, \ldots, a_n) \neq q^{\mathbf{A}}(a_1, \ldots, a_n)$. As $|X| \geq |A|$ we can find a mapping $\alpha: X \to A$ which is onto and $\alpha x_i = a_i$, $1 \leq i \leq n$. Then $\alpha$ can be extended to a surjective homomorphism $\beta: \mathbf{F}_{K^*}(\bar{X}) \to \mathbf{A}$, and $\beta(p) \neq \beta(q)$. Thus $\langle p,q\rangle \notin \ker \beta \in \Psi_{K^*}(X)$, so $\langle p,q\rangle \notin \bigcap \Psi_{K^*}(X)$. Consequently $\bigcap \Psi_{K^*}(X) = \theta_{K^*}(X)$. As $\mathbf{F}_K(\bar{X}) = \mathbf{F}_{K^*}(\bar{X})$ by 11.4, it follows that $\mathbf{F}_K(\bar{X}) = \mathbf{T}(X)/\bigcap \Psi_{K^*}(X)$. Then (see §8 Exercise 11) we see that $\mathbf{F}_K(\bar{X}) \in IP_S(K^*) \subseteq IP_S(K)$. □

**Theorem 11.12.** $V = HP_S$.

PROOF. As

$$P_S \leq SP$$

we have

$$HP_S \leq HSP = V.$$

Given a class $K$ of algebras and sufficiently large $X$, we have

$$\mathbf{F}_{V(K)}(\bar{X}) \in IP_S(K)$$

by 11.11; hence

$$V(K) \subseteq HP_S(K)$$

by 10.11. Thus

$$V = HP_S. \qquad \square$$

REFERENCE

1. G. Birkhoff [1935]

EXERCISES §11

1. Given a type $\mathscr{F}$ and a set of variables $X$ and $p,q \in T(X)$ show that $\mathbf{T}(X) \models p \approx q$ iff $p = q$ (thus $\mathbf{T}(X)$ does not satisfy any interesting identities).
2. If $V$ is a variety and $X$ is infinite, show $V = HSP(\mathbf{F}_V(\bar{X}))$.
3. If $X$ is finite and $\mathrm{Id}_V(X)$ defines $V$ does it follow that $V = HSP(\mathbf{F}_V(\bar{X}))$?
4. Describe free semilattices.
5. Show that if $V = V(\mathbf{A})$ then, given $X \neq \varnothing$, $\mathbf{F}_V(\bar{X})$ can be embedded in $\mathbf{A}^{|A|^{|X|}}$. In particular if $\mathbf{A}$ has no proper subalgebras the embedding is also subdirect.

# §12. Mal'cev Conditions

One of the most fruitful directions of research was initiated by Mal'cev in the 1950's when he showed the connection between permutability of congruences for all algebras in a variety $V$ and the existence of a ternary term $p$ such that $V$ satisfies certain identities involving $p$. The characterization of properties in varieties by the existence of certain terms involved in certain identities we will refer to as *Mal'cev conditions*. This topic has been significantly advanced in recent years by Taylor.

**Lemma 12.1.** *Let $V$ be a variety of type $\mathscr{F}$, and let*

$$p(x_1, \ldots, x_m, y_1, \ldots, y_n),$$
$$q(x_1, \ldots, x_m, y_1, \ldots, y_n)$$

*be terms such that in* $\mathbf{F} = \mathbf{F}_V(\overline{X})$, *where*

$$X = \{x_1, \ldots, x_m, y_1, \ldots, y_n\},$$

*we have*

$$\langle p^{\mathbf{F}}(\overline{x}_1, \ldots, \overline{x}_m, \overline{y}_1, \ldots, \overline{y}_n), q^{\mathbf{F}}(\overline{x}_1, \ldots, \overline{x}_m, \overline{y}_1, \ldots, \overline{y}_n)\rangle \in \Theta(\overline{y}_1, \ldots, \overline{y}_n).$$

*Then*

$$V \models p(x_1, \ldots, x_m, y, \ldots, y) \approx q(x_1, \ldots, x_m, y, \ldots, y).$$

PROOF. The homomorphism

$$\alpha: \mathbf{F}_V(\overline{x}_1, \ldots, \overline{x}_m, \overline{y}_1, \ldots, \overline{y}_n) \to \mathbf{F}_V(\overline{x}_1, \ldots, \overline{x}_m, \overline{y})$$

defined by

$$\alpha(\overline{x}_i) = \overline{x}_i, \qquad 1 \leq i \leq m,$$

and

$$\alpha(\overline{y}_i) = \overline{y}, \qquad 1 \leq i \leq n,$$

is such that

$$\Theta(\overline{y}_1, \ldots, \overline{y}_n) \subseteq \ker \alpha;$$

so

$$\alpha p(\overline{x}_1, \ldots, \overline{x}_m, \overline{y}_1, \ldots, \overline{y}_n) = \alpha q(\overline{x}_1, \ldots, \overline{x}_m, \overline{y}_1, \ldots, \overline{y}_n);$$

thus

$$p(\overline{x}_1, \ldots, \overline{x}_m, \overline{y}, \ldots, \overline{y}) = q(\overline{x}_1, \ldots, \overline{x}_m, \overline{y}, \ldots, \overline{y})$$

in $\mathbf{F}_V(\overline{x}_1, \ldots, \overline{x}_m, \overline{y})$, so by 11.4

$$V \models p(x_1, \ldots, x_m, y, \ldots, y) \approx q(x_1, \ldots, x_m, y, \ldots, y). \qquad \square$$

**Theorem 12.2** (Mal'cev). *Let $V$ be a variety of type $\mathscr{F}$. The variety $V$ is congruence-permutable iff there is a term $p(x,y,z)$ such that*

$$V \models p(x,x,y) \approx y$$

*and*

$$V \models p(x,y,y) \approx x.$$

PROOF. ($\Rightarrow$) If $V$ is congruence-permutable, then in $\mathbf{F}_V(\overline{x},\overline{y},\overline{z})$ we have

$$\langle \overline{x},\overline{z} \rangle \in \Theta(\overline{x},\overline{y}) \circ \Theta(\overline{y},\overline{z})$$

so

$$\langle \overline{x},\overline{z} \rangle \in \Theta(\overline{y},\overline{z}) \circ \Theta(\overline{x},\overline{y}).$$

Hence there is a $p(\overline{x},\overline{y},\overline{z}) \in F_V(\overline{x},\overline{y},\overline{z})$ such that

$$\overline{x}\Theta(\overline{y},\overline{z})p(\overline{x},\overline{y},\overline{z})\Theta(\overline{x},\overline{y})\overline{z}.$$

By 12.1

$$V \models p(x,y,y) \approx x$$

and

$$V \models p(x,x,z) \approx z.$$

($\Leftarrow$) Let $\mathbf{A} \in V$ and suppose $\phi,\psi \in \operatorname{Con} \mathbf{A}$. If

$$\langle a,b \rangle \in \phi \circ \psi,$$

say $a\phi c\psi b$, then

$$b = p(c,c,b)\phi p(a,c,b)\psi p(a,b,b) = a,$$

so

$$\langle b,a \rangle \in \phi \circ \psi.$$

Thus by 5.9

$$\phi \circ \psi = \psi \circ \phi. \qquad \square$$

EXAMPLES. (1) *Groups* $\langle A,\cdot,^{-1},1 \rangle$ are congruence-permutable, for let $p(x,y,z)$ be $x \cdot y^{-1} \cdot z$.

(2) *Rings* $\langle R,+,\cdot,-,0 \rangle$ are congruence-permutable, for let $p(x,y,z)$ be $x - y + z$.

(3) *Quasigroups* $\langle Q,/,\cdot,\backslash \rangle$ are congruence-permutable, for let $p(x,y,z)$ be $(x/(y\backslash y)) \cdot (y\backslash z)$.

**Theorem 12.3.** *Suppose $V$ is a variety for which there is a ternary term $M(x,y,z)$ such that*

$$V \models M(x,x,y) \approx M(x,y,x) \approx M(y,x,x) \approx x.$$

*Then $V$ is congruence-distributive.*

PROOF. Let $\phi,\psi,\chi \in \operatorname{Con} \mathbf{A}$, where $\mathbf{A} \in V$. If

$$\langle a,b \rangle \in \phi \wedge (\psi \vee \chi)$$

then $\langle a,b \rangle \in \phi$ and there exist $c_1, \ldots, c_n$ such that

$$a\psi c_1 \chi c_2 \cdots \psi c_n \chi b.$$

But then as

$$M(a,c_i,b)\phi M(a,c_i,a) = a,$$

for each $i$, we have

$$a = M(a,a,b)(\phi \wedge \psi)M(a,c_1,b)(\phi \wedge \chi)M(a,c_2,b) \cdots M(a,c_n,b)(\phi \wedge \chi)M(a,b,b) = b,$$

so

$$\langle a,b \rangle \in (\phi \wedge \psi) \vee (\phi \wedge \chi).$$

This suffices to show

$$\phi \wedge (\psi \vee \chi) = (\phi \wedge \psi) \vee (\phi \wedge \chi),$$

so $V$ is congruence-distributive. $\square$

EXAMPLE. *Lattices* are congruence-distributive, for let

$$M(x,y,z) = (x \vee y) \wedge (x \vee z) \wedge (y \vee z).$$

**Definition 12.4.** A variety $V$ is *arithmetical* if it is both congruence-distributive and congruence-permutable.

**Theorem 12.5** (Pixley). *A variety $V$ is arithmetical iff it satisfies either of the equivalent conditions*

(a) *There are terms $p$ and $M$ as in* 12.2 *and* 12.3.

(b) *There is a term $m(x,y,z)$ such that*

$$V \models m(x,y,x) \approx m(x,y,y) \approx m(y,y,x) \approx x.$$

PROOF. If $V$ is arithmetical then there is a term $p$ as $V$ is congruence-permutable. Let $\mathbf{F}_V(\bar{x},\bar{y},\bar{z})$ be the free algebra in $V$ freely generated by $\{\bar{x},\bar{y},\bar{z}\}$. Then as

$$\langle \bar{x},\bar{z}\rangle \in \Theta(\bar{x},\bar{z}) \cap [\Theta(\bar{x},\bar{y}) \vee \Theta(\bar{y},\bar{z})]$$

it follows that

$$\langle \bar{x},\bar{z}\rangle \in [\Theta(\bar{x},\bar{z}) \cap \Theta(\bar{x},\bar{y})] \vee [\Theta(\bar{x},\bar{z}) \cap \Theta(\bar{y},\bar{z})];$$

hence

$$\langle \bar{x},\bar{z}\rangle \in [\Theta(\bar{x},\bar{z}) \cap \Theta(\bar{x},\bar{y})] \circ [\Theta(\bar{x},\bar{z}) \cap \Theta(\bar{y},\bar{z})].$$

Choose $M(\bar{x},\bar{y},\bar{z}) \in F_V(\bar{x},\bar{y},\bar{z})$ such that

$$\bar{x}[\Theta(\bar{x},\bar{z}) \cap \Theta(\bar{x},\bar{y})]M(\bar{x},\bar{y},\bar{z})[\Theta(\bar{x},\bar{z}) \cap \Theta(\bar{y},\bar{z})]\,\bar{z}.$$

Then by 12.1,

$$V \models M(x,x,y) \approx M(x,y,x) \approx M(y,x,x) \approx x.$$

If (a) holds then let $m(x,y,z)$ be $p(x,M(x,y,z),z)$. Finally if (b) holds let $p(x,y,z)$ be $m(x,y,z)$ and let $M(x,y,z)$ be $m(x,m(x,y,z),z)$, and use 12.2 and 12.3. □

EXAMPLES. (1) *Boolean algebras* are arithmetical, for let

$$m(x,y,z) = (x \wedge z) \vee (x \wedge y' \wedge z') \vee (x' \wedge y' \wedge z).$$

(2) *Heyting algebras* are arithmetical, for let

$$m(x,y,z) = [(x \to y) \to z] \wedge [(z \to y) \to x] \wedge [x \vee z].$$

Note that 12.3 is not a Mal'cev condition as it is an implication rather than a characterization. Jónsson discovered a Mal'cev condition for congruence-distributive varieties which we will make considerable use of in the last chapter.

**Theorem 12.6** (Jónsson). *A variety $V$ is congruence-distributive iff there is a finite $n$ and terms $p_0(x,y,z), \ldots, p_n(x,y,z)$ such that $V$ satisfies*

$$\begin{aligned}
&p_i(x,y,x) \approx x && 0 \le i \le n\\
&p_0(x,y,z) \approx x, \qquad p_n(x,y,z) \approx z &&\\
&p_i(x,x,y) \approx p_{i+1}(x,x,y) && \textit{for } i \textit{ even}\\
&p_i(x,y,y) \approx p_{i+1}(x,y,y) && \textit{for } i \textit{ odd.}
\end{aligned}$$

PROOF. (⇒) Since

$$\Theta(\overline{x},\overline{z}) \wedge [\Theta(\overline{x},\overline{y}) \vee \Theta(\overline{y},\overline{z})] = [\Theta(\overline{x},\overline{z}) \wedge \Theta(\overline{x},\overline{y})] \vee [\Theta(\overline{x},\overline{z}) \wedge \Theta(\overline{y},\overline{z})]$$

in $\mathbf{F}_V(\overline{x},\overline{y},\overline{z})$ we must have

$$\langle \overline{x},\overline{z} \rangle \in [\Theta(\overline{x},\overline{z}) \wedge \Theta(\overline{x},\overline{y})] \vee [\Theta(\overline{x},\overline{z}) \wedge \Theta(\overline{y},\overline{z})].$$

Thus for some $p_1(\overline{x},\overline{y},\overline{z}), \ldots, p_{n-1}(\overline{x},\overline{y},\overline{z}) \in F_V(\overline{x},\overline{y},\overline{z})$ we have

$$\begin{aligned} &\overline{x}[\Theta(\overline{x},\overline{z}) \wedge \Theta(\overline{x},\overline{y})]p_1(\overline{x},\overline{y},\overline{z}) \\ &p_1(\overline{x},\overline{y},\overline{z})[\Theta(\overline{x},\overline{z}) \wedge \Theta(\overline{y},\overline{z})]p_2(\overline{x},\overline{y},\overline{z}) \\ &\qquad\vdots \\ &p_{n-1}(\overline{x},\overline{y},\overline{z})[\Theta(\overline{x},\overline{z}) \wedge \Theta(\overline{y},\overline{z})]\overline{z}, \end{aligned}$$

and from these the desired equations fall out.

(⇐) For $\phi,\psi,\chi \in \operatorname{Con} \mathbf{A}$, where $\mathbf{A} \in V$, we need to show

$$\phi \wedge (\psi \vee \chi) \subseteq (\phi \wedge \psi) \vee (\phi \wedge \chi),$$

so let

$$\langle a,b \rangle \in \phi \wedge (\psi \vee \chi).$$

Then $\langle a,b \rangle \in \phi$, and for some $c_1, \ldots, c_t$ we have

$$a\psi c_1\chi \ldots c_t\chi b.$$

From this follows, for $0 \leq i \leq n$,

$$p_i(a,a,b)\psi p_i(a,c_1,b)\chi \ldots p_i(a,c_t,b)\chi p_i(a,b,b);$$

hence

$$p_i(a,a,b)(\phi \wedge \psi)p_i(a,c_1,b)(\phi \wedge \chi) \ldots p_i(a,c_t,b)(\phi \wedge \chi)p_i(a,b,b),$$

so

$$p_i(a,a,b)[(\phi \wedge \psi) \vee (\phi \wedge \chi)]p_i(a,b,b),$$

$0 \leq i \leq n$. Then in view of the given equations, $a[(\phi \wedge \psi) \vee (\phi \wedge \chi)]b$, so $V$ is congruence-distributive. □

By looking at the proofs of 12.2 and 12.6 one easily has the following result.

**Theorem 12.7.** *A variety $V$ is congruence-permutable (respectively, congruence-distributive) iff $\mathbf{F}_V(\overline{x},\overline{y},\overline{z})$ has permutable (respectively, distributive) congruences.*

For convenience in future discussions we introduce the following definitions.

**Definition 12.8.** A ternary term $p$ satisfying the conditions in 12.2 for a variety $V$ is called a *Mal'cev term* for $V$, a ternary term $M$ as described in 12.3 is a *majority term* for $V$, and a ternary term $m$ as described in 12.5 is called a $\frac{2}{3}$*-minority term* for $V$.

The reader will find Mal'cev conditions for congruence-modular varieties in Day (1) below.

REFERENCES

1. A. Day [1969]
2. B. Jónsson [1967]
3. A. I. Mal'cev [1954]
4. A. F. Pixley [1963]
5. W. Taylor [1973]

EXERCISES §12

1. Verify the claim that Boolean algebras [Heyting algebras] are arithmetical.
2. Let $V$ be a variety of rings generated by finitely many finite fields. Show that $V$ is arithmetical.
3. Show that the variety of $n$-valued Post algebras is arithmetical.
4. Show that the variety generated by the six-element ortholattice in Figure 19 is arithmetical.

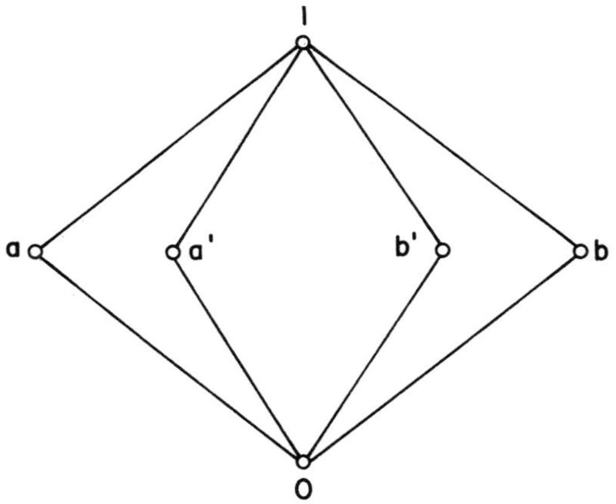


Figure 19

## §13. The Center of an Algebra

Smith (6) introduced a generalization to any algebra in a congruence-permutable variety of the commutator for groups. Hagemann and Herrmann (3) then showed that such commutators exist for any algebra in a congruence-modular variety. Using the commutator one can define the center of such algebras. Another very simple definition of the center, valid for any algebra, was given by Freese and McKenzie (1), and we will use it here.

**Definition 13.1.** Let $\mathbf{A}$ be an algebra of type $\mathscr{F}$. The *center* of $\mathbf{A}$ is the binary relation $Z(\mathbf{A})$ defined by:

$$\langle a,b\rangle \in Z(\mathbf{A})$$

iff for every $p(x,y_1,\ldots,y_n) \in T(x,y_1,\ldots,y_n)$ and for every $c_1,\ldots,c_n,d_1,\ldots,d_n \in A$,

$$p(a,c_1,\ldots,c_n) = p(a,d_1,\ldots,d_n) \quad \text{iff} \quad p(b,c_1,\ldots,c_n) = p(b,d_1,\ldots,d_n).$$

**Theorem 13.2.** *For every algebra* $\mathbf{A}$, *the center* $Z(\mathbf{A})$ *is a congruence on* $\mathbf{A}$.

PROOF. Certainly $Z(\mathbf{A})$ is reflexive, symmetric, and transitive, hence $Z(\mathbf{A})$ is an equivalence relation on $A$. Next let $f$ be an $n$-ary function symbol, and suppose $\langle a_i,b_i\rangle \in Z(\mathbf{A})$, $1 \le i \le n$. Given a term $p(x,y_1,\ldots,y_m)$ and elements $c_1,\ldots,c_m$, $d_1,\ldots,d_m$ of $A$, from the definition of $Z(\mathbf{A})$ we have

$$\begin{aligned} & p(f(a_1,a_2,\ldots,a_n),\vec{c}) = p(f(a_1,a_2,\ldots,a_n),\vec{d}) \\ \text{iff}\quad & p(f(b_1,a_2,\ldots,a_n),\vec{c}) = p(f(b_1,a_2,\ldots,a_n),\vec{d}) \\ & \vdots \\ \text{iff}\quad & p(f(b_1,\ldots,b_{n-1},a_n),\vec{c}) = p(f(b_1,\ldots,b_{n-1},a_n),\vec{d}) \\ \text{iff}\quad & p(f(b_1,\ldots,b_n),\vec{c}) = p(f(b_1,\ldots,b_n),\vec{d}); \end{aligned}$$

hence

$$p(f(\vec{a}),\vec{c}) = p(f(\vec{a}),\vec{d}) \quad \text{iff} \quad p(f(\vec{b}),\vec{c}) = p(f(\vec{b}),\vec{d}),$$

so

$$\langle f(a_1,\ldots,a_n), f(b_1,\ldots,b_n)\rangle \in Z(\mathbf{A}).$$

Thus $Z(\mathbf{A})$ is indeed a congruence. □

Let us actually calculate the above defined center of a group and of a ring.

EXAMPLE. Let $\mathbf{G} = \langle G,\cdot,^{-1},1\rangle$ be a group. If $\langle a,b\rangle \in Z(\mathbf{G})$ then, with the term $p(x,y_1,y_2) = y_1 \cdot x \cdot y_2$ and $c \in G$, we have

$$p(a,a^{-1},c) = p(a,c,a^{-1});$$

hence

$$p(b,a^{-1},c) = p(b,c,a^{-1}),$$

that is,

$$a^{-1} \cdot b \cdot c = c \cdot b \cdot a^{-1}.$$

With $c = 1$ it follows that

$$a^{-1} \cdot b = b \cdot a^{-1};$$

hence for $c \in G$,

$$a^{-1} \cdot b \cdot c = c \cdot a^{-1} \cdot b,$$

consequently $\langle a,b\rangle$ is in the congruence associated with the normal subgroup **N** of **G** which is the usual group-theoretic center of **G**, i.e., $N = \{g \in G : h \cdot g = g \cdot h \text{ for } h \in G\}$.

Conversely, suppose **N** is the usual group-theoretic center of **G**. Then for any term $p(x,y_1,\ldots,y_n)$ and elements $a,b,c_1,\ldots,c_n,d_1,\ldots,d_n \in G$, if $a \cdot b^{-1} \in N$, and if

$$p(a,\vec{c}) = p(a,\vec{d})$$

then

$$p((a \cdot b^{-1}) \cdot b, \vec{c}) = p((a \cdot b^{-1}) \cdot b, \vec{d}),$$

so

$$p(b,\vec{c}) = p(b,\vec{d})$$

as $a \cdot b^{-1}$ is central. So, by symmetry, if $a \cdot b^{-1} \in N$ then

$$p(a,\vec{c}) = p(a,\vec{d}) \quad \text{iff} \quad p(b,\vec{c}) = p(b,\vec{d}),$$

so $\langle a,b\rangle \in Z(\mathbf{G})$.

Thus

$$Z(\mathbf{G}) = \{\langle a,b\rangle \in G^2 : (a \cdot b^{-1}) \cdot c = c \cdot (a \cdot b^{-1}) \text{ for } c \in G\}.$$

Example. Let $\mathbf{R} = \langle R,+,\cdot,-,0\rangle$ be a ring. If $\langle r,s\rangle \in Z(\mathbf{R})$ then, for $t \in R$,

$$(r - \underline{r}) \cdot t = (r - \underline{r}) \cdot 0;$$

hence replacing the underlined $r$ by $s$ we have

$$(r - s) \cdot t = 0.$$

Likewise

$$t \cdot (r - s) = 0,$$

so $r - s \in \operatorname{Ann}(\mathbf{R})$, the annihilator of **R**. Conversely, if $r - s \in \operatorname{Ann}(\mathbf{R})$ and $p(x,y_1,\ldots,y_n)$ is a term and $c_1,\ldots,c_n,d_1,\ldots,d_n \in R$ then from

$$p(r,\vec{c}) = p(r,\vec{d})$$

it follows that

$$p((r - s) + s, \vec{c}) = p((r - s) + s, \vec{d}),$$

and thus

$$p(s,\vec{c}) = p(s,\vec{d}).$$

By symmetry, we have

$$Z(\mathbf{R}) = \{\langle r,s\rangle : r - s \in \operatorname{Ann}(\mathbf{R})\}.$$

Now we return to the fundamental theorem of centrality, namely the characterization of modules up to polynomial equivalence.

**Definition 13.3.** Let **A** be an algebra of type $\mathscr{F}$. To $\mathscr{F}_0$ add symbols $a$ for each $a \in A$, and call the new type $\mathscr{F}_A$, and let $\mathbf{A}_A$ be the algebra of type $\mathscr{F}_A$ which is just **A** with a nullary operation corresponding to each element of $A$. The terms of type $\mathscr{F}_A$ are called the *polynomials* of **A**. We write $p^{\mathbf{A}}$ for $p^{\mathbf{A}_A}$. Two algebras $\mathbf{A}_1 = \langle A, F_1\rangle$ and $\mathbf{A}_2 = \langle A, F_2\rangle$, possibly of different types, on the same universe are said to be *polynomially equivalent* if they have the same set of polynomial functions, i.e., for each polynomial $p(x_1, \ldots, x_n)$ of $\mathbf{A}_1$ there is a polynomial $q(x_1, \ldots, x_n)$ of $\mathbf{A}_2$ such that $p^{\mathbf{A}_1} = q^{\mathbf{A}_2}$, and conversely.

The following proof incorporates elegant arguments due to McKenzie and Taylor.

**Theorem 13.4** [Gumm, Hagemann, Herrmann]. *Let* $\mathbf{A}$ *be an algebra such that* $V(\mathbf{A})$ *is congruence-permutable. Then the following are equivalent*:

(a) $\mathbf{A}$ *is polynomially equivalent to a left* $\mathbf{R}$*-module, for some* $\mathbf{R}$.
(b) $Z(\mathbf{A}) = \nabla_A$.
(c) $\{\langle a,a\rangle : a \in A\}$ *is a coset of a congruence on* $\mathbf{A} \times \mathbf{A}$.

Proof. (a) $\Rightarrow$ (b): If $\mathbf{A}$ is polynomially equivalent to a module $\mathbf{M} = \langle M, +, -, 0, (f_r)_{r\in R}\rangle$, then for every term $p(x, y_1, \ldots, y_n)$ of $\mathbf{A}$ there is a polynomial

$$q(x,y_1, \ldots, y_n) = f_r(x) + f_{r_1}(y_1) + \cdots + f_{r_n}(y_n) + m,$$

of $\mathbf{M}$ such that

$$p^{\mathbf{A}} = q^{\mathbf{M}}.$$

Thus for $a,b,c_1, \ldots, c_n, d_1, \ldots, d_n \in A$, if

$$p(a,c_1, \ldots, c_n) = p(a,d_1, \ldots, d_n)$$

then

$$q(a,c_1, \ldots, c_n) = q(a,d_1, \ldots, d_n);$$

hence if we subtract $f_r(a)$ from both sides,

$$f_{r_1}(c_1) + \cdots + f_{r_n}(c_n) + m = f_{r_1}(d_1) + \cdots + f_{r_n}(d_n) + m,$$

so if we add $f_r(b)$ to both sides,

$$q(b,c_1, \ldots, c_n) = q(b,d_1, \ldots, d_n);$$

consequently

$$p(b,c_1, \ldots, c_n) = p(b,d_1, \ldots, d_n).$$

By symmetry,

$$p(a,\vec{c}) = p(a,\vec{d}) \quad \text{iff} \quad p(b,\vec{c}) = p(b,\vec{d});$$

hence $Z(\mathbf{A}) = \nabla_A$.

(b) $\Leftrightarrow$ (c): First note that $X = \{\langle a,a\rangle : a \in A\}$ is a coset of some congruence on $\mathbf{A} \times \mathbf{A}$ iff it is a coset of $\Theta(X)$, the smallest congruence on $\mathbf{A} \times \mathbf{A}$ obtained by identifying $X$. Now, from §10 Exercise 7,

$$\Theta(\{\langle a,a\rangle : a \in A\}) = t^*(s(\{\langle p^{\mathbf{A}\times\mathbf{A}}(\langle \bar{a},\bar{a}\rangle, \langle c_1,d_1\rangle, \ldots, \langle c_n,d_n\rangle),$$
$$p^{\mathbf{A}\times\mathbf{A}}(\langle \bar{b},\bar{b}\rangle, \langle c_1,d_1\rangle, \ldots, \langle c_n,d_n\rangle)\rangle : \bar{a},\bar{b},c_1, \ldots, c_n,d_1, \ldots, d_n \in A$$

and $p$ is a term$\})) \cup \Delta_{A\times A}$.

Hence $X$ is a coset of $\Theta(X)$ iff for every $\bar{a},\bar{b},c_1,\ldots,c_n,d_1,\ldots,d_n \in A$ and every term $p(x,y_1,\ldots,y_n)$,

$$p^{\mathbf{A}\times\mathbf{A}}(\langle\bar{a},\bar{a}\rangle,\langle c_1,d_1\rangle,\ldots,\langle c_n,d_n\rangle) \in X$$

iff

$$p^{\mathbf{A}\times\mathbf{A}}(\langle\bar{b},\bar{b}\rangle,\langle c_1,d_1\rangle,\ldots,\langle c_n,d_n\rangle) \in X,$$

that is,

$$p^{\mathbf{A}}(\bar{a},\vec{c}) = p^{\mathbf{A}}(\bar{a},\vec{d}) \quad \text{iff} \quad p^{\mathbf{A}}(\bar{b},\vec{c}) = p^{\mathbf{A}}(\bar{b},\vec{d}).$$

Thus $X$ is a coset of $\Theta(X)$ iff $Z(\mathbf{A}) = \nabla_A$.

(b) $\Rightarrow$ (a): Given that $Z(\mathbf{A}) = \nabla_A$, let $p(x,y,z)$ be a Mal'cev term for $V(\mathbf{A})$ Choose any element 0 of $A$ and define, for $a,b \in A$,

$$\begin{aligned} a + b &= p(a,0,b) \\ -a &= p(0,a,0). \end{aligned}$$

Then

$$\begin{aligned} a + 0 &= p(a,0,0) \\ &= a. \end{aligned}$$

Next observe that for $a,b,c,d,e \in A$,

$$p(p(a,a,a),d,p(b,e,\underline{e})) = p(p(a,d,b),e,p(c,c,\underline{e}));$$

hence, as $\langle e,c\rangle \in Z(\mathbf{A})$, we can replace the underlined $e$ by $c$ to obtain

$$p(p(a,a,a),d,p(b,e,c)) = p(p(a,d,b),e,p(c,c,c)),$$

so

$$p(a,d,p(b,e,c)) = p(p(a,d,b),e,c).$$

Setting $d = e = 0$, we have the associative law

$$a + (b + c) = (a + b) + c.$$

Next,

$$\begin{aligned} a + (-a) &= p(a,0,p(0,a,0)) \\ &= p(p(a,0,0),a,0) \\ &= p(a,a,0) \\ &= 0. \end{aligned}$$

By

$$p(a,\underline{b},b) = p(b,\underline{b},a)$$

and the fact that $\langle 0,b\rangle \in Z(\mathbf{A})$, we can replace the underlined $b$ by 0 to obtain

$$p(a,0,b) = p(b,0,a);$$

hence

$$a + b = b + a,$$

so $\langle A,+,-,0\rangle$ is an abelian group.

Next we show that each $n$-ary term function $p^{\mathbf{A}}(x_1,\ldots,x_n)$ of $\mathbf{A}$ is affine for $\langle A,+,-,0\rangle$, i.e., it is a homomorphism from $\langle A,+,-,0\rangle^n$ to $\langle A,+,-,0\rangle$

plus a constant. Let $a_1, \ldots, a_n, b_1, \ldots, b_n \in A$. Then

$$p(a_1 + \underline{0}, \ldots, a_n + 0) + p(0, \ldots, 0) = p(0 + \underline{0}, \ldots, 0 + 0) + p(a_1, \ldots, a_n).$$

As $\langle 0, b_1 \rangle \in Z(\mathbf{A})$ we can replace the underlined 0's by $b_1$ to obtain

$$\begin{aligned} &p(a_1 + b_1, a_2 + 0, \ldots, a_n + 0) + p(0, \ldots, 0) \\ &\qquad = p(0 + b_1, 0 + 0, \ldots, 0 + 0) + p(a_1, \ldots, a_n). \end{aligned}$$

Continuing in this fashion, we obtain

$$\begin{aligned} p(a_1 + b_1, \ldots, a_n + b_n) + p(0, \ldots, 0) &= p(b_1, \ldots, b_n) + p(a_1, \ldots, a_n) \\ &= p(a_1, \ldots, a_n) + p(b_1, \ldots, b_n). \end{aligned}$$

Thus $p^{\mathbf{A}}(x_1, \ldots, x_n) - p^{\mathbf{A}}(0, \ldots, 0)$ is a group homomorphism from $\langle A, +, -, 0 \rangle^n$ to $\langle A, +, -, 0 \rangle$.

To construct the desired module, let $R$ be the set of unary functions $p^{\mathbf{A}}(x, c_1, \ldots, c_n)$ on $A$ obtained by choosing terms $p(x, y_1, \ldots, y_n)$ and elements $c_1, \ldots, c_n \in A$ such that

$$p(0, c_1, \ldots, c_n) = 0.$$

For such unary functions we have

$$p(a + b, c_1, \ldots, c_n) = p(a, c_1, \ldots, c_n) + p(b, 0, \ldots, 0) - p(0, \ldots, 0)$$

and

$$\begin{aligned} p(b, c_1, \ldots, c_n) &= p(b, 0, \ldots, 0) + p(0, c_1, \ldots, c_n) - p(0, \ldots, 0) \\ &= p(b, 0, \ldots, 0) - p(0, \ldots, 0); \end{aligned}$$

hence

$$p(a + b, c_1, \ldots, c_n) = p(a, c_1, \ldots, c_n) + p(b, c_1, \ldots, c_n). \tag{*}$$

Thus each member of $R$ is an endomorphism of $\langle A, +, -, 0 \rangle$.

Clearly $R$ is closed under composition $\circ$, and for $r, s \in R$ define $r + s$ and $-r$ by

$$\begin{aligned} (r + s)(a) &= r(a) + s(a) = p(r(a), 0, s(a)) \\ (-r)(a) &= -r(a) = p(0, r(a), 0). \end{aligned}$$

Then $r + s$, $-r \in R$. Let $\hat{0}$ be the constant function on $A$ with value 0, and let $\hat{1}$ be the identity function on $A$. Then $\hat{0}, \hat{1} \in R$ as well. We claim that $\mathbf{R} = \langle R, +, \cdot, -, \hat{0}, \hat{1} \rangle$ is a ring. Certainly $\langle R, +, -, 0 \rangle$ is an abelian group as the operations are defined pointwise in the abelian group $\langle A, +, -, 0 \rangle$, and $\langle R, \cdot, 1 \rangle$ is a monoid. Thus we only need to look at the distributive laws. If we are given $r, s, t \in R$, then

$$\begin{aligned} [(r + s) \circ t](a) &= (r + s)(t(a)) \\ &= r(t(a)) + s(t(a)) \\ &= (r \circ t)(a) + (s \circ t)(a) \\ &= (r \circ t + s \circ t)(a); \end{aligned}$$

hence

$$(r + s) \circ t = r \circ t + s \circ t.$$

Also

$$\begin{aligned}[r \circ (s + t)](a) &= r((s + t)(a)) \\ &= r(s(a) + t(a)) \\ &= r(s(a)) + r(t(a)) \qquad \text{(by (*) above)} \\ &= (r \circ s)(a) + (r \circ t)(a) \\ &= (r \circ s + r \circ t)(a);\end{aligned}$$

hence

$$r \circ (s + t) = (r \circ s) + (r \circ t).$$

This shows **R** is a ring.

Now to show that $\mathbf{M} = \langle A,+,-,0,(r)_{r \in R}\rangle$ is a left **R**-module, we only need to check the laws concerning scalar multiplication. So let $r,s \in R$, $a,b \in A$. Then

$$\begin{aligned}(r + s)(a) &= r(a) + s(a) \qquad \text{(by definition)} \\ r(a + b) &= r(a) + r(b) \qquad \text{(by (*))} \\ (r \circ s)(a) &= r(s(a)).\end{aligned}$$

Thus **M** is a left **R**-module (indeed a unitary left **R**-module).

The fundamental operations of **M** are certainly expressible by polynomial functions of **A**. Conversely any $n$-ary fundamental operation $f^{\mathbf{A}}(x_1, \ldots, x_n)$ of **A** satisfies, for $a_1, \ldots, a_n \in A$,

$$\begin{aligned}f(a_1, \ldots, a_n) - f(0, \ldots, 0) = {} & (f(a_1, 0, \ldots, 0) - f(0, \ldots, 0)) \\ & + \cdots + (f(0, \ldots, a_n) - f(0, \ldots, 0)).\end{aligned}$$

As

$$\begin{aligned}r_1 &= f^{\mathbf{A}}(x, 0, \ldots, 0) - f^{\mathbf{A}}(0, \ldots, 0) \in R \\ &\ \vdots \\ r_n &= f^{\mathbf{A}}(0, \ldots, 0, x) - f^{\mathbf{A}}(0, \ldots, 0) \in R\end{aligned}$$

it follows that

$$f^{\mathbf{A}}(x_1, \ldots, x_n) = r_1(x_1) + \cdots + r_n(x_n) + f(0, \ldots, 0);$$

hence each fundamental operation of **A** is a polynomial of **M**. This suffices to show that **A** and **M** are polynomially equivalent. □

Actually one only needs to assume $V(\mathbf{A})$ is congruence-modular in Theorem 13.4 (see (4) or (7) below).

## References

1. R. Freese and R. McKenzie [b]
2. H. P. Gumm [a]
3. J. Hagemann and C. Herrmann [1979]
4. C. Herrmann [1979]

5. R. McKenzie [a]
6. J. D. H. Smith [1976]
7. W. Taylor [a]

EXERCISES §13

1. If **A** belongs to an arithmetical variety, show that $Z(\mathbf{A}) = \Delta_A$. [Hint: if $\langle a,b\rangle \in Z(\mathbf{A})$ use $m(a,b,a) = m(b,b,a)$.]

2. Show that $\langle a,b\rangle \in Z(\prod_{i\in I} \mathbf{A}_i)$ iff $\langle a(i),b(i)\rangle \in Z(\mathbf{A}_i)$ for $i \in I$.

3. If $\mathbf{A} \leq \mathbf{B}$ and $Z(\mathbf{B}) = \nabla_B$, show $Z(\mathbf{A}) = \nabla_A$.

4. If $\mathbf{B} \in H(\mathbf{A})$ and **A** is in a congruence-permutable variety, show that $Z(\mathbf{A}) = \nabla_A$ implies $Z(\mathbf{B}) = \nabla_B$. Conclude that in a congruence-permutable variety all members **A** with $Z(\mathbf{A}) = \nabla_A$ constitute a subvariety.

5. Suppose **A** is polynomially equivalent to a module. If $p(x,y,z)$, $q(x,y,z)$ are two Mal'cev terms for **A**, show $p^{\mathbf{A}}(x,y,z) = q^{\mathbf{A}}(x,y,z)$.

6. (Freese and McKenzie). Let $V$ be a congruence permutable variety such that $Z(\mathbf{A}) = \nabla_A$ for every $\mathbf{A} \in V$. Let $p(x,y,z)$ be a Mal'cev term for $V$. Define $R$ by
$$R = \{r(\bar{x},\bar{y}) \in F_V(\bar{x},\bar{y}) : r(\bar{x},\bar{x}) = \bar{x}\}.$$
(Note that if $r(\bar{x},\bar{y}) = s(\bar{x},\bar{y})$, then $r(\bar{x},\bar{x}) = \bar{x}$ iff $s(\bar{x},\bar{x}) = \bar{x}$.) Define the operations $+,\cdot,-,0,1$, on $R$ by
$$\begin{aligned} r(\bar{x},\bar{y}) + s(\bar{x},\bar{y}) &= p(r(\bar{x},\bar{y}),\bar{y}, s(\bar{x},\bar{y}))\\ r(\bar{x},\bar{y}) \cdot s(\bar{x},\bar{y}) &= r(s(\bar{x},\bar{y}), \bar{y})\\ -r(\bar{x},\bar{y}) &= p(\bar{y}, r(\bar{x},\bar{y}), \bar{y})\\ 0 &= \bar{y}\\ 1 &= \bar{x}. \end{aligned}$$
Verify that $\mathbf{R} = \langle R,+,\cdot,-,0,1\rangle$ is a ring with unity. Next, given an algebra $\mathbf{A} \in V$ and $n \in A$, define the operations $+,-,0,(f_r)_{r\in R}$ on $A$ by
$$\begin{aligned} a + b &= p(a,n,b)\\ -a &= p(n,a,n)\\ 0 &= n\\ f_r(a) &= r(a,n). \end{aligned}$$
Now verify that $\langle A,+,-,0,(f_r)_{r\in R}\rangle$ is a unitary **R**-module, and it is polynomially equivalent to **A**.

# §14. Equational Logic and Fully Invariant Congruences

In this section we explore the connections between the identities satisfied by classes of algebras and fully invariant congruences on the term algebra. Using this, we can give a complete set of rules for making deductions of identities from identities. Finally, we show that the possible finite sizes of minimal defining sets of identities of a variety form a convex set.

**Definition 14.1.** A congruence $\theta$ on an algebra **A** is *fully invariant* if for every endomorphism $\alpha$ of **A**,

$$\langle a,b\rangle \in \theta \Rightarrow \langle \alpha a,\alpha b\rangle \in \theta.$$

Let $\mathrm{Con}_{\mathrm{FI}}(\mathbf{A})$ denote *the set of fully invariant congruences on* **A**.

**Lemma 14.2.** $\mathrm{Con}_{\mathrm{FI}}(\mathbf{A})$ *is closed under arbitrary intersection.*

PROOF. (Exercise.) □

**Definition 14.3.** Given an algebra **A** and $S \subseteq A \times A$ let $\Theta_{\mathrm{FI}}(S)$ denote the least fully invariant congruence on **A** containing $S$. The congruence $\Theta_{\mathrm{FI}}(S)$ is called the *fully invariant congruence generated by* $S$.

**Lemma 14.4.** *If we are given an algebra* **A** *of type* $\mathscr{F}$ *then* $\Theta_{\mathrm{FI}}$ *is an algebraic closure operator on* $A \times A$. *Indeed,* $\Theta_{\mathrm{FI}}$ *is 2-ary.*

PROOF. First construct $\mathbf{A} \times \mathbf{A}$, and then to the fundamental operations of $\mathbf{A} \times \mathbf{A}$ add the following:

$$\langle a,a\rangle \quad \text{for } a \in A$$

$$s(\langle a,b\rangle) = \langle b,a\rangle$$

$$t(\langle a,b\rangle,\langle c,d\rangle) = \begin{cases} \langle a,d\rangle & \text{if } b = c \\ \langle a,b\rangle & \text{otherwise} \end{cases}$$

$$e_\sigma(\langle a,b\rangle) = \langle \sigma a,\sigma b\rangle \quad \text{for } \sigma \text{ an endomorphism of } \mathbf{A}.$$

Then it is not difficult to verify that $\theta$ is a fully invariant congruence on **A** iff $\theta$ is a subuniverse of the new algebra we have just constructed. Thus $\Theta_{\mathrm{FI}}$ is an algebraic closure operator.

To see that $\Theta_{\mathrm{FI}}$ is 2-ary let us define a new algebra $\mathbf{A}^*$ by replacing each $n$-ary fundamental operation $f$ of **A** by the set of all unary operations of the form

$$f(a_1,\ldots,a_{i-1},x,a_{i+1},\ldots,a_n)$$

where $a_1,\ldots,a_{i-1},a_{i+1},\ldots,a_n$ are elements of $A$.

**Claim.** Con **A** = Con $\mathbf{A}^*$.

Clearly $\theta \in \mathrm{Con}\,\mathbf{A} \Rightarrow \theta \in \mathrm{Con}\,\mathbf{A}^*$. For the converse suppose that $\theta \in \mathrm{Con}\,\mathbf{A}^*$ and $f \in \mathscr{F}_n$. Then for

$$\langle a_i,b_i\rangle \in \theta, \qquad 1 \le i \le n,$$

we have

$$\langle f(a_1,\ldots,a_{n-1},a_n), f(a_1,\ldots,a_{n-1},b_n)\rangle \in \theta$$
$$\langle f(a_1,\ldots,a_{n-1},b_n), f(a_1,\ldots,b_{n-1},b_n)\rangle \in \theta$$
$$\vdots$$
$$\langle f(a_1,b_2,\ldots,b_n), f(b_1,\ldots,b_n)\rangle \in \theta;$$

hence

$$\langle f(a_1, \ldots, a_n), f(b_1, \ldots, b_n)\rangle \in \theta.$$

Thus

$$\theta \in \operatorname{Con} \mathbf{A}.$$

If now we go back to the beginning of the proof and use $\mathbf{A}^*$ instead of $\mathbf{A}$, but keep the $e_\sigma$'s the same, it follows that $\Theta_{\mathrm{FI}}$ is the closure operator Sg of an algebra all of whose operations are of arity at most 2. Then by 4.2, $\Theta_{\mathrm{FI}}$ is a 2-ary closure operator. □

**Definition 14.5.** Given a set of variables $X$ and a type $\mathscr{F}$, let

$$\tau: \mathrm{Id}(X) \to T(X) \times T(X)$$

be the bijection defined by

$$\tau(p \approx q) = \langle p,q\rangle.$$

**Lemma 14.6.** *For $K$ a class of algebras of type $\mathscr{F}$ and $X$ a set of variables, $\tau(\mathrm{Id}_K(X))$ is a fully invariant congruence on $\mathbf{T}(X)$.*

Proof. As

$$p \approx p \in \mathrm{Id}_K(X) \quad \text{for } p \in T(X)$$
$$p \approx q \in \mathrm{Id}_K(X) \Rightarrow q \approx p \in \mathrm{Id}_K(X)$$
$$p \approx q,\ q \approx r \in \mathrm{Id}_K(X) \Rightarrow p \approx r \in \mathrm{Id}_K(X)$$

it follows that $\tau(\mathrm{Id}_K(X))$ is an equivalence relation on $T(X)$. Now if

$$p_i \approx q_i \in \mathrm{Id}_K(X) \quad \text{for } 1 \le i \le n$$

and if $f \in \mathscr{F}_n$ then it is easily seen that

$$f(p_1, \ldots, p_n) \approx f(q_1, \ldots, q_n) \in \mathrm{Id}_K(X),$$

so $\tau(\mathrm{Id}_K(X))$ is a congruence relation on $\mathbf{T}(X)$. Next, if $\alpha$ is an endomorphism of $\mathbf{T}(X)$ and

$$p(x_1, \ldots, x_n) \approx q(x_1, \ldots, x_n) \in \mathrm{Id}_K(X)$$

then it is again direct to verify that

$$p(\alpha x_1, \ldots, \alpha x_n) \approx q(\alpha x_1, \ldots, \alpha x_n) \in \mathrm{Id}_K(X);$$

hence $\tau(\mathrm{Id}_K(X))$ is fully invariant. □

**Lemma 14.7.** *Given a set of variables $X$ and a fully invariant congruence $\theta$ on $\mathbf{T}(X)$ we have, for $p \approx q \in \mathrm{Id}(X)$,*

$$\mathbf{T}(X)/\theta \models p \approx q \Leftrightarrow \langle p,q\rangle \in \theta.$$

*Thus $\mathbf{T}(X)/\theta$ is free in $V(\mathbf{T}(X)/\theta)$.*

PROOF. (⇒) If

$$p = p(x_1, \ldots, x_n),$$
$$q = q(x_1, \ldots, x_n)$$

then

$$\begin{aligned} &\mathbf{T}(X)/\theta \models p(x_1, \ldots, x_n) \approx q(x_1, \ldots, x_n) \\ \Rightarrow\ & p(x_1/\theta, \ldots, x_n/\theta) = q(x_1/\theta, \ldots, x_n/\theta) \\ \Rightarrow\ & p(x_1, \ldots, x_n)/\theta = q(x_1, \ldots, x_n)/\theta \\ \Rightarrow\ & \langle p(x_1, \ldots, x_n), q(x_1, \ldots, x_n) \rangle \in \theta \\ \Rightarrow\ & \langle p, q \rangle \in \theta. \end{aligned}$$

(⇐) Given $r_1, \ldots, r_n \in T(X)$ we can find an endomorphism $\varepsilon$ of $\mathbf{T}(X)$ with

$$\varepsilon(x_i) = r_i, \qquad 1 \le i \le n;$$

hence

$$\begin{aligned} &\langle p(x_1, \ldots, x_n), q(x_1, \ldots, x_n) \rangle \in \theta \\ \Rightarrow\ & \langle \varepsilon p(x_1, \ldots, x_n), \varepsilon q(x_1, \ldots, x_n) \rangle \in \theta \\ \Rightarrow\ & \langle p(r_1, \ldots, r_n), q(r_1, \ldots, r_n) \rangle \in \theta \\ \Rightarrow\ & p(r_1/\theta, \ldots, r_n/\theta) = q(r_1/\theta, \ldots, r_n/\theta). \end{aligned}$$

Thus

$$\mathbf{T}(X)/\theta \models p \approx q.$$

For the last claim, given $p \approx q \in \mathrm{Id}(X)$,

$$\begin{aligned} \langle p, q \rangle \in \theta &\Leftrightarrow \mathbf{T}(X)/\theta \models p \approx q \\ &\Leftrightarrow V(\mathbf{T}(X)/\theta) \models p \approx q \qquad \text{(by 11.3)}, \end{aligned}$$

so $\mathbf{T}(X)/\theta$ is free in $V(\mathbf{T}(X)/\theta)$ by 11.4. □

**Theorem 14.8.** *Given a subset $\Sigma$ of* Id($X$), *one can find a $K$ such that*

$$\Sigma = \mathrm{Id}_K(X)$$

*iff $\tau(\Sigma)$ is a fully invariant congruence on* $\mathbf{T}(X)$.

PROOF. (⇒) This was proved in 14.6.

(⇐) Suppose $\tau(\Sigma)$ is a fully invariant congruence $\theta$. Let $K = \{\mathbf{T}(X)/\theta\}$. Then by 14.7

$$\begin{aligned} K \models p \approx q &\Leftrightarrow \langle p, q \rangle \in \theta \\ &\Leftrightarrow p \approx q \in \Sigma. \end{aligned}$$

Thus $\Sigma = \mathrm{Id}_K(X)$. □

**Definition 14.9.** A subset $\Sigma$ of Id($X$) is called an *equational theory over $X$* if there is a class of algebras $K$ such that

$$\Sigma = \mathrm{Id}_K(X).$$

**Corollary 14.10.** *The equational theories (of type $\mathscr{F}$) over $X$ form an algebraic lattice which is isomorphic to the lattice of fully invariant congruences on* $\mathbf{T}(X)$.

PROOF. This follows from 14.4 and 14.8. □

**Definition 14.11.** Let $X$ be a set of variables and $\Sigma$ a set of identities of type $\mathscr{F}$ with variables from $X$. For $p,q \in T(X)$ we say

$$\Sigma \models p \approx q$$

(read: "$\Sigma$ yields $p \approx q$") if, given any algebra $\mathbf{A}$,

$$\mathbf{A} \models \Sigma \quad \text{implies} \quad \mathbf{A} \models p \approx q.$$

**Theorem 14.12.** *If $\Sigma$ is a set of identities over $X$ and $p \approx q$ is an identity over $X$, then*

$$\Sigma \models p \approx q \Leftrightarrow \langle p,q \rangle \in \Theta_{\mathrm{FI}}(\tau\Sigma).$$

PROOF. Suppose

$$\mathbf{A} \models \Sigma.$$

Then as $\tau(\mathrm{Id}_{\mathbf{A}}(X))$ is a fully invariant congruence on $\mathbf{T}(X)$ by 14.6, we have

$$\Theta_{\mathrm{FI}}(\tau\Sigma) \subseteq \tau\ \mathrm{Id}_{\mathbf{A}}(X);$$

hence

$$\langle p,q \rangle \in \Theta_{\mathrm{FI}}(\tau\Sigma) \Rightarrow \mathbf{A} \models p \approx q,$$

so

$$\langle p,q \rangle \in \Theta_{\mathrm{FI}}(\tau\Sigma) \Rightarrow \Sigma \models p \approx q.$$

Conversely, by 14.7

$$\mathbf{T}(X)/\Theta_{\mathrm{FI}}(\tau\Sigma) \models \Sigma,$$

so if

$$\Sigma \models p \approx q$$

then

$$\mathbf{T}(X)/\Theta_{\mathrm{FI}}(\tau\Sigma) \models p \approx q;$$

hence by 14.7,

$$\langle p,q \rangle \in \Theta_{\mathrm{FI}}(\tau\Sigma).$$ □

In the proof of 14.4 we gave an explicit description of the operations needed to construct the fully invariant closure $\Theta_{\mathrm{FI}}(S)$ of a set of ordered pairs $S$ from an algebra. This will lead to an elegant set of axioms and rules of inference for working with identities.

**Definition 14.13.** Given a term $p$, the *subterms* of $p$ are defined by:

(1) The term $p$ is a subterm of $p$.

(2) If $f(p_1, \ldots, p_n)$ is a subterm of $p$ and $f \in \mathscr{F}_n$ then each $p_i$ is a subterm of $p$.

**Definition 14.14.** A set of identities $\Sigma$ over $X$ is closed under *replacement* if given any $p \approx q \in \Sigma$ and any term $r \in T(X)$, if $p$ occurs as a subterm of $r$, then letting $s$ be the result of replacing that occurrence of $p$ by $q$, we have $r \approx s \in \Sigma$.

**Definition 14.15.** A set of identities $\Sigma$ over $X$ is closed under *substitution* if for each $p \approx q$ in $\Sigma$ and each $r \in T(X)$, if we replace every occurrence of a given variable $x$ in $p \approx q$ by $r$, then the resulting identity is in $\Sigma$.

**Definition 14.16.** If $\Sigma$ is a set of identities over $X$, then the *deductive closure* $D(\Sigma)$ of $\Sigma$ is the smallest subset of $\mathrm{Id}(X)$ containing $\Sigma$ such that

$p \approx p \in D(\Sigma)$ for $p \in T(X)$
$p \approx q \in D(\Sigma) \Rightarrow q \approx p \in D(\Sigma)$
$p \approx q,\ q \approx r \in D(\Sigma) \Rightarrow p \approx r \in D(\Sigma)$
$D(\Sigma)$ is closed under replacement
$D(\Sigma)$ is closed under substitution.

**Theorem 14.17.** *Given* $\Sigma \subseteq \mathrm{Id}(X)$, $p \approx q \in \mathrm{Id}(X)$,

$$\Sigma \models p \approx q \Leftrightarrow p \approx q \in D(\Sigma).$$

PROOF. The first three closure properties make $\tau D(\Sigma)$ into an equivalence relation containing $\tau\Sigma$, the fourth makes it a congruence, and the last closure property says $\tau D(\Sigma)$ is a fully invariant congruence. Thus

$$\tau D(\Sigma) \supseteq \Theta_{\mathrm{FI}}(\tau\Sigma).$$

However $\tau^{-1}\Theta_{\mathrm{FI}}(\tau\Sigma)$ has all five closure properties and contains $\Sigma$; hence

$$\tau D(\Sigma) = \Theta_{\mathrm{FI}}(\tau\Sigma).$$

Thus

$$\begin{aligned} \Sigma \models p \approx q &\Leftrightarrow \langle p,q\rangle \in \Theta_{\mathrm{FI}}(\tau\Sigma) \qquad \text{(by 14.12)} \\ &\Leftrightarrow p \approx q \in D(\Sigma). \end{aligned}$$

□

Thus we see that using only the most obvious rules for working with identities we can derive all possible consequences. From this we can set up the following equational logic.

**Definition 14.18.** Let $\Sigma$ be a set of identities over $X$. For $p \approx q \in \mathrm{Id}(X)$ we say

$$\Sigma \vdash p \approx q$$

(read "$\Sigma$ proves $p \approx q$") if there is a sequence of identities

$$p_1 \approx q_1, \ldots, p_n \approx q_n$$

from $\mathrm{Id}(X)$ such that each $p_i \approx q_i$ belongs to $\Sigma$, or is of the form $p \approx p$, or is a result of applying any of the last four closure rules of 14.16 to previous identities in the sequence, and the last identity $p_n \approx q_n$ is $p \approx q$. The sequence

$p_1 \approx q_1, \ldots, p_n \approx q_n$ is called a *formal deduction* of $p \approx q$, and $n$ is the *length* of the deduction.

**Theorem 14.19** (Birkhoff: The Completeness Theorem for Equational Logic). *Given* $\Sigma \subseteq \mathrm{Id}(X)$ *and* $p \approx q \in \mathrm{Id}(X)$ *we have*

$$\Sigma \models p \approx q \Leftrightarrow \Sigma \vdash p \approx q.$$

PROOF. Certainly

$$\Sigma \vdash p \approx q \Rightarrow p \approx q \in D(\Sigma)$$

as we have used only properties under which $D(\Sigma)$ is closed in the construction of a formal deduction $p_1 \approx q_1, \ldots, p_n \approx q_n$ of $p \approx q$.

For the converse of this, first it is obvious that

$$\Sigma \vdash p \approx q \quad \text{for } p \approx q \in \Sigma$$

and

$$\Sigma \vdash p \approx p \quad \text{for } p \in T(X).$$

If

$$\Sigma \vdash p \approx q$$

then there is a formal deduction

$$p_1 \approx q_1, \ldots, p_n \approx q_n$$

of $p \approx q$. But then

$$p_1 \approx q_1, \ldots, p_n \approx q_n, \qquad q_n \approx p_n$$

is a formal deduction of $q \approx p$.

If

$$\Sigma \vdash p \approx q, \qquad \Sigma \vdash q \approx r$$

let

$$p_1 \approx q_1, \ldots, p_n \approx q_n$$

be a formal deduction of $p \approx q$ and let

$$\bar{p}_1 \approx \bar{q}_1, \ldots, \bar{p}_k \approx \bar{q}_k$$

be a formal deduction of $q \approx r$.

Then

$$p_1 \approx q_1, \ldots, p_n \approx q_n, \qquad \bar{p}_1 \approx \bar{q}_1, \ldots, \bar{p}_k \approx \bar{q}_k, \qquad p_n \approx \bar{q}_k$$

is a formal deduction of $p \approx r$.

If

$$\Sigma \vdash p \approx q$$

let

$$p_1 \approx q_1, \ldots, p_n \approx q_n$$

be a formal deduction of $p \approx q$. Let

$$r(\ldots, p, \ldots)$$

denote a term with a specific occurrence of the subterm $p$. Then

$$p_1 \approx q_1, \ldots, p_n \approx q_n, \qquad r(\ldots, p_n, \ldots) \approx r(\ldots, q_n, \ldots)$$

is a formal deduction of

$$r(\ldots, p, \ldots) \approx r(\ldots, q, \ldots).$$

Finally, if

$$\Sigma \vdash p_i \approx q_i, \qquad 1 \leq i \leq n,$$

and $f \in \mathscr{F}_n$ then by writing the formal deductions of each $p_i \approx q_i$ in succession and adding the identity $f(p_1, \ldots, p_n) \approx f(q_1, \ldots, q_n)$ at the end we have a formal deduction of the latter, viz.,

$$\ldots, p_1 \approx q_1, \ldots, p_2 \approx q_2, \ldots, \ldots, p_n \approx q_n, f(p_1, \ldots, p_n) \approx f(p_1, \ldots, p_{n-1}, q_n), \ldots.$$

Thus

$$D(\Sigma) \subseteq \{p \approx q : \Sigma \vdash p \approx q\};$$

hence

$$D(\Sigma) = \{p \approx q : \Sigma \vdash p \approx q\},$$

so by 14.17

$$\Sigma \models p \approx q \Leftrightarrow \Sigma \vdash p \approx q. \qquad \square$$

The completeness theorem gives us a two-edged sword for tackling the study of consequences of identities. When using the notion of satisfaction, we look at all the algebras satisfying a given set of identities, whereas when working with $\vdash$ we can use induction arguments on the length of a formal deduction.

Examples. (1) An identity $p \approx q$ is *balanced* if each variable occurs the same number of times in $p$ as in $q$. If $\Sigma$ is a balanced set of identities then using induction on the length of a formal deduction we can show that if $\Sigma \vdash p \approx q$ then $p \approx q$ is balanced. [This is not at all evident if one works with the notion $\models$.]

(2) A famous theorem of Jacobson in ring theory says that if we are given $n \geq 2$, if $\Sigma$ is the set of ring axioms plus $x^n \approx x$, then $\Sigma \models x \cdot y \approx y \cdot x$. However there is no known routine way of writing out a formal deduction, given $n$, of $x \cdot y \approx y \cdot x$. (For special $n$, such as $n = 2,3$, this is a popular exercise.)

Another application of fully invariant congruences in the study of identities is to show the existence of minimal subvarieties.

**Definition 14.20.** A variety $V$ is *trivial* if all algebras in $V$ are trivial. A subclass $W$ of a variety $V$ which is also a variety is called a *subvariety* of $V$. $V$ is a *minimal* (or *equationally complete*) variety if $V$ is not trivial but the only subvariety of $V$ not equal to $V$ is the trivial variety.

**Theorem 14.21.** *Let $V$ be a nontrivial variety. Then $V$ contains a minimal subvariety.*

Proof. Let $V = M(\Sigma)$, $\Sigma \subseteq \mathrm{Id}(X)$ with $X$ infinite (see 11.8). Then $\mathrm{Id}_V(X)$ defines $V$, and as $V$ is nontrivial it follows from 14.6 that $\tau(\mathrm{Id}_V(X))$ is a fully invariant congruence on $\mathbf{T}(X)$ which is not $\nabla$. As

$$\nabla = \Theta_{\mathrm{FI}}(\langle x,y\rangle)$$

for any $x,y \in X$ with $x \neq y$, it follows that $\nabla$ is finitely generated (as a fully invariant congruence). This allows us to use Zorn's lemma to extend $\tau(\mathrm{Id}_V(X))$ to a maximal fully invariant congruence on $\mathbf{T}(X)$, say $\theta$. Then in view of 14.8, $\tau^{-1}\theta$ must define a minimal variety which is a subvariety of $V$. □

Example. The variety of lattices has a unique minimal subvariety, the variety generated by a two-element chain. To see this let $V$ be a minimal subvariety of the variety of lattices. Let $\mathbf{L}$ be a nontrivial lattice in $V$. As $\mathbf{L}$ contains a two-element sublattice, we can assume $\mathbf{L}$ is a two-element lattice. Now $V(\mathbf{L})$ is not trivial, and $V(\mathbf{L}) \subseteq V$, hence $V(\mathbf{L}) = V$. [We shall see in V§8 Exercise 2 that $V$ is the variety of all distributive lattices.]

We close this section with a look at an application of Tarski's irredundant basis theorem to sizes of minimal defining sets of identities.

**Definition 14.22.** Given a variety $V$ and a set of variables $X$ let

$$\mathrm{IrB}(\mathrm{Id}_V(X)) = \{|\Sigma| : \Sigma \text{ is a minimal finite set of identities over } X \text{ defining } V\}.$$

**Theorem 14.23** (Tarski). *Given a variety $V$ and a set of variables $X$, $\mathrm{IrB}(\mathrm{Id}_V(X))$ is a convex set.*

Proof. For $\Sigma \subseteq \mathrm{Id}_V(X)$, $\Sigma \vDash \mathrm{Id}_V(X)$ implies

$$\Theta_{\mathrm{FI}}(\tau\Sigma) = \tau\mathrm{Id}_V(X).$$

As $\Theta_{\mathrm{FI}}$ is 2-ary by 14.4, from 4.4 we have the result. □

## References

1. G. Birkhoff [1935]
2. A. Tarski [1975]

## Exercises §14

1. Show that the fully invariant congruences on an algebra $\mathbf{A}$ form a complete sublattice of $\mathbf{Con\ A}$.
2. Show that every variety of mono-unary algebras is defined by a single identity.

3. Verify the claim that consequences of balanced identities are again balanced.

4. Given a type $\mathscr{F}$ and a maximal fully invariant congruence $\theta$ on $\mathbf{T}(x,y)$ show that $V(\mathbf{T}(x,y)/\theta)$ is a minimal variety, and every minimal variety is of this form.

5. If $V$ is a minimal variety of groups show that $\mathbf{F}_V(\bar{x})$ is nontrivial, hence $V = V(\mathbf{F}_V(\bar{x}))$. Determine all minimal varieties of groups.

6. Determine all minimal varieties of semigroups.

7. If $p(x)$ is a term and $\Sigma$ is a set of identities such that $\Sigma \models p(x) \approx x$ and $\Sigma \models p(x) \approx p(y)$, show that $\Sigma \models x \approx y$; hence $M(\Sigma)$ is a trivial variety.

8. Let $f,g$ be two unary operation symbols. Let $N$ be the set of natural numbers, and for $I \subseteq N$ let

$$\Sigma_I = \{fgf^n g^2(x) \approx x : n \in I\} \cup \{fgf^n g^2(x) \approx fgf^n g^2(y) : n \notin I\}.$$

Show that $M(\Sigma_I)$ is not a trivial variety, but for $I \neq J$, $M(\Sigma_I) \cap M(\Sigma_J)$ is trivial. Conclude that there are $2^\omega$ minimal varieties of bi-unary algebras; hence some variety of bi-unary algebras is not defined by a finite set of identities.

9. If a variety $V$ is defined by an infinite minimal set of identities show that $V$ has at least continuum many varieties above it.

10. (The compactness theorem for equational logic) If a variety $V$ is defined by a finite set of identities, then for any other set $\Sigma$ of identities defining $V$ show that there is a finite subset $\Sigma_0$ of $\Sigma$ which defines $V$.

11. Given $\Sigma \subseteq \mathrm{Id}(X)$ let an *elementary deduction* from $\Sigma$ be one of the form

$$r(\ldots,\varepsilon p,\ldots) \approx r(\ldots,\varepsilon q,\ldots),$$

which is an identity obtained from $p \approx q$, where $p \approx q$ or $q \approx p \in \Sigma$, by first substituting for some variable $x$ the term $\varepsilon p$, where $\varepsilon$ is an endomorphism of $\mathbf{T}(X)$, and then replacing some occurrence of $\varepsilon p$ in a term by $\varepsilon q$. Show that $D(\Sigma)$ is the set of $r \approx s$ such that $r = s$ or there exist elementary deductions $r_i \approx s_i$, $1 \leq i \leq n$, with $r = r_1$, $s_i = r_{i+1}$, $1 \leq i < n$, and $s_n = s$, provided $X$ is infinite.

12. Write out a formal deduction of $x \cdot y \approx y \cdot x$ from the ring axioms plus $x \cdot x \approx x$.

# CHAPTER III
# Selected Topics

Now that we have covered the most basic aspects of universal algebra, let us take a brief look at how universal algebra relates to two other popular areas of mathematics. First we discuss two topics from combinatorics which can conveniently be regarded as algebraic systems, namely Steiner triple systems and mutually orthogonal Latin squares. In particular we will show how to refute Euler's conjecture. Then we treat finite state acceptors as partial unary algebras and look at the languages they accept—this will include the famous Kleene theorem on regular languages.

## §1. Steiner Triple Systems, Squags, and Sloops

**Definition 1.1.** A *Steiner triple system* on a set $A$ is a family $\mathscr{S}$ of three-element subsets of $A$ such that each pair of distinct elements from $A$ is contained in exactly one member of $\mathscr{S}$. $|A|$ is called the *order* of the Steiner triple system.

If $|A| = 1$ then $\mathscr{S} = \varnothing$, and if $|A| = 3$ then $\mathscr{S} = \{A\}$. Of course there are no Steiner triple systems on $A$ if $|A| = 2$. The following result gives some constraints on $|A|$ and $|\mathscr{S}|$. (Actually they are the best possible, but we will not prove this fact.)

**Theorem 1.2.** *If $\mathscr{S}$ is a Steiner triple system on a finite set $A$, then*

(a) $|\mathscr{S}| = |A| \cdot (|A| - 1)/6$

(b) $|A| \equiv 1$ or $3 \pmod 6$.

**Proof.** For (a) note that each member of $\mathscr{S}$ contains three distinct pairs of elements of $A$, and as each pair of elements appears in only one member of $\mathscr{S}$, it follows that the number of pairs of elements from $A$ is exactly $3|\mathscr{S}|$, i.e.,

$$\binom{|A|}{2} = 3|\mathscr{S}|.$$

To show that (b) holds, fix $a \in A$ and let $T_1, \ldots, T_k$ be the members of $\mathscr{S}$ to which $a$ belongs. Then the doubletons $T_1 - \{a\}, \ldots, T_k - \{a\}$ are mutually disjoint as no pair of elements of $A$ is contained in two disjoint triples of $\mathscr{S}$; and $A - \{a\} = (T_1 - \{a\}) \cup \cdots \cup (T_k - \{a\})$ as each member of $A - \{a\}$ is in some triple along with the element $a$. Thus $2 \,\big|\, |A| - 1$, so $|A| \equiv 1 \pmod 2$. From (a) we see that $|A| \equiv 0$ or $1 \pmod 3$; hence we have $|A| \equiv 1$ or $3 \pmod 6$. □

Thus after $|A| = 3$ the next possible size $|A|$ is 7. Figure 20 shows a Steiner triple system of order 7, where we require that three numbers be in a triple iff they lie on one of the lines drawn or on the circle. The reader will quickly convince himself that this is the only Steiner triple system of order 7 (up to a relabelling of the elements).

Are there some easy ways to construct new Steiner triple systems from old ones? If we convert to an algebraic system it will become evident that our standard constructions in universal algebra apply. A natural way of introducing a binary operation on $A$ is to require

$$a \cdot b = c \quad \text{if } \{a,b,c\} \in \mathscr{S}. \tag{*}$$

Unfortunately this leaves $a \cdot a$ undefined. We conveniently get around this by defining

$$a \cdot a = a. \tag{**}$$

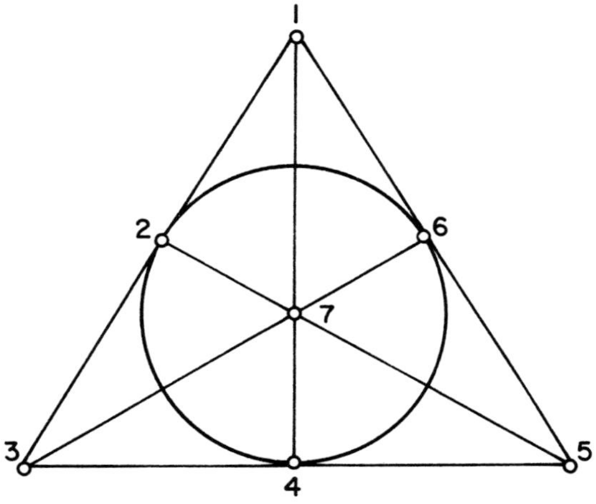


Figure 20

Although the associative law for $\cdot$ fails already in the system of order 3, nonetheless we have the identities

(Sq1) $x \cdot x \approx x$
(Sq2) $x \cdot y \approx y \cdot x$
(Sq3) $x \cdot (x \cdot y) \approx y$.

**Definition 1.3.** A groupoid satisfying the identities (Sq1)–(Sq3) above is called a *squag* (or *Steiner quasigroup*).

Now we will show that the variety of squags precisely captures the Steiner triple systems.

**Theorem 1.4.** *If* $\langle A,\cdot\rangle$ *is a squag define* $\mathcal{S}$ *to be the set of three-element subsets* $\{a,b,c\}$ *of* $A$ *such that the product of any two elements gives the third. Then* $\mathcal{S}$ *is a Steiner triple system on* $A$.

PROOF. Suppose $a \cdot b = c$ holds. Then

$$a \cdot (a \cdot b) = a \cdot c,$$

so by (Sq3)

$$b = a \cdot c.$$

Continuing, we see that the product of any two of $a,b,c$ gives the third. Thus in view of (Sq1), if any two are equal, all three are equal. Consequently for any two distinct elements of $A$ there is a unique third element (distinct from the two) such that the product of any two gives the third. Thus $\mathcal{S}$ is indeed a Steiner triple system on $A$. □

Another approach to converting a Steiner triple system $\mathcal{S}$ on $A$ to an algebra is to adjoin a new element, called 1, and replace (**) by

$$a \cdot a = 1 \qquad (**')$$

$$a \cdot 1 = 1 \cdot a = a. \qquad (**'')$$

This leads to a groupoid with identity $\langle A \cup \{1\},\cdot,1\rangle$ satisfying the identities

(Sl1) $x \cdot x \approx 1$
(Sl2) $x \cdot y \approx y \cdot x$
(Sl3) $x \cdot (x \cdot y) \approx y$.

**Definition 1.5.** A groupoid with a distinguished element $\langle A,\cdot,1\rangle$ is called a *sloop* (or *Steiner loop*) if the identities (Sl1)–(Sl3) hold.

**Theorem 1.6.** *If* $\langle A,\cdot,1\rangle$ *is a sloop and* $|A| \geq 2$, *define* $\mathcal{S}$ *to be the three-element subsets of* $A - \{1\}$ *such that the product of any two distinct elements gives the third. Then* $\mathcal{S}$ *is a Steiner triple system on* $A - \{1\}$.

PROOF. (Similar to 1.4.) □

## §2. Quasigroups, Loops, and Latin Squares

A quasigroup is usually defined to be a groupoid $\langle A,\cdot\rangle$ such that for any elements $a,b \in A$ there are unique elements $c,d$ satisfying

$$a \cdot c = b$$
$$d \cdot a = b.$$

The definition of quasigroups we adopted in II§1 has two extra binary operations \ and /, left division and right division respectively, which allow us to consider quasigroups as an equational class. Recall that the axioms for quasigroups $\langle A,/,\cdot,\backslash\rangle$ are given by

$$x\backslash(x \cdot y) \approx y \qquad (x \cdot y)/y \approx x$$
$$x \cdot (x\backslash y) \approx y \qquad (x/y) \cdot y \approx x.$$

To convert a quasigroup $\langle A,\cdot\rangle$ in the usual definition to one in our definition let $a/b$ be the unique solution $c$ of $c \cdot b = a$, and let $a\backslash b$ be the unique solution $d$ of $a \cdot d = b$. The four equations above are then easily verified. Conversely, given a quasigroup $\langle A,/,\cdot,\backslash\rangle$ by our definition and $a,b \in A$, suppose $c$ is such that $a \cdot c = b$. Then $a\backslash(a \cdot c) = a\backslash b$; hence $c = a\backslash b$, so only one such $c$ is possible. However, $a \cdot (a\backslash b) = b$, so there is one such $c$. Similarly, we can show that there is exactly one $d$ such that $d \cdot a = b$, namely $d = b/a$. Thus the two definitions of quasigroups are, in an obvious manner, equivalent.

A loop is usually defined to be a quasigroup with an identity element $\langle A,\cdot,1\rangle$. In our definition we have an algebra $\langle A,/,\cdot,\backslash,1\rangle$; and such loops form an equational class.

Returning to a Steiner triple system $\mathscr{S}$ on $A$ we see that the associated squag $\langle A,\cdot\rangle$ is indeed a quasigroup, for if $a \cdot c = b$ then $a \cdot (a \cdot c) = a \cdot b$, so $c = a \cdot b$, and furthermore $a \cdot (a \cdot b) = b$; hence if we are given $a,b$ there is a unique $c$ such that $a \cdot c = b$. Similarly, there is a unique $d$ such that $d \cdot a = b$. In the case of squags we do not need to introduce the additional operations / and \ to obtain an equational class, for in this case / and \ are the same as $\cdot$. Squags are sometimes called *idempotent totally symmetric quasigroups*.

Given any finite groupoid $\langle A,\cdot\rangle$ we can write out the multiplication table of $\langle A,\cdot\rangle$ in a square array, giving the *Cayley table* of $\langle A,\cdot\rangle$ (see Figure 21).

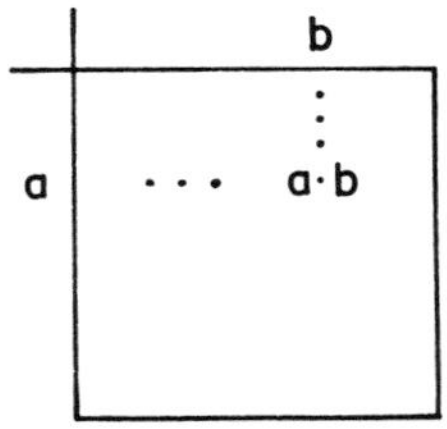


Figure 21

If we are given the Cayley table for a finite groupoid $\langle A, \cdot \rangle$, it is quite easy to check whether or not $\langle A, \cdot \rangle$ is actually a quasigroup.

**Theorem 2.1.** *A finite groupoid* $\mathbf{A}$ *is a quasigroup iff every element of* $A$ *appears exactly once in each row and in each column of the Cayley table of* $\langle A, \cdot \rangle$.

PROOF. If we are given $a,b \in A$, then there is exactly one $c$ satisfying $a \cdot c = b$ iff $b$ occurs exactly once in the $a$th row of the Cayley table of $\langle A, \cdot \rangle$, and there is exactly one $d$ such that $d \cdot a = b$ iff $b$ occurs exactly once in the $a$th column of the Cayley table. □

**Definition 2.2.** A *Latin square* of *order n* is an $n \times n$ matrix $(a_{ij})$ of elements from an $n$ element set $A$ such that each member of $A$ occurs exactly once in each row and each column of the matrix. (See Figure 22 for a Latin square of order 4.)

| | | | |
|---|---|---|---|
| a | b | c | d |
| d | c | a | b |
| b | a | d | c |
| c | d | b | a |

Figure 22

From Theorem 2.1 it is clear that Latin squares are in an obvious one-to-one correspondence with quasigroups by using Cayley tables.

REFERENCES

1. R. H. Bruck [6], [1963]
2. R. W. Quackenbush [1976]

## §3. Orthogonal Latin Squares

**Definition 3.1.** If $(a_{ij})$ and $(b_{ij})$ are two Latin squares of order $n$ with entries from $A$ with the property that for each $\langle a,b \rangle \in A \times A$ there is exactly one index $ij$ such that $\langle a,b \rangle = \langle a_{ij}, b_{ij} \rangle$, then we say that $(a_{ij})$ and $(b_{ij})$ are *orthogonal Latin squares.*

Figure 23 shows an example of orthogonal Latin squares of order 3. In the late 1700's Euler was asked if there were orthogonal Latin squares of

| | | |
|---|---|---|
| a | b | c |
| b | c | a |
| c | a | b |

| | | |
|---|---|---|
| a | b | c |
| c | a | b |
| b | c | a |

Figure 23

order 6. Euler conjectured: *if* $n \equiv 2 \pmod 4$ *then there do not exist orthogonal Latin squares of order n*. However he was unable to prove even a single case of this conjecture for $n > 2$. In 1900 Tarry verified the conjecture for $n = 6$ (this is perhaps surprising if one considers that there are more than 800 million Latin squares on a set of six elements). Later Macneish gave a construction of orthogonal Latin squares of all orders $n$ where $n \not\equiv 2 \pmod 4$. Then in 1959–60, Bose, Parker, and Shrikhande showed that $n = 2,6$ are the only values for which Euler's conjecture is actually true! Following the elegant presentation of Evans we will show, by converting orthogonal Latin squares into algebras, how to construct a pair of orthogonal Latin squares of order 54, giving a counterexample to Euler's conjecture.

In view of §2, two orthogonal Latin squares on a set $A$ correspond to two quasigroups $\langle A,/,\cdot,\backslash\rangle$ and $\langle A,\oslash,\circ,\obslash\rangle$ such that the map $\langle a,b\rangle \mapsto \langle a \cdot b, a \circ b\rangle$ is a permutation of $A \times A$. For a finite set $A$ this will be a bijection iff there exist functions $*_l$ and $*_r$ from $A \times A$ to $A$ such that

$$*_l(a \cdot b, a \circ b) = a$$
$$*_r(a \cdot b, a \circ b) = b.$$

Thus we are led to the following algebraic structures.

**Definition 3.2** (Evans). A *pair of orthogonal Latin squares* is an algebra

$$\langle A,/,\cdot,\backslash,\oslash,\circ,\obslash,*_l,*_r\rangle$$

with eight binary operations such that

(i) $\langle A,/,\cdot,\backslash\rangle$ is a quasigroup
(ii) $\langle A,\oslash,\circ,\obslash\rangle$ is a quasigroup
(iii) $*_l(x \cdot y, x \circ y) \approx x$
(iv) $*_r(x \cdot y, x \circ y) \approx y$.

The *order* of such an algebra is the cardinality of its universe. Let POLS be the *variety of pairs of orthogonal Latin squares.*

Now let us show how to construct a pair of orthogonal Latin squares of order $n$ for any $n$ which is not congruent to 2 (mod 4).

**Lemma 3.3.** *If $q$ is a prime power and $q \geq 3$, then there is a member of* POLS *of order q.*

Proof. Let $\langle K,+,\cdot\rangle$ be a finite field of order $q$, and let $e_1,e_2$ be two distinct nonzero elements of $K$. Then define two binary operations $\square_1$ and $\square_2$ on $K$ by

$$a \,\square_i\, b = e_i \cdot a + b.$$

Note that the two groupoids $\langle K,\square_1\rangle$ and $\langle K,\square_2\rangle$ are actually quasigroups, for $a \,\square_i\, c = b$ holds iff $c = b - e_i \cdot a$, and $d \,\square_i\, a = b$ holds iff $d = e_i^{-1} \cdot (b - a)$. Also we have that

$$\langle a \,\square_1\, b, a \,\square_2 b\rangle = \langle c \,\square_1\, d, c \,\square_2\, d\rangle$$

implies

$$e_1 \cdot a + b = e_1 \cdot c + d$$
$$e_2 \cdot a + b = e_2 \cdot c + d;$$

hence

$$e_1 \cdot (a - c) = d - b$$
$$e_2 \cdot (a - c) = d - b$$

and thus, as $e_1 \neq e_2$,

$$a = c \quad \text{and} \quad b = d.$$

Thus the Cayley tables of $\langle K,\square_1\rangle$ and $\langle K,\square_2\rangle$ give rise to orthogonal Latin squares of order $q$. □

**Theorem 3.4.** *If* $n \equiv 0$, 1, *or* 3 (mod 4), *then there is a pair of orthogonal Latin squares of order* $n$.

Proof. Note that $n \equiv 0,1$ or 3 (mod 4) iff $n = 2^\alpha p_1^{\alpha_1} \cdots p_k^{\alpha_k}$ with $\alpha \neq 1$, $\alpha_i \geq 1$, and each $p_i$ is an odd prime. The case $n = 1$ is trivial, and for $n \geq 3$ use 3.3 to construct $\mathbf{A}_0,\mathbf{A}_1, \ldots ,\mathbf{A}_k$ in POLS of order $2^\alpha, p_1^{\alpha_1}, \ldots ,p_k^{\alpha_k}$ respectively. Then $\mathbf{A}_0 \times \mathbf{A}_1 \times \ldots \times \mathbf{A}_k$ is the desired algebra. □

To refute Euler's conjecture we need to be more clever.

**Definition 3.5.** An algebra $\langle A,F\rangle$ is a *binary algebra* if each of the fundamental operations is binary. A binary algebra $\langle A,F\rangle$ is *idempotent* if

$$f(x,x) \approx x$$

holds for each function symbol $f$.

**Definition 3.6.** Let IPOLS be the variety of idempotent algebras in POLS.

Our goal is to show that there is an idempotent pair of orthogonal Latin squares of order 54. We construct this algebra by using a block design obtained from the projective plane of order 7 to paste together some small members of IPOLS which come from finite fields.

**Definition 3.7.** A variety $V$ of algebras is *binary idempotent* if
(i) the members of $V$ are binary idempotent algebras, and
(ii) $V$ can be defined by identities involving at most two variables.

Note that IPOLS is a binary idempotent variety.

**Definition 3.8.** A *2-design* is a tuple $\langle B, B_1, \ldots, B_k \rangle$ where
(i) $B$ is a finite set,
(ii) each $B_i$ is a subset of $B$ (called a *block*),
(iii) $|B_i| \geq 2$ for all $i$, and
(iv) each two-element subset of $B$ is contained in exactly one block.

The crucial idea is contained in the following.

**Lemma 3.9.** *Let $V$ be a binary idempotent variety and let $\langle B, B_1, \ldots, B_k \rangle$ be a 2-design. Let $n = |B|$, $n_i = |B_i|$. If $V$ has members of size $n_i$, $1 \leq i \leq k$, then $V$ has a member of size $n$.*

Proof. Let $\mathbf{A}_i \in V$ with $|A_i| = n_i$. We can assume $A_i = B_i$. Then for each binary function symbol $f$ in the type of $V$ we can find a binary function $f^B$ on $B$ such that when we restrict $f^B$ to $B_i$ it agrees with $f^{\mathbf{A}_i}$ (essentially we let $f^B$ be the union of the $f^{\mathbf{A}_i}$). As $V$ can be defined by two variable identities $p(x,y) \approx q(x,y)$ which hold on each $\mathbf{A}_i$, it follows that we have constructed an algebra $\mathbf{B}$ in $V$ with $|B| = n$. ☐

**Lemma 3.10.** *If $q$ is a prime power and $q \geq 4$, then there is a member of IPOLS of size $q$. In particular, there are members of sizes 5, 7, and 8.*

Proof. Again let $\mathbf{K}$ be a field of order $q$, let $e_1, e_2$ be two distinct elements of $K - \{0,1\}$, and define two binary operations $\Box_1, \Box_2$ on $K$ by

$$a \Box_i b = e_i \cdot a + (1 - e_i) \cdot b.$$

We leave it to the reader to verify that the Cayley tables of $\langle K, \Box_1 \rangle$ and $\langle K, \Box_2 \rangle$ give rise to an idempotent pair of orthogonal Latin squares. ☐

Now we need a construction from finite projective geometry. Given a finite field $F$ of cardinality $n + 1$ we form the *projective plane* $\mathscr{P}_n$ *of order* $n$ by letting the *points* be the subsets of $F^3$ of the form $a + U$ where $a \in F^3$ and $U$ is a one-dimensional subspace of $F^3$ (as a vector space over $F$), and by letting the *lines* be the subsets of $F^3$ of the form $a + V$ where $a \in F^3$ and $V$ is a two-dimensional subspace of $F^3$. One can readily verify that every line of $\mathscr{P}_n$ contains $n + 1$ points, and every point of $\mathscr{P}_n$ "belongs to" (i.e., is contained in) $n + 1$ lines; and there are $n^2 + n + 1$ points and $n^2 + n + 1$ lines. Furthermore, any two distinct points belong to exactly one line and any two distinct lines meet in exactly one point.

**Lemma 3.11.** *There is a* 2-*design* $\langle B,B_1,\ldots,B_k\rangle$ *with* $|B| = 54$ *and* $|B_i| \in \{5,7,8\}$ *for* $1 \leq i \leq k$.

PROOF. Let $\pi$ be the projective plane of order 7. This has 57 points and each line contains 8 points. Choose three points on one line and remove them. Let $B$ be the set of the remaining 54 points, and let the $B_i$ be the sets obtained by intersecting the lines of $\pi$ with $B$. Then $\langle B,B_1,\ldots,B_k\rangle$ is easily seen to be a 2-design since each pair of points from $B$ lies on a unique line of $\pi$, and $|B_i| \in \{5,7,8\}$. □

**Theorem 3.12.** *There is an idempotent pair of orthogonal Latin squares of order* 54.

PROOF. Just combine 3.9, 3.10, and 3.11. □

REFERENCE

1. T. Evans [1979]

## §4. Finite State Acceptors

In 1943 McCulloch and Pitts developed a model of nerve nets which was later formalized as various types of finite state machines. The idea is quite simple. One considers the nervous system as a finite collection of internal neurons and sensory neurons and considers time as divided into suitably small subintervals such that in each subinterval each neuron either fires once or is inactive. The firing of a given neuron during any one subinterval will send impulses to certain other internal neurons during that subinterval. Such impulses are either activating or deactivating. If an internal neuron receives sufficiently many (the threshold of the neuron) activating impulses and no deactivating impulses in a given subinterval, then it fires during the next subinterval of time. The sensory neurons can only be excited to fire by external stimuli. In any given subinterval of time, the state of the network of internal neurons is defined by noting which neurons are firing and which are not, and the input during any given subinterval to the network is determined by which sensory neurons are firing and which are not. We call an input during a subinterval of time a letter, the totality of letters constituting the alphabet. A sequence of inputs (in consecutive subintervals) is a word. A word is accepted (or recognized) by the neural network if after the sensory neurons proceed through the sequence of inputs given by the letters of the word the internal neurons at some specified number of subintervals later are in some one of the so-called accepting states.

In his 1956 paper, Kleene analyzed the possibilities for the set of all words which could be accepted by a neural network and showed that they are precisely the regular languages. Later Myhill showed the connection between these languages and certain congruences on the monoid of words. Let us now abstract from the nerve nets, where we consider the states as points and the letters of the alphabet as functions acting on the states, i.e., if we are in a given state and read a given letter, the resulting state describes the action of the letter on the given state.

**Definition 4.1.** A *finite state acceptor* (abbreviated f.s.a.) of type $\mathscr{F}$ (where the type is finite) is a 4-tuple $\mathbf{A} = \langle A,F,a_0,A_0\rangle$, where $\langle A,F\rangle$ is a finite unary algebra of type $\mathscr{F}$, $a_0 \in A$, and $A_0 \subseteq A$. The set $A$ is the set of *states* of $\mathbf{A}$, $a_0$ is the *initial state*, and $A_0$ is the *set of final states.*

**Definition 4.2.** If we are given a finite type $\mathscr{F}$ of unary algebras, let $\langle \mathscr{F}^*,\cdot,1\rangle$ be the monoid of strings on $\mathscr{F}$. Given a string $w \in \mathscr{F}^*$, an f.s.a. $\mathbf{A}$ of type $\mathscr{F}$, and an element $a \in A$, let $w(a)$ be the element resulting from applying the "term" $w(x)$ to $a$; for example if $w = fg$ then $w(a) = f(g(a))$, and $1(a) = a$.

**Definition 4.3.** A *language* of type $\mathscr{F}$ is a subset of $\mathscr{F}^*$. A string $w$ from $\mathscr{F}^*$ is *accepted* by an f.s.a. $\mathbf{A} = \langle A,F,a_0,A_0\rangle$ of type $\mathscr{F}$ if $w(a_0) \in A_0$. The *language accepted by* $\mathbf{A}$, written $\mathscr{L}(\mathbf{A})$, is the set of strings from $\mathscr{F}^*$ accepted by $\mathbf{A}$. ("Language" has a different meaning in this section from that given in II§1.)

**Definition 4.4.** Given languages $L$,$L_1$, and $L_2$ of type $\mathscr{F}$ let

$$L_1 \cdot L_2 = \{w_1 \cdot w_2 : w_1 \in L_1, w_2 \in L_2\} \text{ and}$$
$$L^* = \text{the subuniverse of } \langle \mathscr{F}^*,\cdot,1\rangle \text{ generated by } L.$$

The set of *regular languages* of types $\mathscr{F}$ is the smallest collection of subsets of $\mathscr{F}^*$ which contains the singleton languages $\{f\}$, $f \in \mathscr{F} \cup \{1\}$, and is closed under the set-theoretic operations, $\cup$, $\cap$, $'$, and the operations $\cdot$ and $*$ defined above.

To prove that the languages accepted by f.s.a.'s form precisely the class of regular languages it is convenient to introduce partial algebras.

**Definition 4.5.** A *partial unary algebra* of type $\mathscr{F}$ is a pair $\langle A,F\rangle$ where $F$ is a family of partially defined unary functions on $A$ indexed by $\mathscr{F}$, i.e., the domain and range of each function $f$ are contained in $A$.

**Definition 4.6.** A *partial finite state acceptor* (partial f.s.a.) $\mathbf{A} = \langle A,F,a_0,A_0\rangle$ of type $\mathscr{F}$ has the same definition as an f.s.a. of type $\mathscr{F}$, except that we only require that $\langle A,F\rangle$ be a partial unary algebra of type $\mathscr{F}$. Also the *language accepted by* **A**, $\mathscr{L}(\mathbf{A})$, is defined as in 4.3. (Note that for a given $w \in \mathscr{F}^*$, $w(a)$ might not be defined for some $a \in A$.)

**Lemma 4.7.** *Every language accepted by a partial* f.s.a. *is accepted by some* f.s.a.

PROOF. Given a partial f.s.a. $\mathbf{A} = \langle A,F,a_0,A_0\rangle$ choose $b \notin A$ and let $B = A \cup \{b\}$. For $f \in \mathscr{F}$ and $a \in A \cup \{b\}$, if $f(a)$ is not defined in **A** let $f(a) = b$. This gives an f.s.a. which accepts the same language as **A**. □

**Definition 4.8.** If $\langle A,F,a_0, A_0\rangle$ is a partial f.s.a. then, for $a \in A$ and $w \in \mathscr{F}^*$, the *range of w applied to a*, written $\mathrm{Rg}(w,a)$, is the set

$$\{f_n(a), f_{n-1} f_n(a), \ldots, f_1 \ldots f_n(a)\}$$

where $w = f_1 \ldots f_n$; and it is $\{a\}$ if $w = 1$.

**Lemma 4.9.** *The language accepted by any* f.s.a. *is regular.*

PROOF. Let $L$ be the language of the partial f.s.a. $\mathbf{A} = \langle A,F,a_0,A_0\rangle$. We will prove the lemma by induction on $|A|$. First note that $\varnothing$ is a regular language as $\varnothing = \{f\} \cap \{f\}'$ for any $f \in \mathscr{F}$. For the ground case suppose $|A| = 1$. If $A_0 = \varnothing$ then $\mathscr{L}(\mathbf{A}) = \varnothing$, a regular language. If $A_0 = \{a_0\}$ let

$$\mathscr{G} = \{f \in \mathscr{F} : f(a_0) \text{ is defined}\}.$$

Then

$$\mathscr{L}(\mathbf{A}) = \mathscr{G}^* = \left(\bigcup_{f \in \mathscr{G}} \{f\}\right)^*,$$

also a regular language.

For the induction step assume that $|A| > 1$, and for any partial f.s.a. $\mathbf{B} = \langle B,F,b_0, B_0\rangle$ with $|B| < |A|$ the language $\mathscr{L}(\mathbf{B})$ is regular. If $A_0 = \varnothing$ then, as before, $\mathscr{L}(\mathbf{A}) = \varnothing$, a regular language. So assume $A_0 \neq \varnothing$. The crux of the proof is to decompose any acceptable word into a product of words which one can visualize as giving a sequence of cycles when applied to $a_0$, followed by a noncycle, mapping from $a_0$ to a member of $A_0$ if $a_0 \notin A_0$. Let

$$C = \{\langle f_1,f_2\rangle \in \mathscr{F} \times \mathscr{F} : f_1 w f_2(a_0) = a_0 \text{ for some } w \in \mathscr{F}^*,$$
$$f_2(a_0) \neq a_0, \text{ and } \mathrm{Rg}(w; f_2(a_0)) \subseteq A - \{a_0\}\}$$

which we picture as in Figure 24. Now, for $\langle f_1,f_2\rangle \in C$ let

$$C_{f_1 f_2} = \{w \in \mathscr{F}^* : f_1 w f_2(a_0) = a_0, \mathrm{Rg}(w; f_2(a_0)) \subseteq A - \{a_0\}\}.$$

Then $C_{f_1 f_2}$ is the language accepted by

$$\langle A - \{a_0\}, F, f_2(a_0), f_1^{-1}(a_0) - \{a_0\}\rangle \;;$$

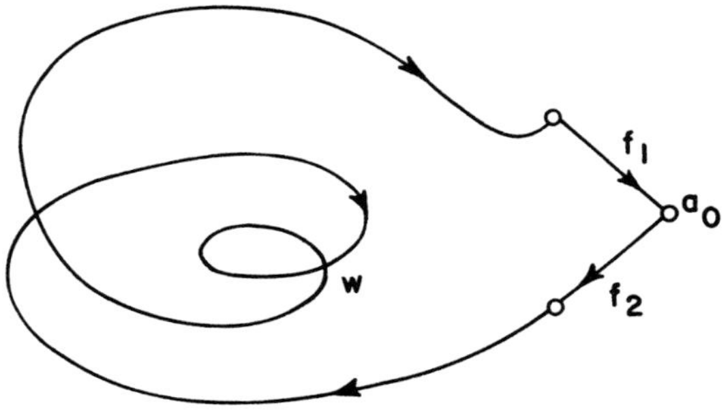


Figure 24

hence, by the induction hypothesis, $C_{f_1 f_2}$ is regular. Let

$$\mathscr{H} = \{f \in \mathscr{F}: f(a_0) = a_0\} \cup \{1\}$$

and

$$\mathscr{D} = \{f \in \mathscr{F}: f(a_0) \neq a_0\}.$$

For $f \in \mathscr{D}$ let

$$E_f = \{w \in \mathscr{F}^*: wf(a_0) \in A_0, \operatorname{Rg}(w, f(a_0)) \subseteq A - \{a_0\}\}.$$

We see that $E_f$ is the language accepted by

$$\langle A - \{a_0\}, F, f(a_0), A_0 - \{a_0\}\rangle;$$

hence by the induction hypothesis, it is also regular. Let

$$E = \begin{cases} \bigcup_{f \in \mathscr{D}} E_f \cdot \{f\} & \text{if } a_0 \notin A_0 \\ (\bigcup_{f \in \mathscr{D}} E_f \cdot \{f\}) \cup \{1\} & \text{if } a_0 \in A_0. \end{cases}$$

Then

$$L = E \cdot \left(\mathscr{H} \cup \bigcup_{\langle f_1, f_2\rangle \in C} \{f_1\} \cdot C_{f_1 f_2} \cdot \{f_2\}\right)^*,$$

a regular language. □

**Definition 4.10.** Given a type $\mathscr{F}$ and $t \notin \mathscr{F}$ let the *deletion homomorphism*

$$\delta_t : (\mathscr{F} \cup \{t\})^* \to \mathscr{F}^*$$

be the homomorphism defined by

$$\delta_t(f) = f \quad \text{for } f \in \mathscr{F}$$
$$\delta_t(t) = 1.$$

**Lemma 4.11.** *If $L$ is a language of type $\mathscr{F} \cup \{t\}$, where $t \notin \mathscr{F}$, which is also the language accepted by some* f.s.a., *then $\delta_t(L)$ is a language of type $\mathscr{F}$ which is the language accepted by some* f.s.a.

Proof. Let $\mathbf{A} = \langle A, F \cup \{t\}, a_0, A_0\rangle$ be an f.s.a. with $\mathscr{L}(\mathbf{A}) = L$. For $w \in \mathscr{F}^*$ define

$$S_w = \{\overline{w}(a_0) : \overline{w} \in (\mathscr{F} \cup \{t\})^*, \delta_t(\overline{w}) = w\}$$

and let

$$B = \{S_w : w \in \mathscr{F}^*\}.$$

This is of course finite as $A$ is finite. For $f \in \mathscr{F}$ define

$$f(S_w) = S_{fw}.$$

This makes sense as $S_{fw}$ depends only on $S_w$, not on $w$. Next let

$$b_0 = S_1,$$

and let

$$B_0 = \{S_w : S_w \cap A_0 \neq \varnothing\}.$$

Then

$$\begin{aligned} &\langle B, F, b_0, B_0\rangle \text{ accepts } w \\ \text{iff}\quad & w(S_1) \in B_0 \\ \text{iff}\quad & S_w \cap A_0 \neq \varnothing \\ \text{iff}\quad & \overline{w}(a_0) \in A_0 \text{ for some } \overline{w} \in \delta_t^{-1}(w) \\ \text{iff}\quad & \overline{w} \in L \text{ for some } \overline{w} \in \delta_t^{-1}(w) \\ \text{iff}\quad & w \in \delta_t(L). \end{aligned}$$

□

**Theorem 4.12** (Kleene). *Let L be a language. Then L is the language accepted by some* f.s.a. *iff L is regular.*

Proof. We have already proved (⇒) in 4.9. For the converse we proceed by induction. If $L = \{f\}$ then we can use the partial f.s.a. in Figure 25, where all functions not drawn are undefined, and $A_0 = \{a\}$. If $L = \{1\}$ use $A = A_0 = \{a_0\}$ with all $f$'s undefined.

Next suppose $L_1$ is the language of $\langle A, F, a_0, A_0\rangle$ and $L_2$ is the language of $\langle B, F, b_0, B_0\rangle$. Then $L_1 \cap L_2$ is the language of $\langle A \times B, F, \langle a_0, b_0\rangle, A_0 \times B_0\rangle$, where $f(\langle a,b\rangle)$ is defined to be $\langle f(a), f(b)\rangle$; and $L_1'$ is the language of $\langle A, F, a_0, A - A_0\rangle$ (we are assuming $\langle A, F, a_0, A_0\rangle$ is an f.s.a.). Combining these we see by De Morgan's law that $L_1 \cup L_2$ is the language of a suitable f.s.a.

To handle $L_1 \cdot L_2$ we first expand our type to $\mathscr{F} \cup \{t\}$. Then mapping each member of $B_0$ to the input of a copy of $\mathbf{A}$ as in Figure 26 we see that $L_1 \cdot \{t\} \cdot L_2$ is the language of some f.s.a.; hence if we use 4.11 it follows that $L_1 \cdot L_2$ is the language of some f.s.a.

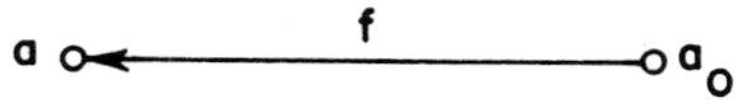


Figure 25

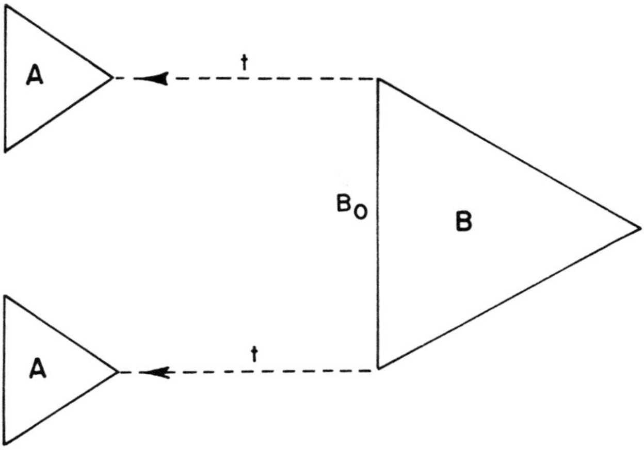


Figure 26

Similarly for $L_1^*$, let $t$ map each element of $A_0$ to $a_0$ as in Figure 27. Then $(L_1 \cdot \{t\})^* \cdot L_1$ is the language of this partial f.s.a.; hence

$$L_1^* = \delta_t[(L_1 \cdot \{t\})^* \cdot L_1 \cup \{1\}]$$

is the language of some f.s.a. This proves Kleene's theorem. □

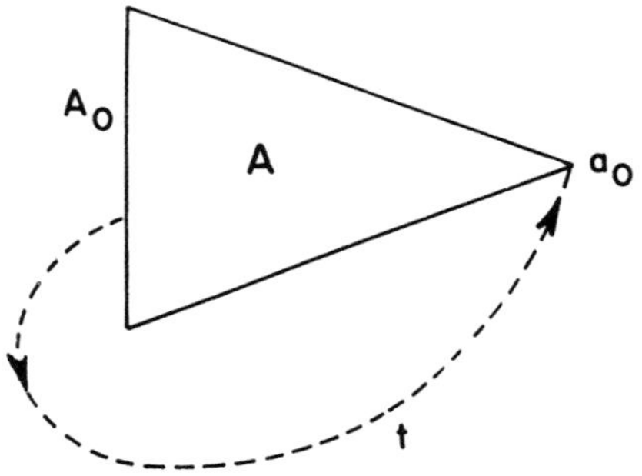


Figure 27

Another approach to characterizing languages accepted by f.s.a.'s of type $\mathscr{F}$ uses congruences on $\langle \mathscr{F}^*, \cdot, 1 \rangle$.

**Definition 4.13.** Let $\tau$ be the mapping from $\mathscr{F}^*$ to $T(x)$, the set of terms of type $\mathscr{F}$ over $x$, defined by $\tau(w) = w(x)$.

**Lemma 4.14.** *The mapping $\tau$ is an isomorphism between the monoid $\langle \mathscr{F}^*, \cdot, 1 \rangle$ and the monoid $\langle T(x), \circ, x \rangle$, where $\circ$ is "composition."*

Proof. (Exercise.) □

**Definition 4.15.** For $\theta \in \text{Con}\langle \mathscr{F}^*,\cdot,1\rangle$ let

$$\theta(x) = \{\langle w_1(x),w_2(x)\rangle : \langle w_1,w_2\rangle \in \theta\}.$$

**Lemma 4.16.** *The map $\theta \mapsto \theta(x)$ is a lattice isomorphism from the lattice of congruences of $\langle \mathscr{F}^*,\cdot,1\rangle$ to the lattice of fully invariant congruences on $\mathbf{T}(x)$. (See* II§14.)

Proof. Suppose $\theta \in \text{Con}\langle \mathscr{F}^*,\cdot,1\rangle$ and $\langle w_1,w_2\rangle \in \theta$. Then for $u \in \mathscr{F}^*$, $\langle uw_1,uw_2\rangle \in \theta$ suffices to show that $\theta(x)$ is a congruence on $\mathbf{T}(x)$, and $\langle w_1u,w_2u\rangle \in \theta$ shows that $\theta(x)$ is fully invariant. The remaining details we leave to the reader. □

**Lemma 4.17.** *If $L$ is a language of type $\mathscr{F}$ accepted by some* f.s.a., *then there is a $\theta \in \text{Con}\langle \mathscr{F}^*,\cdot,1\rangle$ such that $\theta$ is of finite index (i.e., $\langle \mathscr{F}^*,\cdot,1\rangle/\theta$ is finite) and $L^\theta = L$ (see* II§6.16), *i.e., $L$ is a union of cosets of $\theta$.*

Proof. Choose $\mathbf{A}$ an f.s.a. of type $\mathscr{F}$ such that $\mathscr{L}(\mathbf{A}) = L$. Let $\mathbf{F}_A(\bar{x})$ be the free algebra freely generated by $\bar{x}$ in the variety $V(\langle A,F\rangle)$. Let

$$\alpha : \mathbf{T}(x) \to \mathbf{F}_A(\bar{x})$$

be the natural homomorphism defined by $\alpha(x) = \bar{x}$, and let

$$\beta : \mathbf{F}_A(\bar{x}) \to \langle A,F\rangle$$

be the homomorphism defined by $\beta(\bar{x}) = a_0$. Then, with

$$L(x) = \{w(x) : w \in L\},$$
$$L(x) = \alpha^{-1}\beta^{-1}(A_0)$$
$$= \bigcup_{p \in \beta^{-1}(A_0)} p/\ker \alpha;$$

hence

$$L(x) = L(x)^{\ker \alpha}.$$

As $\ker \alpha$ is a fully invariant congruence on $\mathbf{T}(x)$ we have $\ker \alpha = \theta(x)$ for some $\theta \in \text{Con}\langle \mathscr{F}^*,\cdot,1\rangle$. Thus

$$L(x) = L(x)^{\theta(x)};$$

hence

$$L = L^\theta.$$

As $\ker \alpha$ is of finite index it follows that $\theta$ is also of finite index. □

**Theorem 4.18** (Myhill). *Let $L$ be a language of type $\mathscr{F}$. Then $L$ is the language of some* f.s.a. *iff there is a $\theta \in \text{Con}\langle \mathscr{F}^*,\cdot,1\rangle$ of finite index such that $L^\theta = L$.*

PROOF. ($\Rightarrow$) This was handled in 4.17.

($\Leftarrow$) Suppose $\theta$ is a congruence of finite index on $\mathscr{F}^*$ such that $L^\theta = L$. Let

$$\begin{aligned} A &= \{w/\theta : w \in \mathscr{F}\} \\ f(w/\theta) &= fw/\theta \quad \text{for } f \in \mathscr{F} \\ a_0 &= 1/\theta \\ A_0 &= \{w/\theta : w \in L\}. \end{aligned}$$

Then

$$\begin{aligned} &\langle A,F,a_0,A_0\rangle \text{ accepts } w \\ \text{iff}\quad & w(1/\theta) \in A_0 \\ \text{iff}\quad & w/\theta \in A_0 \\ \text{iff}\quad & w/\theta = u/\theta \text{ for some } u \in L \\ \text{iff}\quad & w \in L. \end{aligned}$$

□

**Definition 4.19.** Given a language $L$ of type $\mathscr{F}$ define the binary relation $\equiv_L$ on $\mathscr{F}^*$ by

$$w_1 \equiv_L w_2 \quad \text{iff} \quad (uw_1v \in L \Leftrightarrow uw_2v \in L \text{ for } u,v \in \mathscr{F}^*).$$

**Lemma 4.20.** *If we are given $L$, a language of type $\mathscr{F}$, then $\equiv_L$ is the largest congruence $\theta$ on $\langle \mathscr{F}^*,\cdot,1\rangle$ such that $L^\theta = L$.*

PROOF. Suppose $L^\theta = L$. Then for $\langle w_1,w_2\rangle \in \theta$ and $u,v \in \mathscr{F}^*$, $\langle uw_1v,uw_2v\rangle \in \theta$; hence $uw_1v \in L \Leftrightarrow uw_2v \in L$ as $uw_1v/\theta = uw_2v/\theta$ and $L = \bigcup_{w \in L} w/\theta$. Thus $\theta \subseteq \equiv_L$.

Next $\equiv_L$ is easily seen to be an equivalence relation on $\mathscr{F}^*$. If $w_1 \equiv_L w_2$ and $\hat{w}_1 \equiv_L \hat{w}_2$ then for $u,v \in \mathscr{F}^*$,

$$\begin{aligned} & uw_1\hat{w}_1v \in L \\ \text{iff}\quad & uw_1\hat{w}_2v \in L \\ \text{iff}\quad & uw_2\hat{w}_2v \in L; \end{aligned}$$

hence

$$w_1\hat{w}_1 \equiv_L w_2\hat{w}_2,$$

so $\equiv_L$ is indeed a congruence on $\langle \mathscr{F}^*,\cdot,1\rangle$.

If now $w \in L$ and $w \equiv_L \hat{w}$ then

$$1\cdot w\cdot 1 \in L \Leftrightarrow 1\cdot \hat{w}\cdot 1 \in L$$

implies $\hat{w} \in L$; hence $w/\equiv_L \subseteq L$. Thus $L^{\equiv_L} = L$. □

**Definition 4.21.** If we are given a language $L$ of type $\mathscr{F}$, then the *syntactic monoid* $M_L$ of $L$ is defined by

$$M_L = \langle \mathscr{F}^*,\cdot, 1\rangle/\equiv_L.$$

**Theorem 4.22.** *A language $L$ is accepted by some* f.s.a. *iff $M_L$ is finite.*

PROOF. Just combine 4.18 and 4.20. □

REFERENCES

1. J. Brzozowski [7], [7a]
2. S. C. Kleene [1956]
3. J. Myhill [1957]
4. J. von Neumann [37]

# CHAPTER IV
# Starting from Boolean Algebras . . .

Boolean algebras, essentially introduced by Boole in the 1850's to codify the laws of thought, have been a popular topic of research since then. A major breakthrough was the duality between Boolean algebras and Boolean spaces discovered by Stone in the 1930's. Stone also proved that Boolean algebras and Boolean rings are essentially the same in that one can convert via terms from one to the other. Following Stone's papers numerous results appeared which generalized or used his results to obtain structure theorems—these include the work of McCoy (rings), Rosenbloom (Post algebras), Arens and Kaplansky (rings), Foster (Boolean powers), Foster and Pixley (various notions of primality), Dauns and Hofmann (biregular rings), Pierce (rings), Comer (cylindric algebras and general algebras), and Bulman–Fleming, Keimel, and Werner (discriminator varieties).

Since every Boolean algebra can be represented as a field of sets, the class of Boolean algebras is sometimes regarded as being rather uncomplicated. However, when one starts to look at basic questions concerning decidability, rigidity, direct products, etc., they are associated with some of the most challenging results. Our major goal in this chapter will be representation theorems based on Boolean algebras, with some fascinating digressions.

## §1. Boolean Algebras

Let us repeat our definition from II§1.

**Definition 1.1.** A *Boolean algebra* is an algebra $\langle B, \vee, \wedge, ', 0, 1\rangle$ with two binary operations, one unary operation (called *complementation*), and two nullary operations which satisfies:

B1: $\langle B, \vee, \wedge \rangle$ is a distributive lattice
B2: $x \wedge 0 \approx 0$, $x \vee 1 \approx 1$
B3: $x \wedge x' \approx 0$, $x \vee x' \approx 1$.

Thus Boolean algebras form an equational class, hence a variety. Some useful properties of Boolean algebras follow.

**Lemma 1.2.** *Let* **B** *be a Boolean algebra. Then* **B** *satisfies*

B4: $a \wedge b = 0$ and $a \vee b = 1$ *imply* $a = b'$
B5: $(x')' \approx x$
B6: $(x \vee y)' \approx x' \wedge y'$, $(x \wedge y)' \approx x' \vee y'$ (*DeMorgan's Laws*).

PROOF. If

$$a \wedge b = 0$$

then

$$\begin{aligned} a' &= a' \vee (a \wedge b) \\ &= (a' \vee a) \wedge (a' \vee b) \\ &= 1 \wedge (a' \vee b) \\ &= a' \vee b; \end{aligned}$$

hence $a' \geq b$, and if

$$a \vee b = 1$$

then

$$\begin{aligned} a' &= a' \wedge (a \vee b) \\ &= (a' \wedge a) \vee (a' \wedge b) \\ &= 0 \vee (a' \wedge b) \\ &= a' \wedge b. \end{aligned}$$

Thus $a' \leq b$; hence

$$b = a'.$$

This proves B4.

Now

$$a' \wedge a = 0 \quad \text{and} \quad a' \vee a = 1;$$

hence

$$a = (a')'$$

by B4, so B5 is established. Finally

$$\begin{aligned} (x \vee y) \vee (x' \wedge y') &\approx x \vee [y \vee (x' \wedge y')] \\ &\approx x \vee [(y \vee x') \wedge (y \vee y')] \\ &\approx x \vee y \vee x' \\ &\approx 1 \end{aligned}$$

and

$$\begin{aligned} (x \vee y) \wedge (x' \wedge y') &\approx [x \wedge (x' \wedge y')] \vee [y \wedge (x' \wedge y')] \\ &\approx 0 \vee 0 \\ &\approx 0. \end{aligned}$$

Thus by B4

$$x' \wedge y' \approx (x \vee y)'.$$

Similarly (interchanging $\vee$ and $\wedge$, 0 and 1), we establish

$$x' \vee y' \approx (x \wedge y)'. \qquad \square$$

Perhaps the best known Boolean algebras are the following.

**Definition 1.3.** Let $X$ be a set. The *Boolean algebra of subsets of* $X$, $\mathbf{Su}(X)$, has as its universe $\mathrm{Su}(X)$ and as operations $\cup,\cap,',\varnothing,X$. The Boolean algebra $\mathbf{2}$ is given by $\langle 2,\vee,\wedge,',0,1\rangle$ where $\langle 2,\vee,\wedge\rangle$ is a two element lattice with $0 < 1$, and where $0' = 1$, $1' = 0$; also $\mathbf{1} = \langle\{\varnothing\},\vee,\wedge,',\varnothing,\varnothing\rangle$.

It is an easy exercise to see that if $|X| = 1$ then $\mathbf{Su}(X) \cong \mathbf{2}$; and $\mathbf{Su}(\varnothing) = \mathbf{1}$.

**Lemma 1.4.** *Let $X$ be a set. Then* $\mathbf{Su}(X) \cong \mathbf{2}^X$.

Proof. Let $\alpha\colon \mathrm{Su}(X) \to 2^X$ be such that

$$\alpha(Y)(x) = 1 \quad \text{iff} \quad x \in Y.$$

Then $\alpha$ is a bijection, and both $\alpha$ and $\alpha^{-1}$ are order-preserving maps between $\langle \mathrm{Su}(X),\subseteq\rangle$ and $\langle 2^X,\leq\rangle$; hence we have a lattice isomorphism. Also for $Y \subseteq X$

$$\begin{aligned}\alpha(Y')(x) = 1 \quad &\text{iff} \quad x \notin Y\\ &\text{iff} \quad \alpha(Y)(x) = 0;\end{aligned}$$

hence

$$\alpha(Y')(x) = (\alpha(Y)(x))',$$

so

$$\alpha(Y') = (\alpha(Y))'.$$

As

$$\alpha(\varnothing) = 0 \quad \text{and} \quad \alpha(X) = 1$$

we have an isomorphism. $\square$

**Definition 1.5.** If $\mathbf{B}$ is a Boolean algebra and $a \in B$, let $\mathbf{B}\restriction a$ be the algebra

$$\langle[0,a],\vee,\wedge,^*,0,a\rangle$$

where $[0,a]$ is the interval $\{x \in B: 0 \leq x \leq a\}$, $\vee$ and $\wedge$ are the same as in $\mathbf{B}$ except restricted to $[0,a]$, and $x^*$ is defined to be $a \wedge x'$.

**Lemma 1.6.** *If $\mathbf{B}$ is a Boolean algebra and $a \in B$ then $\mathbf{B}\restriction a$ is also a Boolean algebra.*

PROOF. Clearly $\langle [0,a], \vee, \wedge \rangle$ is a distributive lattice, as it is a sublattice of $\langle B, \vee, \wedge \rangle$. For $b \in [0,a]$ we have

$$b \wedge 0 = 0, \qquad b \vee a = a,$$

$$\begin{aligned} b \wedge b^* &= b \wedge (a \wedge b') \\ &= 0, \end{aligned}$$

$$\begin{aligned} b \vee b^* &= b \vee (a \wedge b') \\ &= (a \wedge b) \vee (a \wedge b') \\ &= a \wedge (b \vee b') \\ &= a \wedge 1 \\ &= a. \end{aligned}$$

Thus $\mathbf{B}\restriction a$ is a Boolean algebra. □

**Lemma 1.7.** *If* $\mathbf{B}$ *is a Boolean algebra and* $a \in B$ *then the map*

$$\alpha_a : B \to B\restriction a$$

*defined by*

$$\alpha_a(b) = a \wedge b$$

*is a surjective homomorphism from* $\mathbf{B}$ *to* $\mathbf{B}\restriction a$.

PROOF. If $b,c \in B$ then

$$\begin{aligned} \alpha_a(b \vee c) &= a \wedge (b \vee c) \\ &= (a \wedge b) \vee (a \wedge c) \\ &= \alpha_a(b) \vee \alpha_a(c), \end{aligned}$$

$$\begin{aligned} \alpha_a(b \wedge c) &= a \wedge (b \wedge c) \\ &= (a \wedge b) \wedge (a \wedge c) \\ &= \alpha_a(b) \wedge \alpha_a(c), \end{aligned}$$

$$\begin{aligned} \alpha_a(b') &= a \wedge b' \\ &= (a \wedge a') \vee (a \wedge b') \\ &= a \wedge (a' \vee b') \\ &= a \wedge (a \wedge b)' \\ &= (\alpha_a(b))^*, \end{aligned}$$

$$\alpha_a(0) = 0 \quad \text{and} \quad \alpha_a(1) = a.$$

Thus $\alpha_a$ is indeed a homomorphism. □

**Theorem 1.8.** *If* $\mathbf{B}$ *is a Boolean algebra and* $a \in B$, *then*

$$\mathbf{B} \cong \mathbf{B}\restriction a \times \mathbf{B}\restriction a'.$$

PROOF. Let

$$\alpha: B \to B\restriction a \times B\restriction a'$$

be defined by

$$\alpha(b) = \langle \alpha_a(b), \alpha_{a'}(b) \rangle.$$

It is easily seen that $\alpha$ is a homomorphism, and for

$$\langle b,c \rangle \in B\restriction a \times B\restriction a'$$

we have

$$\begin{aligned} \alpha(b \vee c) &= \langle a \wedge (b \vee c), a' \wedge (b \vee c) \rangle \\ &= \langle b,c \rangle \end{aligned}$$

as

$$\begin{aligned} a \wedge (b \vee c) &= (a \wedge b) \vee (a \wedge c) \\ &= b \vee 0 \\ &= b, \end{aligned}$$

etc. Thus $\alpha$ is surjective. Now if

$$\alpha(b) = \alpha(c)$$

for any $b,c \in B$ then

$$a \wedge b = a \wedge c \quad \text{and} \quad a' \wedge b = a' \wedge c$$

so

$$(a \wedge b) \vee (a' \wedge b) = (a \wedge c) \vee (a' \wedge c);$$

hence

$$(a \vee a') \wedge b = (a \vee a') \wedge c,$$

and thus

$$b = c.$$

This guarantees that $\alpha$ is the desired isomorphism. □

**Corollary 1.9** (Stone). **2** *is, up to isomorphism, the only directly indecomposable Boolean algebra which is nontrivial.*

PROOF. If **B** is a Boolean algebra and $|B| > 2$, let $a \in B$, $0 < a < 1$. Then $0 < a' < 1$, and hence both $\mathbf{B}\restriction a$ and $\mathbf{B}\restriction a'$ are nontrivial. From 1.8 it follows that **B** is not directly indecomposable. □

**Corollary 1.10** (Stone). *Every finite Boolean algebra is isomorphic to the Boolean algebra of all subsets of some finite set X.*

PROOF. Every finite Boolean algebra **B** is isomorphic to a direct product of directly indecomposable Boolean algebras by II§7.10; hence $\mathbf{B} \cong \mathbf{2}^n$ for some finite $n$. Now 1.4 applies. □

**Definition 1.11.** A *field of subsets* of the set $X$ is a subalgebra of $\mathbf{Su}(X)$, i.e., a family of subsets of $X$ closed under union, intersection, and complementation and containing $\varnothing$ and $X$, with the operations of $\mathbf{Su}(X)$.

**Corollary 1.12.** *Every Boolean algebra is isomorphic to a subdirect power of* **2**, *hence* (Stone) *every Boolean algebra is isomorphic to a field of sets.*

Proof. The only nontrivial subdirectly irreducible Boolean algebra is **2**, in view of 1.9. Thus Birkhoff's theorem guarantees that for every Boolean algebra **B** there is an $X$ and a subdirect embedding $\alpha:\mathbf{B} \to \mathbf{2}^X$, hence by 1.4 there is an embedding $\beta:\mathbf{B} \to \mathbf{Su}(X)$. □

**Definition 1.13.** Let $BA$ be the class of Boolean algebras.

The next result is immediate from 1.12.

**Corollary 1.14.** $BA = V(\mathbf{2}) = ISP(S) = IP_s(S)$, *where* $S = \{\mathbf{1},\mathbf{2}\}$.

References


1. P. R. Halmos [18]
2. L. Henkin, J. D. Monk, A. Tarski [19]
3. R. Sikorski [32]


Exercises §1

1. A subset $J$ of a set $I$ is a *cofinite subset of* $I$ if $I - J$ is finite. Show that the collection of finite and cofinite subsets of $I$ form a subuniverse of $\mathbf{Su}(I)$.

2. If $\mathbf{B}_1$ and $\mathbf{B}_2$ are two finite Boolean algebras with $|\mathbf{B}_1| = |\mathbf{B}_2|$, show $\mathbf{B}_1 \cong \mathbf{B}_2$.

3. Let **B** be a Boolean algebra. An element $b \in B$ is called an *atom* of **B** if $b$ covers 0 (see I§1). Show that an isomorphism between two Boolean algebras maps atoms to atoms.

4. Show that an infinite free Boolean algebra is *atomless* (i.e., has no atoms).

5. Show that any two denumerable atomless Boolean algebras are isomorphic. [Hint: Let $\mathbf{B}_0,\mathbf{B}_1$ be two such Boolean algebras. Given an isomorphism $\alpha:\mathbf{B}_0' \to \mathbf{B}_1'$, $\mathbf{B}_i'$ a finite subalgebra of $\mathbf{B}_i$, and $\mathbf{B}_0' \le \mathbf{B}_0'' \le \mathbf{B}_0$ with $\mathbf{B}_0''$ finite, show there is a $\mathbf{B}_1''$ with $\mathbf{B}_1' \le \mathbf{B}_1'' \le \mathbf{B}_1$ and an isomorphism $\beta:\mathbf{B}_0'' \to \mathbf{B}_1''$ extending $\alpha$. Iterate this procedure, alternately choosing the domain from $\mathbf{B}_0$, then from $\mathbf{B}_1$.]

6. If **B** is a (nontrivial) finite Boolean algebra show that the subalgebra of $\mathbf{B}^{B^B}$ generated by the projection maps $\pi_b:B^B \to B$, where $\pi_b(f) = f(b)$, has cardinality $2^{2^{|B|}}$.

7. Let $\mathbf{F}(n)$ denote the free Boolean algebra freely generated by $\{\bar{x}_1, \ldots, \bar{x}_n\}$. Show $\mathbf{F}(n) \cong \mathbf{2}^{2^n}$. [Hint: Use Exercise 6 above and II§11 Exercise 5.]

8. If **B** is a Boolean algebra and $a,b \in B$ with $a \wedge b = 0$ are such that $\mathbf{B}\restriction a \cong \mathbf{B}\restriction b$, show that there is an automorphism $\alpha$ of **B** such that $\alpha(a) = b$ and $\alpha(b) = a$.

9. If **A** is an algebra such that **Con A** is distributive, show that the factor congruences on **A** form a Boolean lattice which is a sublattice of **Con A**.

10. Let **B** be a subalgebra of $\mathbf{Su}(X)$. Show that $\hat{B} = \{Y \subseteq X : (Y \cap Z') \cup (Z \cap Y')$ is finite for some $Z \in B\}$ is a subuniverse of $\mathbf{Su}(X)$, and $\hat{B}$ contains all the atoms of $\mathbf{Su}(X)$.

11. Given a cardinal $\kappa \geq \omega$ and a set $X$ show that $\{Y \subseteq X: |Y| < \kappa \text{ or } |X - Y| < \kappa\}$ is a subuniverse of $\mathbf{Su}(X)$.

The study of cylindric algebras (see II§1) has parallels with the study of Boolean algebras. Let $CA_n$ denote *the class of cylindric algebras of dimension n*, and let $c(x)$ be the term $c_0(c_1(\ldots(c_{n-1}(x))\ldots))$. We will characterize the directly indecomposable members of $CA_n$ below.

12. Show $CA_n$ satisfies the following:

    (a) $c_i(x) \approx 0 \leftrightarrow x \approx 0$
    (b) $c_i(c_i(x)) \approx c_i(x)$
    (c) $x \wedge c_i(y) \approx 0 \leftrightarrow c_i(x) \wedge y \approx 0$
    (d) $c_i(x \vee y) \approx c_i(x) \vee c_i(y)$
    (e) $c(x \vee y) \approx c(x) \vee c(y)$
    (f) $c_i(c_i(x) \wedge c_i(y)) \approx c_i(x) \wedge c_i(y)$
    (g) $c_i((c_i x)') \approx (c_i x)'$
    (h) $c_i(x) \leq c(x)$
    (i) $c((cx)') \approx (cx)'$.

13. For $\mathbf{A} \in CA_n$ and $a \in A$ with $c(a) = a$ define $\mathbf{A}\restriction a$ to be the algebra $\langle [0,a], \vee, \wedge, {}^*, c_0, \ldots, c_{n-1}, 0, a, d_{00} \wedge a, \ldots, d_{n-1,n-1} \wedge a\rangle$, where the operations $\vee, \wedge, c_0, \ldots, c_{n-1}$ are the restrictions of the corresponding operations of $\mathbf{A}$ to $[0,a]$, and $x^* = a \wedge x'$. Show $\mathbf{A}\restriction a \in CA_n$, and the map $\alpha: A \to A\restriction a$ defined by $\alpha(b) = b \wedge a$ is a surjective homomorphism from $\mathbf{A}$ to $\mathbf{A}\restriction a$.

14. If $\mathbf{A} \in CA_n$, $a \in A$, show that $c(a) = a$ implies $c(a') = a'$. Hence show that if $c(a) = a$ then the natural map from $\mathbf{A}$ to $\mathbf{A}\restriction a \times \mathbf{A}\restriction a'$ is an isomorphism. Conclude that $\mathbf{A} \in CA_n$ is directly indecomposable iff it satisfies $a \neq 0 \to c(a) = 1$ for $a \in A$.

15. A member of $CA_1$ is called a *monadic algebra*. Show that the following construction describes all finite monadic algebras. Given finite Boolean algebras $\mathbf{B}_1, \ldots, \mathbf{B}_k$ define $c_0$ on each $\mathbf{B}_i$ by $c_0(0) = 0$ and $c_0(a) = 1$ if $a \neq 0$, and let $d_{00} = 1$. Call the resulting monadic algebras $\mathbf{B}_i^*$. Now form the product $\mathbf{B}_1^* \times \cdots \times \mathbf{B}_k^*$.

## §2. Boolean Rings

The observation that Boolean algebras could be regarded as rings is due to Stone.

**Definition 2.1.** A ring $\mathbf{R} = \langle R, +, \cdot, -, 0, 1\rangle$ is *Boolean* if $\mathbf{R}$ satisfies

$$x^2 \approx x.$$

**Lemma 2.2.** *If* $\mathbf{R}$ *is a Boolean ring then* $\mathbf{R}$ *satisfies*

$$x + x \approx 0 \quad \text{and} \quad x \cdot y \approx y \cdot x.$$

PROOF. Let $a,b \in R$. Then

$$(a + a)^2 = a + a$$

implies

$$a^2 + a^2 + a^2 + a^2 = a + a;$$

hence

$$a + a + a + a = a + a,$$

so

$$a + a = 0.$$

Thus

$$\mathbf{R} \models x + x \approx 0.$$

Now

$$(a + b)^2 = a + b,$$

so

$$a^2 + a \cdot b + b \cdot a + b^2 = a + b;$$

hence

$$a + a \cdot b + b \cdot a + b = a + b,$$

yielding

$$a \cdot b + b \cdot a = 0.$$

As

$$a \cdot b + a \cdot b = 0$$

this says

$$a \cdot b + a \cdot b = a \cdot b + b \cdot a,$$

so

$$a \cdot b = b \cdot a.$$

Thus

$$\mathbf{R} \models x \cdot y \approx y \cdot x. \qquad \square$$

**Theorem 2.3** (Stone). (a) *Let* $\mathbf{B} = \langle B, \vee, \wedge, ', 0, 1 \rangle$ *be a Boolean algebra. Define* $\mathbf{B}^{\otimes}$ *to be the algebra* $\langle B, +, \cdot, -, 0, 1 \rangle$, *where*

$$a + b = (a \wedge b') \vee (a' \wedge b)$$
$$a \cdot b = a \wedge b$$
$$-a = a.$$

*Then* $\mathbf{B}^{\otimes}$ *is a Boolean ring.*

(b) *Let* $\mathbf{R} = \langle R, +, \cdot, -, 0, 1 \rangle$ *be a Boolean ring. Define* $\mathbf{R}^{\otimes}$ *to be the algebra* $\langle R, \vee, \wedge, ', 0, 1 \rangle$ *where*

$$a \vee b = a + b + a \cdot b$$
$$a \wedge b = a \cdot b$$
$$a' = 1 + a.$$

*Then* $\mathbf{R}^{\otimes}$ *is a Boolean algebra.*

(c) *Given* $\mathbf{B}$ *and* $\mathbf{R}$ *as above we have* $\mathbf{B}^{\otimes\otimes} = \mathbf{B}$, $\mathbf{R}^{\otimes\otimes} = \mathbf{R}$.

PROOF. (a) Let $a,b,c \in B$. Then

(i) $a + 0 = (a \wedge 0') \vee (a' \wedge 0)$
$\quad = a \wedge 1$
$\quad = a$

(ii) $a + b = (a \wedge b') \vee (a' \wedge b)$
$= (b \wedge a') \vee (b' \wedge a)$
$= b + a$

(iii) $a + a = (a \wedge a') \vee (a' \wedge a)$
$= 0$

(iv) $a + (b + c) = [a \wedge (b + c)'] \vee [a' \wedge (b + c)]$
$= \{a \wedge [(b \wedge c') \vee (b' \wedge c)]'\} \vee \{a' \wedge [(b \wedge c') \vee (b' \wedge c)]\}$
$= \{a \wedge [(b' \vee c) \wedge (b \vee c')]\} \vee \{(a' \wedge b \wedge c') \vee (a' \wedge b' \wedge c)\}$
$= \{a \wedge [(b' \wedge c') \vee (c \wedge b)]\} \vee \{(a' \wedge b \wedge c') \vee (a' \wedge b' \wedge c)\}$
$= (a \wedge b' \wedge c') \vee (a \wedge b \wedge c) \vee (a' \wedge b \wedge c') \vee (a' \wedge b' \wedge c)$
$= (a \wedge b \wedge c) \vee (a \wedge b' \wedge c') \vee (b \wedge c' \wedge a') \vee (c \wedge a' \wedge b')$.

The value of this last expression does not change if we permute $a$,$b$ and $c$ in any manner; hence $c + (a + b) = a + (b + c)$, so by (ii) $(a + b) + c = a + (b + c)$.

(v) $a \cdot 1 = 1 \cdot a = a$

(vi) $a \cdot (b \cdot c) = a \wedge (b \wedge c)$
$= (a \wedge b) \wedge c$
$= (a \cdot b) \cdot c$

(vii) $a \cdot (b + c) = a \wedge [(b \wedge c') \vee (b' \wedge c)]$
$= (a \wedge b \wedge c') \vee (a \wedge b' \wedge c)$

and

$$\begin{aligned}(a \cdot b) + (a \cdot c) &= [(a \wedge b) \wedge (a \wedge c)'] \vee [(a \wedge b)' \wedge (a \wedge c)] \\ &= [(a \wedge b) \wedge (a' \vee c')] \vee [(a' \vee b') \wedge (a \wedge c)] \\ &= (a \wedge b \wedge c') \vee (b' \wedge a \wedge c)\end{aligned}$$

so

$$a \cdot (b + c) = (a \cdot b) + (a \cdot c).$$

(viii) $a \cdot a = a \wedge a$
$= a$.

Thus $\mathbf{B}^{\otimes}$ is a Boolean ring.

(b) Let $a,b,c \in R$. Then

(i) $a \vee b = a + b + a \cdot b$
$= b + a + b \cdot a$
$= b \vee a$

(ii) $a \wedge b = a \cdot b$
$= b \cdot a$
$= b \wedge a$

(iii) $a \vee (b \vee c) = a + (b \vee c) + a \cdot (b \vee c)$
$= a + (b + c + b \cdot c) + a \cdot (b + c + b \cdot c)$
$= a + b + c + a \cdot b + a \cdot c + b \cdot c + a \cdot b \cdot c$.

The value of this last expression does not change if we permute $a$, $b$ and $c$, so

$$a \vee (b \vee c) = c \vee (a \vee b);$$

hence by (i) above

$$a \vee (b \vee c) = (a \vee b) \vee c.$$

(iv) $a \wedge (b \wedge c) = a \cdot (b \cdot c)$
$= (a \cdot b) \cdot c$
$= (a \wedge b) \wedge c$

(v) $a \vee a = a + a + a \cdot a$
$= 0 + a$
$= a$

(vi) $a \wedge a = a \cdot a$
$= a$

(vii) $a \vee (a \wedge b) = a + (a \wedge b) + a \cdot (a \wedge b)$
$= a + a \cdot b + a \cdot (a \cdot b)$
$= a + a \cdot b + a \cdot b$
$= a$

(viii) $a \wedge (a \vee b) = a \cdot (a + b + a \cdot b)$
$= a^2 + a \cdot b + a^2 \cdot b$
$= a + a \cdot b + a \cdot b$
$= a$

(ix) $a \wedge (b \vee c) = a \cdot (b + c + b \cdot c)$
$= a \cdot b + a \cdot c + a \cdot b \cdot c$
$= a \cdot b + a \cdot c + a \cdot b \cdot a \cdot c$
$= (a \wedge b) \vee (a \wedge c)$

(x) $a \wedge 0 = a \cdot 0$
$= 0$

(xi) $a \vee 1 = a + 1 + a \cdot 1$
$= 1$

(xii) $a \wedge a' = a \cdot (1 + a)$
$= a + a^2$
$= a + a$
$= 0$

(xiii) $a \vee a' = a + (1 + a) + a \cdot (1 + a)$
$= a + 1 + a + a + a^2$
$= 1.$

Thus $\mathbf{R}^{\otimes}$ is a Boolean algebra.

(c) Suppose $\mathbf{B} = \langle B, \vee, \wedge, ', 0, 1 \rangle$ is a Boolean algebra and $a, b \in B$. Then with $\mathbf{B}^{\otimes} = \langle B, +, \cdot, -, 0, 1 \rangle$

(i) $a \cdot b = a \wedge b$

(ii) $1 + a = (1 \wedge a') \vee (1' \wedge a) = a'$

(iii) $a + b + a \cdot b = a + (b + a \cdot b)$
$= a + b \cdot (1 + a)$
$= a + b \cdot a'$
$= [a \wedge (b \wedge a')'] \vee [a' \wedge (b \wedge a')]$
$= [a \wedge (b' \vee a)] \vee [a' \wedge b]$
$= a \vee (a' \wedge b)$
$= (a \vee a') \wedge (a \vee b)$
$= a \vee b.$

Thus $\mathbf{B}^{\otimes\otimes} = \mathbf{B}$.

Next suppose $\mathbf{R} = \langle R, +, \cdot, -, 0, 1\rangle$ is a Boolean ring. Then with $\mathbf{R}^{\otimes} = \langle R, \vee, \wedge, ', 0, 1\rangle$ and $a,b \in R$,

(i) $$\begin{aligned}(a \wedge b') \vee (a' \wedge b) &= [a \cdot (1 + b)] + [(1 + a) \cdot b] \\ &\quad + [a \cdot (1 + b) \cdot (1 + a) \cdot b] \\ &= [a + a \cdot b] + [b + a \cdot b] + 0 \\ &= a + b\end{aligned}$$

(ii) $a \wedge b = a \cdot b$.

Thus $\mathbf{R}^{\otimes\otimes} = \mathbf{R}$. □

The reader should verify that $\otimes$ has nice properties with respect to $H,S$, and $P$; for example if $\mathbf{B}_1, \mathbf{B}_2 \in BA$ then

(i) If $\alpha: \mathbf{B}_1 \to \mathbf{B}_2$ is a homomorphism then $\alpha: \mathbf{B}_1^{\otimes} \to \mathbf{B}_2^{\otimes}$ is a homomorphism between Boolean rings.

(ii) If $\mathbf{B}_1 \leq \mathbf{B}_2$ then $\mathbf{B}_1^{\otimes} \leq \mathbf{B}_2^{\otimes}$.

(iii) If $\mathbf{B}_i \in BA$ for $i \in I$ then $(\prod_{i \in I} \mathbf{B}_i)^{\otimes} = \prod_{i \in I} \mathbf{B}_i^{\otimes}$.

## References

1. P. R. Halmos [18]
2. M. H. Stone [1936]

## Exercises §2

1. If $\mathbf{A}$ is a Boolean algebra [Boolean ring] and $\mathbf{A}_0$ is a subalgebra of $\mathbf{A}$, show $\mathbf{A}_0^{\otimes}$ is a subalgebra of $\mathbf{A}^{\otimes}$.

2. If $\mathbf{A}_1, \mathbf{A}_2$ are Boolean algebras [Boolean rings] and $\alpha: \mathbf{A}_1 \to \mathbf{A}_2$ is a homomorphism then $\alpha$ is also a homomorphism from $\mathbf{A}_1^{\otimes}$ to $\mathbf{A}_2^{\otimes}$.

3. If $(\mathbf{A}_i)_{i \in I}$ is an indexed family of Boolean algebras [Boolean rings], then $(\prod_{i \in I} \mathbf{A}_i)^{\otimes} = \prod_{i \in I} \mathbf{A}_i^{\otimes}$.

4. If we are given an arbitrary ring $\mathbf{R}$, then an element $a \in R$ is a *central idempotent* if $a \cdot b = b \cdot a$ for all $b \in R$, and $a^2 = a$. If $\mathbf{R}$ is a ring with identity show that the central idempotents of $\mathbf{R}$ form a Boolean algebra using the operations: $a \vee b = a + b - a \cdot b$, $a \wedge b = a \cdot b$, and $a' = 1 - a$.

5. If $\theta$ is a congruence on a ring $\mathbf{R}$ with identity, show that $\theta$ is a factor congruence iff $0/\theta$ is a principal ideal of $\mathbf{R}$ generated by a central idempotent. Hence the factor congruences on $\mathbf{R}$ form a sublattice of $\mathbf{Con\,R}$ which is a Boolean lattice.

An *ordered basis* (Mostowski/Tarski) for a Boolean algebra $\mathbf{B}$ is a subset $A$ of $B$ which is a chain under the ordering of $\mathbf{B}$, $0 \notin A$, and every member of $B$ can be expressed in the form $a_1 + \cdots + a_n$, $a_i \in A$.

6. If $A$ is an ordered basis of $\mathbf{B}$, show that $1 \in A$, $A$ is a basis for the vector space $\langle B, +\rangle$ over the two-element field, and each nonzero element of $B$ can be uniquely expressed in the form $a_1 + \cdots + a_n$ with $a_i \in A$, $a_1 < a_2 < \cdots < a_n$.

7. Show that every countable Boolean algebra has an ordered basis.

# §3. Filters and Ideals

Since congruences on rings are associated with ideals it follows that the same must hold for Boolean rings. The translation of Boolean rings to Boolean algebras, namely $\mathbf{R} \mapsto \mathbf{R}^{\otimes}$, gives rise to ideals of Boolean algebras. The image of an ideal under $'$ gives a filter.

**Definition 3.1.** Let $\mathbf{B} = \langle B, \vee, \wedge, ', 0, 1\rangle$ be a Boolean algebra. A subset $I$ of $B$ is called an *ideal* of $\mathbf{B}$ if

(i) $0 \in I$

(ii) $a,b \in I \Rightarrow a \vee b \in I$

(iii) $a \in I$ and $b \leq a \Rightarrow b \in I$.

A subset $F$ of $B$ is called a *filter* of $\mathbf{B}$ if

(i) $1 \in F$

(ii) $a,b \in F \Rightarrow a \wedge b \in F$

(iii) $a \in F$ and $b \geq a \Rightarrow b \in F$.

**Theorem 3.2.** *Let* $\mathbf{B} = \langle B, \vee, \wedge, ', 0, 1\rangle$ *be a Boolean algebra. Then* $I$ *is an ideal of* $\mathbf{B}$ *iff* $I$ *is an ideal of* $\mathbf{B}^{\otimes}$.

Proof. Recall that $I$ is an ideal of a ring $\mathbf{B}^{\otimes}$ iff

$$0 \in I,$$
$$a,b \in I \Rightarrow a + b \in I$$

as $-b = b$, and

$$a \in I,\ b \in R \Rightarrow a \cdot b \in I.$$

So suppose $I$ is an ideal of $\mathbf{B}$. Then

$$0 \in I,$$

and if $a,b \in I$ then

$$a \wedge b' \leq a,$$
$$a' \wedge b \leq b,$$

so

$$a \wedge b',\ a' \wedge b \in I;$$

hence

$$a + b = (a \wedge b') \vee (a' \wedge b) \in I.$$

Now if $a \in I$ and $b \in B$ then

$$a \wedge b \leq a,$$

so

$$a \cdot b = a \wedge b \in I.$$

Thus $I$ is an ideal of $\mathbf{B}^{\otimes}$.

Next suppose $I$ is an ideal of $\mathbf{B}^{\otimes}$. Then $0 \in I$, and if $a,b \in I$ then

$$a \cdot b \in I;$$

hence

$$a \vee b = a + b + a \cdot b \in I.$$

If $a \in I$ and $b \leq a$ then

$$b = a \wedge b = a \cdot b \in I,$$

so $I$ is an ideal of $\mathbf{B}$. □

**Definition 3.3.** If $X \subseteq B$, where $\mathbf{B}$ is a Boolean algebra, let

$$X' = \{a' : a \in X\}.$$

The next result shows that ideals and filters come in pairs.

**Lemma 3.4.** *Given a Boolean algebra* $\mathbf{B}$, *then*

(a) *For* $I \subseteq B$, *$I$ is an ideal iff $I'$ is a filter,*

(b) *For* $F \subseteq B$, *$F$ is a filter iff $F'$ is an ideal.*

Proof. First

$$0 \in I \quad \text{iff} \quad 1 = 0' \in I'.$$

If $a,b \in I$ then

$$a \vee b \in I \quad \text{iff} \quad (a \vee b)' = a' \wedge b' \in I'.$$

For $a \in I$ we have $b \leq a$ iff $a' \leq b'$; hence $b \in I$ iff $b' \in I'$. This proves (a), and (b) is handled similarly. □

The following is now an easy consequence of results about rings, but we will give a direct proof.

**Theorem 3.5.** *Let* $\mathbf{B}$ *be a Boolean algebra. If $\theta$ is a binary relation on $B$, then $\theta$ is a congruence on* $\mathbf{B}$ *iff $0/\theta$ is an ideal, and for $a,b \in B$ we have*

$$\langle a,b \rangle \in \theta \quad \textit{iff} \quad a + b \in 0/\theta.$$

Proof. ($\Rightarrow$) Suppose $\theta$ is a congruence on $\mathbf{B}$. Then

$$0 \in 0/\theta,$$

and if $a,b \in 0/\theta$ then

$$\langle a,0 \rangle \in \theta,$$
$$\langle b,0 \rangle \in \theta,$$

so

$$\langle a \vee b, 0 \vee 0 \rangle \in \theta,$$

i.e.,

$$a \vee b \in 0/\theta.$$

Now if $a \in 0/\theta$ and $b \leq a$ then

$$\langle a,0 \rangle \in \theta$$

so

$$\langle a \wedge b, 0 \wedge b \rangle \in \theta,$$

i.e.,

$$\langle b,0 \rangle \in \theta;$$

hence

$$b \in 0/\theta.$$

This shows that $0/\theta$ is an ideal. Now

$$\langle a,b \rangle \in \theta$$

implies

$$\langle a + b, b + b \rangle \in \theta,$$

i.e.,

$$\langle a + b, 0 \rangle \in \theta,$$

so

$$a + b \in 0/\theta.$$

Conversely,

$$a + b \in 0/\theta$$

implies

$$\langle a + b, 0 \rangle \in \theta,$$

so

$$\langle (a + b) + b, 0 + b \rangle \in \theta,$$

thus

$$\langle a,b \rangle \in \theta.$$

($\Leftarrow$) For this direction first note that $\theta$ is an equivalence relation on $B$ as

$$\langle a,a \rangle \in \theta,$$

since

$$a + a = 0,$$

for $a \in B$; if

$$\langle a,b \rangle \in \theta$$

then

$$\langle b,a \rangle \in \theta$$

as

$$a + b = b + a;$$

and if

$$\langle a,b \rangle \in \theta,$$
$$\langle b,c \rangle \in \theta$$

then

$$a + c = (a + b) + (b + c) \in 0/\theta;$$

hence

$$\langle a,c \rangle \in \theta.$$

Next to show that $\theta$ is compatible with the operations of **B**, let $a_1,a_2,b_1,b_2 \in B$ with

$$\langle a_1,a_2\rangle \in \theta,$$
$$\langle b_1,b_2\rangle \in \theta.$$

Then

$$\begin{aligned}(a_1 \wedge b_1) + (a_2 \wedge b_2) &= (a_1 \cdot b_1) + (a_2 \cdot b_2)\\ &= (a_1 \cdot b_1) + [(a_1 \cdot b_2) + (a_1 \cdot b_2)] + (a_2 \cdot b_2)\\ &= a_1 \cdot (b_1 + b_2) + (a_1 + a_2) \cdot b_2 \in 0/\theta,\end{aligned}$$

so

$$\langle a_1 \wedge b_1, a_2 \wedge b_2\rangle \in \theta.$$

Next

$$\begin{aligned}(a_1 \vee b_1) + (a_2 \vee b_2) &= (a_1 + b_1 + a_1 \cdot b_1) + (a_2 + b_2 + a_2 \cdot b_2)\\ &= (a_1 + a_2) + (b_1 + b_2) + (a_1 \wedge b_1 + a_2 \wedge b_2) \in 0/\theta\end{aligned}$$

as each of the three summands belongs to $0/\theta$, so

$$\langle a_1 \vee b_1, a_2 \vee b_2\rangle \in \theta.$$

From

$$a_1 + a_2 \in 0/\theta$$

it follows that

$$(1 + a_1) + (1 + a_2) \in 0/\theta,$$

so

$$\langle a_1',a_2'\rangle \in \theta.$$

This suffices to show that $\theta$ is a congruence. □

**Definition 3.6.** If $I$ is an ideal of a Boolean algebra **B**, let **B**$/I$ denote the quotient algebra **B**$/\theta$, where

$$\langle a,b\rangle \in \theta \quad \text{iff} \quad a + b \in I.$$

Let $b/I$ denote the equivalence class $b/\theta$ for $b \in B$. If $F$ is a filter of **B** let **B**$/F$ denote **B**$/F'$ and let $b/F$ denote $b/F'$ (see 3.3).

Since we have established a correspondence between ideals, filters, and congruences of Boolean algebras it is natural to look at the corresponding lattices.

**Lemma 3.7.** *The set of ideals and the set of filters of a Boolean algebra are closed under arbitrary intersection.*

PROOF. (Exercise.) □

**Definition 3.8.** Given a Boolean algebra **B** and a set $X \subseteq B$ let $I(X)$ denote the least ideal containing $X$, called the *ideal generated by* $X$, and let $F(X)$ denote the least filter containing $X$, called the *filter generated by* $X$.

**Lemma 3.9.** *For* **B** *a Boolean algebra and* $X \subseteq B$, *we have*

(a) $I(X) = \{b \in B : b \leq b_1 \vee \cdots \vee b_n$ *for some* $b_1, \ldots, b_n \in X\} \cup \{0\}$

(b) $F(X) = \{b \in B : b \geq b_1 \wedge \cdots \wedge b_n$ *for some* $b_1, \ldots, b_n \in X\} \cup \{1\}$.

Proof. For (a) note that

$$0 \in I(X),$$

and for $b_1, \ldots, b_n \in X$ we must have

$$b_1 \vee \cdots \vee b_n \in I(X),$$

so

$$I(X) \supseteq \{b \in B : b \leq b_1 \vee \cdots \vee b_n \text{ for some } b_1, \ldots, b_n \in X\} \cup \{0\}.$$

All we need to do is show that the latter set is an ideal as it certainly contains $X$, and for this it suffices to show that it is closed under join. If

$$b \leq b_1 \vee \cdots \vee b_n,$$
$$c \leq c_1 \vee \cdots \vee c_m$$

with $b_1, \ldots, b_n, c_1, \ldots, c_m \in X$ then

$$b \vee c \leq b_1 \vee \cdots \vee b_n \vee c_1 \vee \cdots \vee c_m$$

so $I(X)$ is as described. The discussion of $F(X)$ parallels the above. □

**Lemma 3.10.** *Let* **B** *be a Boolean algebra.*

(a) *The set of ideals of* **B** *forms a distributive lattice (under* $\subseteq$*) where, for ideals* $I_1, I_2$,

$$I_1 \wedge I_2 = I_1 \cap I_2,$$
$$I_1 \vee I_2 = \{a \in B : a \leq a_1 \vee a_2 \quad \textit{for some } a_1 \in I_1,\ a_2 \in I_2\}.$$

(b) *The set of filters of* **B** *forms a distributive lattice (under* $\subseteq$*) where, for filters* $F_1, F_2$,

$$F_1 \wedge F_2 = F_1 \cap F_2,$$
$$F_1 \vee F_2 = \{a \in B : a \geq a_1 \wedge a_2 \quad \textit{for some } a_1 \in F_1,\ a_2 \in F_2\}.$$

(c) *Both of these lattices are isomorphic to* **Con B**.

Proof. From 3.5 it is evident that the mapping

$$\theta \mapsto 0/\theta$$

from congruences on **B** to ideals of **B** is a bijection such that

$$\theta_1 \subseteq \theta_2 \quad \text{iff} \quad 0/\theta_1 \subseteq 0/\theta_2.$$

Thus the ideals form a lattice isomorphic to **Con B**. The calculations given for $\wedge$ and $\vee$ follow from 3.7 and 3.9. The filters are handled similarly. The distributivity of these lattices follows from the fact that Boolean algebras form a congruence-distributive variety, see II§12, or one can verify this directly. □

A remarkable role will be played in this text by maximal filters, the so-called ultrafilters.

**Definition 3.11.** A filter $F$ of a Boolean algebra $\mathbf{B}$ is an *ultrafilter* if $F$ is maximal with respect to the property that $0 \notin F$. A *maximal ideal* of $\mathbf{B}$ is an ideal which is maximal with respect to the property that $1 \notin I$. (Thus only nontrivial Boolean algebras can have ultrafilters or maximal ideals.)

In view of 3.4 $F$ is an ultrafilter of $\mathbf{B}$ iff $F'$ is a maximal ideal of $\mathbf{B}$, and $I$ is a maximal ideal of $\mathbf{B}$ iff $I'$ is an ultrafilter. The following simple criterion is most useful.

**Theorem 3.12.** *Let $F$ be a filter [$I$ be an ideal] of* $\mathbf{B}$. *Then $F$ is an ultrafilter [$I$ is a maximal ideal] of* $\mathbf{B}$ *iff for any $a \in B$, exactly one of $a,a'$ belongs to $F$ [belongs to $I$].*

Proof. Suppose $F$ is a filter of $\mathbf{B}$.

($\Rightarrow$) If $F$ is an ultrafilter then

$$\mathbf{B}/F \cong \mathbf{2}$$

by 1.9 as $\mathbf{B}/F$ is simple by II§8.9. Let

$$\nu : \mathbf{B} \to \mathbf{B}/F$$

be the natural homomorphism. For $a \in B$,

$$\nu(a') = \nu(a)'$$

so

$$\nu(a) = 1/F \quad \text{or} \quad \nu(a') = 1/F,$$

as $\mathbf{B}/F \cong \mathbf{2}$; hence

$$a \in F \quad \text{or} \quad a' \in F.$$

If we are given $a \in B$ then exactly one of $a$, $a'$ is in $F$ as

$$a \wedge a' = 0 \notin F.$$

($\Leftarrow$) For $a \in B$ suppose exactly one of $a,a' \in F$. Then if $F_1$ is a filter of $\mathbf{B}$ with

$$F \subseteq F_1 \quad \text{and} \quad F \neq F_1$$

let $a \in F_1 - F$. As $a' \in F$ we have

$$0 = a \wedge a' \in F_1;$$

hence $F_1 = B$. Thus $F$ is an ultrafilter. The ideals are handled in the same manner.

□

**Corollary 3.13.** *Let* **B** *be a Boolean algebra.*

(a) *Let $F$ be a filter of* **B**. *Then $F$ is an ultrafilter of* **B** *iff $0 \notin F$ and for $a,b \in B$, $a \vee b \in F$ iff $a \in F$ or $b \in F$.*

(b) (*Stone*) *Let $I$ be an ideal of* **B**. *Then $I$ is a maximal ideal of* **B** *iff $1 \notin I$ and for $a,b \in B$, $a \wedge b \in I$ iff $a \in I$ or $b \in I$.*

PROOF. We will prove the case of filters.

($\Rightarrow$) Suppose $F$ is an ultrafilter with

$$a \vee b \in F.$$

As

$$(a \vee b) \wedge (a' \wedge b') = 0 \notin F$$

we have

$$a' \wedge b' \notin F;$$

hence

$$a' \notin F \quad \text{or} \quad b' \notin F.$$

By 3.12 either

$$a \in F \quad \text{or} \quad b \in F.$$

($\Leftarrow$) Since $1 \in F$, given $a \in B$ we have

$$a \vee a' \in F;$$

hence

$$a \in F \quad \text{or} \quad a' \in F.$$

Both $a,a'$ cannot belong to $F$ as

$$a \wedge a' = 0 \notin F.$$

□

**Definition 3.14.** An ideal $I$ of a Boolean algebra is called a *prime ideal* if $1 \notin I$ and

$$a \wedge b \in I \quad \text{implies} \quad a \in I \text{ or } b \in I.$$

Thus we have just seen that the prime ideals of a Boolean algebra are precisely the maximal ideals.

**Theorem 3.15.** *Let* **B** *be a Boolean algebra.*

(a) (*Stone*) *If $a \in B - \{0\}$, then there is a prime ideal $I$ such that $a \notin I$.*

(b) *If $a \in B - \{1\}$, then there is an ultrafilter $U$ of* **B** *with $a \notin U$.*

PROOF. (a) If $a \in B - \{0\}$ let

$$\alpha : \mathbf{B} \to \mathbf{2}^J$$

be any subdirect embedding of **B** into $\mathbf{2}^J$ for some $J$ (see 1.12). Then

$$\alpha(a) \neq \alpha(0),$$

so for some $j \in J$ we have

$$(\pi_j \circ \alpha)(a) \neq (\pi_j \circ \alpha)(0).$$

As

$$\pi_j \circ \alpha : \mathbf{B} \to \mathbf{2}$$

is onto it follows that

$$\theta = \ker(\pi_j \circ \alpha)$$

is a maximal congruence on $\mathbf{B}$; hence

$$I = 0/\theta$$

is a maximal ideal, thus a prime ideal, and $a \notin I$.

(b) is handled similarly. □

**Lemma 3.16.** *Let $\mathbf{B}_1$ and $\mathbf{B}_2$ be Boolean algebras and suppose*

$$\alpha : \mathbf{B}_1 \to \mathbf{B}_2$$

*is a homomorphism. If $U$ is an ultrafilter of $\mathbf{B}_2$, then $\alpha^{-1}(U)$ is an ultrafilter of $\mathbf{B}_1$.*

PROOF. Let $U$ be an ultrafilter of $\mathbf{B}_2$ and $\beta$ the natural map from $\mathbf{B}_2$ to $\mathbf{B}_2/U$. Then

$$\alpha^{-1}(U) = (\beta \circ \alpha)^{-1}(1),$$

hence $\alpha^{-1}(U)$ is an ultrafilter of $\mathbf{B}_1$ (as the ultrafilters of $\mathbf{B}_1$ are just the preimages of 1 under homomorphisms to $\mathbf{2}$). □

**Theorem 3.17.** *Let $\mathbf{B}$ be a Boolean algebra.*

(a) *If $F$ is a filter of $\mathbf{B}$ and $a \in B - F$, then there is an ultrafilter $U$ with $F \subseteq U$ and $a \notin U$.*

(b) (*Stone*) *If $I$ is an ideal of $\mathbf{B}$ and $a \in B - I$, then there is a maximal ideal $M$ with $I \subseteq M$ and $a \notin M$.*

PROOF. For (a) choose an ultrafilter $U^*$ of $\mathbf{B}/F$ by 3.15 with

$$a/F \notin U^*.$$

Then let $U$ be the inverse image of $U^*$ under the canonical map from $\mathbf{B}$ to $\mathbf{B}/F$. (b) is handled similarly. □

EXERCISES §3

1. If $\mathbf{B}$ is a Boolean algebra and $a,b,c,d \in B$, show that

$$\langle a,b\rangle \in \Theta(c,d) \Leftrightarrow a + b \leq c + d.$$

2. If $\mathbf{B}$ is a Boolean algebra, show that the mapping $\alpha$ from $B$ to the lattice of ideals of $\mathbf{B}$ defined by $\alpha(b) = I(b)$ embeds the Boolean lattice $\langle B, \vee, \wedge\rangle$ into the lattice of ideals of $\mathbf{B}$.

3. If $U$ is an ultrafilter of a Boolean algebra **B** show that $\bigwedge U$ exists, and is an atom $b$ or equals 0. In the former case show $U = F(b)$. (Such an ultrafilter is called a *principal ultrafilter*.)

4. If **B** is the Boolean algebra of finite and cofinite subsets of an infinite set $I$ show that there is exactly one nonprincipal ultrafilter of **B**.

5. If $\mathbf{H} = \langle H, \vee, \wedge, \rightarrow, 0, 1\rangle$ is a Heyting algebra, a *filter* of **H** is a nonempty subset $F$ of $H$ such that (i) $a,b \in F \Rightarrow a \wedge b \in F$ and (ii) $a \in F$, $a \leq b \Rightarrow b \in F$. Show (1) if $\theta \in \operatorname{Con} \mathbf{H}$ then $1/\theta$ is a filter, and $\langle a,b\rangle \in \theta$ iff $(a \rightarrow b) \wedge (b \rightarrow a) \in 1/\theta$, and (2) if $F$ is a filter of **H** then $\theta = \{\langle a,b\rangle \in H^2 : (a \rightarrow b) \wedge (b \rightarrow a) \in F\}$ is a congruence and $F = 1/\theta$.

6. If $\mathbf{A} = \langle A, \vee, \wedge, ', c_0, \ldots, c_{n-1}, 0, 1, d_{00}, \ldots, d_{n-1,n-1}\rangle$ is a cylindric algebra, a *cylindric ideal* of **A** is a subset $I$ of $A$ which is an ideal of the Boolean algebra $\langle A, \vee, \wedge, ', 0, 1\rangle$ and is such that $c(a) \in I$ whenever $a \in I$. Using the exercises of §1 show (1) if $\theta \in \operatorname{Con} \mathbf{A}$ then $0/\theta$ is a cylindric ideal and $\langle a,b\rangle \in \theta$ iff $a + b \in 0/\theta$, and (2) if $I$ is a cylindric ideal of **A** then $\theta = \{\langle a,b\rangle \in A^2 : a + b \in I\}$ is a congruence on **A** with $I = 0/\theta$.

7. Show that a finite-dimensional cylindric algebra **A** is subdirectly irreducible iff $a \in A$ and $a \neq 0$ imply $c(a) = 1$.

## §4. Stone Duality

We will refer to the duality Stone established between Boolean algebras and certain topological spaces as Stone duality. In the following when we speak of a "clopen" set we will mean of course a closed and open set.

**Definition 4.1.** A topological space is a *Boolean space* if it (i) is Hausdorff, (ii) is compact, and (iii) has a basis of clopen subsets.

**Definition 4.2.** Let **B** be a Boolean algebra. Define $\mathbf{B}^*$ to be the topological space whose underlying set is the collection $\mathbf{B}^*$ of ultrafilters of **B**, and whose topology has a subbasis consisting of all sets of the form

$$N_a = \{U \in \mathbf{B}^* : a \in U\},$$

for $a \in B$.

**Lemma 4.3.** *If* **B** *is a Boolean algebra and* $a,b \in B$ *then*

$$N_a \cup N_b = N_{a \vee b},$$
$$N_a \cap N_b = N_{a \wedge b},$$

*and*

$$N_{a'} = (N_a)'.$$

*Thus in particular the* $N_a$*'s form a basis for the topology of* $\mathbf{B}^*$.

PROOF.

$$\begin{aligned} U \in N_a \cup N_b \quad &\text{iff} \quad a \in U \text{ or } b \in U \\ &\text{iff} \quad a \vee b \in U \\ &\text{iff} \quad U \in N_{a \vee b}. \end{aligned}$$

Thus

$$N_a \cup N_b = N_{a \vee b}.$$

The other two identities can be derived similarly. □

**Lemma 4.4.** *Let $X$ be a topological space. Then the clopen subsets of $X$ form a subuniverse of* $\mathbf{Su}(X)$.

PROOF. (Exercise.) □

**Definition 4.5.** If $X$ is a topological space, let $X^*$ be the subalgebra of $\mathbf{Su}(X)$ with universe the collection of clopen subsets of $X$.

**Theorem 4.6** (Stone). (a) *Let* $\mathbf{B}$ *be a Boolean algebra. Then* $\mathbf{B}^*$ *is a Boolean space, and* $\mathbf{B}$ *is isomorphic to* $\mathbf{B}^{**}$ *under the mapping*

$$a \mapsto N_a.$$

(b) *Let $X$ be a Boolean space. Then $X^*$ is a Boolean algebra, and $X$ is homeomorphic to $X^{**}$ under the mapping*

$$x \mapsto \{N \in X^* : x \in N\}.$$

PROOF. (a) To show that $\mathbf{B}^*$ is compact let $(N_a)_{a \in J}$ be a basic open cover of $\mathbf{B}^*$, where $J \subseteq B$. Now suppose no finite subset of $J$ has 1 as its join in $\mathbf{B}$. Then $J$ is contained in a maximal ideal $M$, and

$$U = M'$$

is an ultrafilter with

$$U \cap J = \varnothing.$$

But then

$$U \notin N_a$$

for $a \in J$, which is impossible. Hence for some finite subset $J_0$ of $J$ we have

$$\bigvee J_0 = 1.$$

As

$$\bigvee J_0 \in U$$

for every ultrafilter $U$ we must have

$$U \in N_a$$

for some $a \in J_0$ by 3.13, so $(N_a)_{a \in J_0}$ is a cover of $\mathbf{B}^*$. Thus $\mathbf{B}^*$ is compact. It clearly has a basis of clopen sets as each $N_a$ is clopen since

$$N_a \cap N_{a'} = \varnothing,$$
$$N_a \cup N_{a'} = \mathbf{B}^*.$$

Now if

$$U_1 \neq U_2$$

in $\mathbf{B}^*$ let

$$a \in U_1 - U_2.$$

Then

$$U_1 \in N_a,$$
$$U_2 \in N_{a'},$$

so $\mathbf{B}^*$ is Hausdorff. Thus $\mathbf{B}^*$ is a Boolean space.

The mapping $a \mapsto N_a$ is clearly a homomorphism from $\mathbf{B}$ to $\mathbf{B}^{**}$ in view of 4.3. If $a,b \in B$ and $a \neq b$ then

$$(a \vee b) \wedge (a \wedge b)' \neq 0,$$

so by 3.15(a) there is a prime ideal $I$ such that

$$(a \vee b) \wedge (a \wedge b)' \notin I,$$

so there is an ultrafilter $U$ $(= I')$ such that

$$(a \vee b) \wedge (a \wedge b)' \in U.$$

But then

$$(a \wedge b)' \in U$$

so

$$a \wedge b \notin U;$$

hence

$$a \notin U \quad \text{or} \quad b \notin U;$$

but as

$$a \vee b \in U$$

we have

$$a \in U \quad \text{or} \quad b \in U$$

so exactly one of $a,b$ is in $U$; hence

$$N_a \neq N_b.$$

Thus the mapping is injective. If now $N$ is any clopen subset of $\mathbf{B}^*$ then, being open, $N$ is a union of basic open subsets $N_a$, and being a closed subset of a compact space, $N$ is compact. Thus $N$ is a finite union of basic open sets, so $N$ is equal to some $N_a$, by 4.3.

Thus $\mathbf{B} \cong \mathbf{B}^{**}$ under the above mapping.

(b) $X^*$ is a Boolean algebra by 4.4. Let

$$\alpha : X \to X^{**}$$

be the mapping

$$\alpha(x) = \{N \in X^* : x \in N\}.$$

(Note that $\alpha(x)$ is indeed an ultrafilter of $X^*$). If

$$x, y \in X \quad \text{and} \quad x \neq y$$

then

$$\alpha(x) \neq \alpha(y)$$

as $X$ is Hausdorff and has a basis of clopen subsets. If $U$ is an ultrafilter of $X^*$ then $U$ is a family of closed subsets of $X$ with the finite intersection property, so as $X$ is compact we must have

$$\bigcap U \neq \varnothing.$$

It easily follows that for $x \in \bigcap U$,

$$U \subseteq \alpha(x);$$

thus

$$U = \alpha(x)$$

by the maximality of $U$. Thus $\alpha$ is a bijection.

A clopen subset of $X^{**}$ looks like

$$\{U \in X^{**} : N \in U\}$$

for $N \in X^*$, i.e., for $N$ a clopen subset of $X$. Now

$$\begin{aligned}\alpha(N) &= \{U \in X^{**} : \alpha(x) = U \text{ for some } x \in N\}\\ &= \{U \in X^{**} : N \in U\},\end{aligned}$$

so $\alpha$ is an open map. Also

$$\begin{aligned}\alpha^{-1}\{U \in X^{**} : N \in U\} &= \{x \in X : \alpha(x) \in \{U \in X^{**} : N \in U\}\}\\ &= \{x \in X : x \in N\}\\ &= N,\end{aligned}$$

so $\alpha$ is continuous. Thus $\alpha$ is the desired homeomorphism. □

**Definition 4.7.** Given two disjoint topological spaces, $X_1$, $X_2$, define the *union* of $X_1$, $X_2$ to be the topological space whose underlying set is $X_1 \cup X_2$ and whose open sets are precisely the subsets of the form $O_1 \cup O_2$ where $O_i$ is open in $X_i$.

Given two topological spaces $X_1$, $X_2$, let

$$X_1 \mathbin{\dot\cup} X_2$$

denote the topological space whose underlying set is

$$\{1\} \times X_1 \cup \{2\} \times X_2$$

and whose open subsets are precisely the subsets of the form

$$\{1\} \times O_1 \cup \{2\} \times O_2$$

where $O_i$ is open in $X_i$, $i = 1, 2$. $X_1 \cup X_2$ is called the *disjointed union* of $X_1, X_2$.

The next result is used in the next section.

**Lemma 4.8.** *Given two Boolean algebras $\mathbf{B}_1$ and $\mathbf{B}_2$, the Boolean spaces $(\mathbf{B}_1 \times \mathbf{B}_2)^*$ and $\mathbf{B}_1^* \cup \mathbf{B}_2^*$ are homomorphic.*

PROOF. The case that $|B_1| = |B_2| = 1$ is trivial, so we assume $|B_1 \times B_2| \geq 2$. Given an ultrafilter $U$ in $(\mathbf{B}_1 \times \mathbf{B}_2)^*$, let $\pi_i(U)$ be the image of $U$ under the projection homomorphism

$$\pi_i: \mathbf{B}_1 \times \mathbf{B}_2 \to \mathbf{B}_i.$$

**Claim:** $U = \pi_1(U) \times B_2$ or $U = B_1 \times \pi_2(U)$.

To see this note that

$$\langle 1,0 \rangle \vee \langle 0,1 \rangle = \langle 1,1 \rangle \in U$$

implies

$$\langle 1,0 \rangle \in U \quad \text{or} \quad \langle 0,1 \rangle \in U.$$

If $\langle 1,0 \rangle \in U$ then

$$\langle b_1,b_2 \rangle \in U \Rightarrow \langle b_1,0 \rangle = \langle b_1,b_2 \rangle \wedge \langle 1,0 \rangle \in U;$$

hence

$$\pi_1(U) \times \{0\} \subseteq U,$$

so

$$\pi_1(U) \times B_2 \subseteq U.$$

As

$$U \subseteq \pi_1(U) \times \pi_2(U)$$

we have

$$U = \pi_1(U) \times B_2.$$

Likewise we handle the case $\langle 0,1 \rangle \in U$. This finishes the proof of the claim.

From the claim it is easy to verify that either $\pi_1(U)$ or $\pi_2(U)$ is a filter, and then an ultrafilter. So let us define the map

$$\beta: (\mathbf{B}_1 \times \mathbf{B}_2)^* \to \mathbf{B}_1^* \cup \mathbf{B}_2^*$$

by

$$\beta(U) = \{i\} \times \pi_i(U)$$

for $i$ such that $\pi_i(U)$ is an ultrafilter of $\mathbf{B}_i$. The map $\beta$ is easily seen to be injective in view of the claim. If $\bar{U} \in \mathbf{B}_1^*$ then $\bar{U} \times B_2 \in (\mathbf{B}_1 \times \mathbf{B}_2)^*$, so

$$\beta(\bar{U} \times B_2) = \{1\} \times \bar{U},$$

and a similar argument for $\bar{U} \in \mathbf{B}_2^*$ shows $\beta$ is also surjective. Finally, we have

$$\begin{aligned}
\beta(N_{\langle b_1,b_2 \rangle}) &= \{\beta(U): U \in (\mathbf{B}_1 \times \mathbf{B}_2)^*, \langle b_1,b_2 \rangle \in U\} \\
&= \{\beta(U): U \in (\mathbf{B}_1 \times \mathbf{B}_2)^*, \langle b_1,0 \rangle \in U \text{ or } \langle 0,b_2 \rangle \in U\} \\
&= \{\beta(U): U \in (\mathbf{B}_1 \times \mathbf{B}_2)^*, b_1 \in \pi_1(U) \text{ or } b_2 \in \pi_2(U)\} \\
&= \{1\} \times N_{b_1} \cup \{2\} \times N_{b_2}.
\end{aligned}$$

□

Actually Stone goes on to establish relationships between the following pairs:

| | | |
|---|---|---|
| Boolean algebras | $\longleftrightarrow$ | Boolean spaces |
| filters | $\longleftrightarrow$ | closed subsets |
| ideals | $\longleftrightarrow$ | open subsets |
| homomorphisms | $\longleftrightarrow$ | continuous maps. |

However what we have done above suffices for our goals, so we leave the other relationships for the reader to establish in the exercises.

## REFERENCES

1. P. R. Halmos [18]
2. M. H. Stone [1937]

## EXERCISES §4

1. Show that a finite topological space is a Boolean space iff it is *discrete* (i.e., every subset is open).

2. If $X$ is a Boolean space and $I$ is any set, show that the Tychonoff product $X^I$ is a Boolean space and if $I$ is infinite and $|X| > 1$ then $(X^I)^*$ is an atomless Boolean algebra.

3. Show that a countably infinite free Boolean algebra $\mathbf{B}$ has a Boolean space homeomorphic to $2^\omega$, where 2 is the discrete space $\{0, 1\}$; hence $\mathbf{B}$ is isomorphic to the Boolean algebra of closed and open subsets of the *Cantor* discontinuum. Conclude also that $\mathbf{B}$ has continuum many nonprincipal ultrafilters.

4. Given any set $I$ show that $(\mathbf{Su}(I))^*$ is the Stone–Čech compactification of the discrete space $I$.

5. Give a topological description of the Boolean space of the algebra of finite and cofinite subsets of an infinite set $I$.

6. For $\mathbf{B}$ a Boolean algebra and $U \in \mathbf{B}^{***}$ show that there is an $x \in \mathbf{B}^*$ with $\bigcap U = \{x\}$, and $U = \{N \in B^{**} : x \in N\}$.

7. If $\mathbf{B}$ is a Boolean algebra and $F$ is a filter of $\mathbf{B}$ show that $F^* = \bigcap\{N_b : b \in F\}$ is a closed subset of $\mathbf{B}^*$ and the map $F \mapsto F^*$ is an isomorphism from the lattice of filters of $\mathbf{B}$ to the lattice of closed subsets of $\mathbf{B}^*$ and $b \in F$ iff $N_b \supseteq F^*$.

8. If $\mathbf{B}$ is a Boolean algebra and $I$ is an ideal of $\mathbf{B}$ show $I^* = \bigcup\{N_b : b \in I\}$ is an open subset of $\mathbf{B}^*$ such that the map $I \mapsto I^*$ is an isomorphism from the lattice of ideals of $\mathbf{B}$ to the lattice of open subsets of $\mathbf{B}^*$ with $b \in I$ iff $N_b \subseteq I^*$.

9. If $\alpha : \mathbf{B}_1 \to \mathbf{B}_2$ is a Boolean algebra homomorphism, let $\alpha^* : \mathbf{B}_2^* \to \mathbf{B}_1^*$ be defined by $\alpha^*(U) = \alpha^{-1}(U)$. Show $\alpha^*$ is a continuous mapping from $\mathbf{B}_2^*$ to $\mathbf{B}_1^*$ which is injective if $\alpha$ is surjective, and surjective if $\alpha$ is injective.

10. If $\alpha : X_1 \to X_2$ is a continuous map between Boolean spaces, let $\alpha^* : X_2^* \to X_1^*$ be defined by $\alpha^*(N) = \alpha^{-1}(N)$. Then show $\alpha^*$ is a Boolean algebra homomorphism such that $\alpha^*$ is injective if $\alpha$ is surjective, and surjective if $\alpha$ is injective.

11. Show that the atoms of a Boolean algebra **B** correspond to the *isolated points* of $\mathbf{B}^*$ (a point $x \in \mathbf{B}^*$ is isolated if $\{x\}$ is a clopen subset of $\mathbf{B}^*$).

12. Given a chain $\langle C, \leq \rangle$ define the *interval topology* on $C$ to be the topology generated by the open sets $\{c \in C : c > a\}$ and $\{c \in C : c < a\}$, for $a \in C$. Show that this gives a Boolean space iff $\langle C, \leq \rangle$ is an algebraic lattice (see I§4 Ex. 4).

13. If $\lambda$ is an ordinal, show that the interval topology on $\lambda$ gives a Boolean space iff $\lambda$ is not a limit ordinal.

14. Given Boolean spaces $X_1, \ldots, X_n$ such that $X_i \cap X_j = \{x\}$ for $i \neq j$, show that the space $Y = \bigcup_{1 \leq i \leq n} X_i$ with open sets $\{\bigcup_{1 \leq i \leq n} U_i : U_i$ open in $X_i$, $x$ belongs to all or none of the $U_i$'s$\}$ is again a Boolean space.

## §5. Boolean Powers

The Boolean power construction goes back at least to a paper of Arens and Kaplansky in 1948, and it has parallels in earlier work of Gelfand. Arens and Kaplansky were concerned with rings, and in 1953 Foster generalized Boolean powers to arbitrary algebras. This construction provides a method for translating numerous fascinating properties of Boolean algebras into other varieties, and, as we shall see in §7, provides basic representation theorems.

**Definition 5.1.** If **B** is a Boolean algebra and **A** an arbitrary algebra, let $A[\mathbf{B}]^*$ be the set of continuous functions from $\mathbf{B}^*$ to $A$, giving $A$ the discrete topology.

**Lemma 5.2.** *If we are given* **A**,**B** *as in* 5.1, $A[\mathbf{B}]^*$ *is a subuniverse of* $\mathbf{A}^X$, *where* $X = \mathbf{B}^*$.

Proof. Let $c_1, \ldots, c_n \in A[\mathbf{B}]^*$. As $X$ is compact, each $c_i$ has a finite range, and, for $a \in A$, $c_i^{-1}(a)$ is a clopen subset of $X$. Thus we can visualize a typical member of $A[\mathbf{B}]^*$ as in Figure 28, namely a step function with finitely many

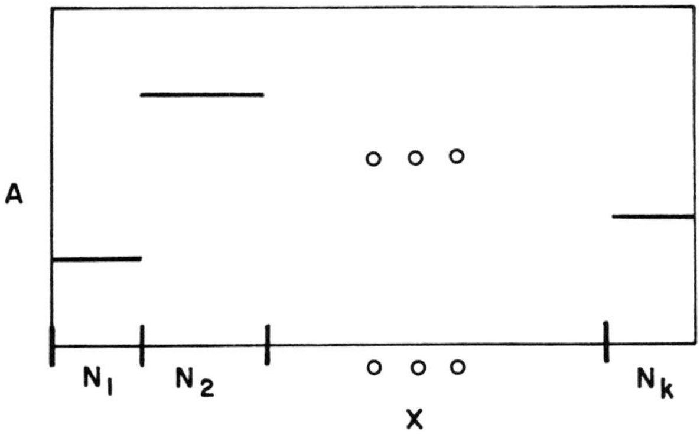


Figure 28

steps, each step occuring over a clopen subset of $X$. If $\mathbf{A}$ is of type $\mathscr{F}$ and $f \in \mathscr{F}_n$ then if we choose clopen subsets $N_1, \ldots, N_k$ which partition $X$ such that each $c_i$ is constant on each $N_j$, $i = 1, \ldots, n$, $j = 1, \ldots, k$, it is clear that $f(c_1, \ldots, c_n)$ is constant on each $N_j$. Consequently, $f(c_1, \ldots, c_n) \in A[\mathbf{B}]^*$. □

**Definition 5.3.** Given $\mathbf{A}$,$\mathbf{B}$ as in 5.1, let $\mathbf{A}[\mathbf{B}]^*$ denote the subalgebra of $\mathbf{A}^X$, $X = \mathbf{B}^*$, with universe $A[\mathbf{B}]^*$. $\mathbf{A}[\mathbf{B}]^*$ is called the (*bounded*) *Boolean power* of $\mathbf{A}$ by $\mathbf{B}$. (Note that $\mathbf{A}[\mathbf{1}]^*$ is a trivial algebra.)

**Theorem 5.4.** *The following results hold for Boolean powers*:

(a) $\mathbf{A}[\mathbf{B}]^*$ *is a subdirect power of* $\mathbf{A}$.
(b) $\mathbf{A}$ *can be embedded in* $\mathbf{A}[\mathbf{B}]^*$ *if* $\mathbf{B}$ *is not trivial.*
(c) $\mathbf{A}[\mathbf{2}]^* \cong \mathbf{A}$.
(d) $\mathbf{A}[\mathbf{B}_1 \times \mathbf{B}_2]^* \cong \mathbf{A}[\mathbf{B}_1]^* \times \mathbf{A}[\mathbf{B}_2]^*$.
(e) $(\mathbf{A}_1 \times \mathbf{A}_2)[\mathbf{B}]^* \cong \mathbf{A}_1[\mathbf{B}]^* \times \mathbf{A}_2[\mathbf{B}]^*$.

Proof. For (a) and (b) note that the constant functions of $A^X$ are in $A[\mathbf{B}]^*$. (c) follows from noting that $\mathbf{2}^*$ is a one-element space, so the only functions in $A^X$ are constant functions.

Let $C(X,\mathbf{A})$ denote the set of continuous functions from $X$ to $A$, for $X$ a Boolean space, and let $\mathbf{C}(X,\mathbf{A})$ denote the subalgebra of $\mathbf{A}^X$ with universe $C(X,\mathbf{A})$. Given two disjoint Boolean spaces $X_1$,$X_2$ define

$$\alpha: C(X_1 \cup X_2,\mathbf{A}) \to C(X_1,\mathbf{A}) \times C(X_2,\mathbf{A})$$

by

$$\alpha c = \langle c\restriction_{X_1}, c\restriction_{X_2}\rangle.$$

As $X_1$,$X_2$ are clopen in $X_1 \cup X_2$ it is not difficult to see that $\alpha$ is a bijection, and if $c_1, \ldots, c_n \in C(X_1 \cup X_2,\mathbf{A})$ and $f$ is a fundamental operation of arity $n$, then

$$\begin{aligned}\alpha f(c_1, \ldots, c_n) &= \langle f(c_1, \ldots, c_n)\restriction_{X_1}, f(c_1, \ldots, c_n)\restriction_{X_2}\rangle \\ &= \langle f(c_1\restriction_{X_1}, \ldots, c_n\restriction_{X_1}), f(c_1\restriction_{X_2}, \ldots, c_n\restriction_{X_2})\rangle \\ &= f(\langle c_1\restriction_{X_1}, c_1\restriction_{X_2}\rangle, \ldots, \langle c_n\restriction_{X_1}, c_n\restriction_{X_2}\rangle) \\ &= f(\alpha c_1, \ldots, \alpha c_n),\end{aligned}$$

so $\alpha$ is an isomorphism. As

$$\mathbf{A}[\mathbf{B}]^* = \mathbf{C}(\mathbf{B}^*,\mathbf{A})$$

it follows from 4.8 that

$$\mathbf{A}[\mathbf{B}_1 \times \mathbf{B}_2]^* \cong \mathbf{A}[\mathbf{B}_1]^* \times \mathbf{A}[\mathbf{B}_2]^*.$$

This proves (d).

Next define

$$\alpha: A_1[\mathbf{B}]^* \times A_2[\mathbf{B}]^* \to (A_1 \times A_2)[\mathbf{B}]^*$$

by

$$\alpha(\langle c_1, c_2\rangle)(x) = \langle c_1 x, c_2 x\rangle.$$

Clearly this is a well-defined injective map. If $c \in (A_1 \times A_2)[\mathbf{B}]^*$ let $N_1,\ldots,N_k$ be a partition of $\mathbf{B}^*$ into clopen subsets such that $c$ is constant on each $N_j$. Then let

$$c_i(x) = (\pi_i c)(x),$$

$i = 1,2$. Then $c_i \in A_i[\mathbf{B}]^*$ as $c_i$ is constant on each $N_j$, and

$$\alpha(\langle c_1,c_2\rangle) = c,$$

so $\alpha$ is surjective. If $\langle c_1^j,c_2^j\rangle \in A_1[\mathbf{B}]^* \times A_2[\mathbf{B}]^*$, $1 \le j \le n$, and if $f$ is a fundamental $n$-ary operation then

$$\begin{aligned}\alpha f(\langle c_1^1,c_2^1\rangle,\ldots,\langle c_1^n,c_2^n\rangle)(x) &= \alpha(\langle f(c_1^1,\ldots,c_1^n), f(c_2^1,\ldots,c_2^n)\rangle)(x)\\ &= \langle f(c_1^1,\ldots,c_1^n)(x), f(c_2^1,\ldots,c_2^n)(x)\rangle\\ &= \langle f(c_1^1x,\ldots,c_1^nx), f(c_2^1x,\ldots,c_2^nx)\rangle\\ &= f(\langle c_1^1x,c_2^1x\rangle,\ldots,\langle c_1^nx,c_2^nx\rangle)\\ &= f(\alpha(\langle c_1^1,c_2^1\rangle)(x),\ldots,\alpha(\langle c_1^n,c_2^n\rangle)(x))\\ &= f(\alpha\langle c_1^1,c_2^1\rangle,\ldots,\alpha\langle c_1^n,c_2^n\rangle)(x);\end{aligned}$$

hence

$$\alpha f(\langle c_1^1,c_2^1\rangle,\ldots,\langle c_1^n,c_2^n\rangle) = f(\alpha\langle c_1^1,c_2^1\rangle,\ldots,\alpha\langle c_1^n,c_2^n\rangle).$$

This proves

$$\mathbf{A}_1[\mathbf{B}]^* \times \mathbf{A}_2[\mathbf{B}]^* \cong (\mathbf{A}_1 \times \mathbf{A}_2)[\mathbf{B}]^*$$

as $\alpha$ is an isomorphism. □

The next result is used in §7, and provides the springboard for the generalization of Boolean powers given in §8.

**Definition 5.5.** If $a,b \in \prod_{i\in I} A_i$ the *equalizer* of $a$ and $b$ is

$$[\![a = b]\!] = \{i \in I : a(i) = b(i)\},$$

and if $J_1,\ldots,J_n$ partition $I$ and $a_1,\ldots,a_n \in \prod_{i\in I} A_i$ then

$$a_1\restriction_{J_1} \cup \cdots \cup a_n\restriction_{J_n}$$

denotes the function $a$ where

$$a(i) = a_k(i) \quad \text{if} \quad i \in J_k.$$

**Theorem 5.6.** *Let* $\mathbf{B}$ *be a Boolean algebra and* $\mathbf{A}$ *any algebra. With* $X = \mathbf{B}^*$, *a subset* $S$ *of* $A^X$ *is* $A[\mathbf{B}]^*$ *iff* $S$ *satisfies*

(a) *the constant functions of* $A^X$ *are in* $S$,

(b) *for* $c_1,c_2 \in S$, $[\![c_1 = c_2]\!]$ *is a clopen subset of* $X$, *and*

(c) *for* $c_1,c_2 \in S$ *and* $N$ *a clopen subset of* $X$,

$$c_1\restriction_N \cup c_2\restriction_{X-N} \in S.$$

PROOF. ($\Rightarrow$) We have already noted that the constant functions are in $A[\mathbf{B}]^*$. For part (b) note that $c \in A[\mathbf{B}]^*$ implies $c^{-1}(a)$ is clopen for $a \in A$ as $c$ is continuous. Also as $c$ has finite range,

$$[\![c_1 = c_2]\!] = \bigcup_{a \in A} c_1^{-1}(a) \cap c_2^{-1}(a)$$

is a clopen subset of $X$. Finally

$$c = c_1 \restriction_N \cup c_2 \restriction_{X-N}$$

is in $A[\mathbf{B}]^*$ as

$$c^{-1}(a) = (c_1^{-1}(a) \cap N) \cup (c_2^{-1}(a) \cap (X - N)),$$

a clopen subset of $X$ for $a \in A$.

($\Leftarrow$) For $a \in A$ let $c_a \in A^X$ be the constant function with value $a$. From (b) we have, for $c \in S$,

$$c^{-1}(a) = [\![c = c_a]\!],$$

a clopen subset of $X$; hence $c$ is continuous, so $c \in A[\mathbf{B}]^*$. Finally, if $c \in A[\mathbf{B}]^*$ let

$$N_a = [\![c = c_a]\!]$$

for $a \in A$. Then

$$c = \bigcup_{a \in A} c_a \restriction_{N_a},$$

so by (c) $c \in S$. □

REFERENCES

1. B. Banaschewski and E. Nelson [1980]
2. S. Burris [1975b]
3. A. L. Foster [1953a], [1953b]

EXERCISES §5

1. Given Boolean algebras $\mathbf{B}_1$, $\mathbf{B}_2$ define $\mathbf{B}_1 * \mathbf{B}_2$ to be $(\mathbf{B}_1^* \times \mathbf{B}_2^*)^*$. Show that for any $\mathbf{A}$, $(\mathbf{A}[\mathbf{B}_1]^*)[\mathbf{B}_2]^* \cong \mathbf{A}[\mathbf{B}_1 * \mathbf{B}_2]^*$; hence $(\mathbf{A}[\mathbf{B}_1]^*)[\mathbf{B}_2]^* \cong (\mathbf{A}[\mathbf{B}_2]^*)[\mathbf{B}_1]^*$.

2. If $F$ is a filter of $\mathbf{B}$ define $\theta_F$ on $\mathbf{A}[\mathbf{B}]^*$ by $\langle a,b \rangle \in \theta_F$ iff $[\![a = b]\!] \supseteq F^*$ (see §4 Ex. 7). Show that $\mathbf{A}[\mathbf{B}]^*/\theta_F \cong \mathbf{A}[\mathbf{B}/F]^*$.

3. Show that $|A[\mathbf{B}]^*| = |A| \cdot |B|$ if either $|A|$ or $|B|$ is infinite, and the other is nontrivial.

4. (Bergman). Let $\mathbf{M}$ be a module. Given two countably infinite Boolean algebras $\mathbf{B}_1$, $\mathbf{B}_2$ show that $\mathbf{M}[\mathbf{B}_1]^* \cong \mathbf{M}[\mathbf{B}_2]^*$. (Hint: (Lawrence) Let $Q_i$ be an ordered basis (see §2 Ex. 7) for $\mathbf{B}_i$, $i = 1, 2$, and let $\alpha: Q_1 \to Q_2$ be a bijection. For $a \in M$ and $q \in Q_i$ let $C_a \restriction_q$ denote the member of $M[\mathbf{B}_i]^*$ with

$$C_a \restriction_q(x) = \begin{cases} a & \text{if } x \in N_q \\ 0 & \text{if } x \notin N_q. \end{cases}$$

Then each member of $M[\mathbf{B}_i]^*$ can be uniquely written in the form $C_{a_1}\restriction_{q_1} + \cdots + C_{a_n}\restriction_{q_n}$, where $q_1 < \cdots < q_n$, $q_j \in Q_i$. Define $\beta: M[\mathbf{B}_1]^* \to M[\mathbf{B}_2]^*$ by

$$C_{a_1}\restriction_{q_1} + \cdots + C_{a_n}\restriction_{q_n} \mapsto C_{a_1}\restriction_{\alpha q_1} + \cdots + C_{a_n}\restriction_{\alpha q_n},$$

where $q_1 < \cdots < q_n$. Then $\beta$ is the desired isomorphism.)

Show that we can replace **M** by any algebra **A** which is polynomially equivalent to a module.

# §6. Ultraproducts and Congruence–distributive Varieties

One of the most popular constructions, first introduced by Łoś (pronounced "wash") in 1955, is the ultraproduct. We will make good use of it in both this and the next chapter. The main result in this section is a new description due to Jónsson, using ultraproducts, of congruence-distributive varieties generated by a class $K$.

**Definition 6.1.** For any set $I$, members of $\mathbf{Su}(I)^*$ are called *ultrafilters over $I$*. Let $\mathbf{A}_i$, $i \in I$, be a family of algebras of a given type, and let $U$ be an ultrafilter over $I$. Define $\theta_U$ on $\prod_{i\in I} A_i$ by

$$\langle a,b\rangle \in \theta_U \quad \text{iff} \quad [\![a = b]\!] \in U,$$

where $[\![a = b]\!]$ is as defined in 5.5.

**Lemma 6.2.** *With $\mathbf{A}_i$, $i \in I$, and $U$ as above, $\theta_U$ is a congruence on $\prod_{i\in I} \mathbf{A}_i$.*

Proof. Obviously, $\theta_U$ is reflexive and symmetric. If

$$\langle a,b\rangle \in \theta_U \quad \text{and} \quad \langle b,c\rangle \in \theta_U$$

then

$$[\![a = c]\!] \supseteq [\![a = b]\!] \cap [\![b = c]\!]$$

implies

$$[\![a = c]\!] \in U,$$

so

$$\langle a,c\rangle \in \theta_U.$$

If

$$\langle a_1,b_1\rangle, \ldots, \langle a_n,b_n\rangle \in \theta_U$$

and $f$ is a fundamental $n$-ary operation then

$$[\![f(a_1, \ldots, a_n) = f(b_1, \ldots, b_n)]\!] \supseteq [\![a_1 = b_1]\!] \cap \cdots \cap [\![a_n = b_n]\!]$$

implies

$$\langle f(a_1, \ldots, a_n), f(b_1, \ldots, b_n)\rangle \in \theta_U.$$

Thus $\theta_U$ is a congruence. □

**Definition 6.3.** With $\mathbf{A}_i$, $i \in I$, and $U$ an ultrafilter over $I$ we define the *ultraproduct*

$$\prod_{i \in I} \mathbf{A}_i/U$$

to be

$$\prod_{i \in I} \mathbf{A}_i/\theta_U.$$

The elements of $\prod_{i \in I} A_i/U$ are denoted by $a/U$, where $a \in \prod_{i \in I} A_i$.

**Lemma 6.4.** *For $a/U$, $b/U$ in an ultraproduct $\prod_{i \in I} \mathbf{A}_i/U$, we have*

$$a/U = b/U \quad \text{iff} \quad [\![a = b]\!] \in U.$$

PROOF. This is an immediate consequence of the definition. □

**Lemma 6.5.** *If $\{\mathbf{A}_i : i \in I\}$ is a finite set of finite algebras, say $\{\mathbf{B}_1, \ldots, \mathbf{B}_k\}$, (I can be infinite), and $U$ is an ultrafilter over $I$ then $\prod_{i \in I} \mathbf{A}_i/U$ is isomorphic to one of the algebras $\mathbf{B}_1, \ldots, \mathbf{B}_k$, namely to that $\mathbf{B}_j$ such that*

$$\{i \in I : \mathbf{A}_i = \mathbf{B}_j\} \in U.$$

PROOF. Let

$$S_j = \{i \in I : \mathbf{A}_i = \mathbf{B}_j\}.$$

Then

$$I = S_1 \cup \cdots \cup S_m$$

implies (by 3.13) that for some $j$,

$$S_j \in U.$$

Let $B_j = \{b_1, \ldots, b_k\}$, where the $b$'s are all distinct, and choose $a_1, \ldots, a_k \in \prod_{i \in I} A_i$ such that

$$a_1(i) = b_1, \ldots, a_k(i) = b_k$$

if $i \in S_j$. Then if we are given $a \in \prod_{i \in I} A_i$,

$$[\![a = a_1]\!] \cup \cdots \cup [\![a = a_k]\!] \supseteq S_j,$$

so

$$[\![a = a_1]\!] \in U \text{ or} \ldots \text{or } [\![a = a_k]\!] \in U;$$

hence

$$a/U = a_1/U \qquad \text{or} \qquad \ldots \qquad \text{or } a/U = a_k/U.$$

Also it should be evident that $a_1/U, \ldots, a_k/U$ are all distinct. Thus $\prod_{i \in I} A_i/U$ has exactly $k$ elements, $a_1/U, \ldots, a_k/U$. Now for $f$ a fundamental $n$-ary operation and for

$$\{b_{i_1}, \ldots, b_{i_n}, b_{i_{n+1}}\} \subseteq \{b_1, \ldots, b_k\}$$

with

$$f(b_{i_1}, \ldots, b_{i_n}) = b_{i_{n+1}},$$

we have

$$[\![f(a_{i_1},\ldots,a_{i_n}) = a_{i_{n+1}}]\!] \supseteq S_j;$$

hence

$$f(a_{i_1}/U,\ldots,a_{i_n}/U) = a_{i_{n+1}}/U.$$

Thus the map

$$\alpha\colon \prod_{i\in I} A_i/U \to B_j$$

defined by

$$\alpha(a_t/U) = b_t,$$

$1 \le t \le k$, is an isomorphism. □

**Lemma 6.6** (Jónsson). *Let $W$ be a family of subsets of $I$ ($\neq \emptyset$) such that*
(i) *$I \in W$,*
(ii) *if $J \in W$ and $J \subseteq K \subseteq I$ then $K \in W$, and*
(iii) *if $J_1 \cup J_2 \in W$ then $J_1 \in W$ or $J_2 \in W$. Then there is an ultrafilter $U$ over $I$ with*

$$U \subseteq W.$$

PROOF. If $\emptyset \in W$ then $W = \mathrm{Su}(I)$, so any ultrafilter will do. If $\emptyset \notin W$, then $\mathrm{Su}(I) - W$ is a proper ideal; extend it to a maximal ideal and take the complementary ultrafilter. □

**Definition 6.7.** We denote the class of ultraproducts of members of $K$ by $P_U(K)$.

**Theorem 6.8** (Jónsson). *Let $V(K)$ be a congruence-distributive variety. If $\mathbf{A}$ is a subdirectly irreducible algebra in $V(K)$, then*

$$\mathbf{A} \in HSP_U(K);$$

hence

$$V(K) = IP_SHSP_U(K).$$

PROOF. Suppose $\mathbf{A}$ is a nontrivial subdirectly irreducible algebra in $V(K)$. Then for some choice of $\mathbf{A}_i \in K$, $i \in I$, and for some $\mathbf{B} \le \prod_{i\in I} \mathbf{A}_i$ there is a surjective homomorphism

$$\alpha\colon \mathbf{B} \to \mathbf{A},$$

as $V(K) = HSP(K)$. Let

$$\theta = \ker \alpha.$$

For $J \subseteq \mathrm{I}$ let

$$\theta_J = \left\{\langle a,b\rangle \in \left(\prod_{i\in I} A_i\right)^2 : J \subseteq [\![a = b]\!]\right\}.$$

One easily verifies that $\theta_J$ is a congruence on $\prod_{i\in I} \mathbf{A}_i$. Let

$$\theta_J\restriction_B = \theta_J \cap B^2$$

be the restriction of $\theta_J$ to $\mathbf{B}$, and define $W$ to be

$$\{J \subseteq I : \theta_J \restriction_B \subseteq \theta\}.$$

Clearly

$$I \in W, \qquad \varnothing \notin W$$

and if

$$J \in W \quad \text{and} \quad J \subseteq K \subseteq I$$

then

$$\theta_K \restriction_B \subseteq \theta,$$

as

$$\theta_K \restriction_B \subseteq \theta_J \restriction_B.$$

Now suppose

$$J_1 \cup J_2 \in W,$$

i.e.,

$$\theta_{J_1 \cup J_2} \restriction_B \subseteq \theta.$$

As

$$\theta_{J_1 \cup J_2} = \theta_{J_1} \cap \theta_{J_2},$$

it follows that

$$(\theta_{J_1 \cup J_2}) \restriction_B = \theta_{J_1} \restriction_B \cap \theta_{J_2} \restriction_B.$$

Since

$$\theta = \theta \vee (\theta_{J_1} \restriction_B \cap \theta_{J_2} \restriction_B)$$

it follows that

$$\theta = (\theta \vee \theta_{J_1} \restriction_B) \cap (\theta \vee \theta_{J_2} \restriction_B)$$

by distributivity, and as Theorem II§6.20 gives

$$\begin{aligned} \mathbf{Con\ B}/\theta &\cong [\theta, \nabla] \\ &\leq \mathbf{Con\ B} \end{aligned}$$

we must have from the fact that $\mathbf{B}/\theta$ is subdirectly irreducible (it is isomorphic to $\mathbf{A}$)

$$\theta = \theta \vee \theta_{J_i} \restriction_B$$

for $i = 1$ or 2; hence

$$\theta_{J_i} \restriction_B \subseteq \theta$$

for $i = 1$ or 2, so either $J_1$ or $J_2$ is in $W$. By 6.6, there is an ultrafilter $U$ contained in $W$. From the definition of $W$ we have

$$\theta_U \restriction_B \subseteq \theta$$

as

$$\theta_U = \bigcup \{\theta_J : J \in U\}.$$

Let $v$ be the natural homomorphism from $\prod_{i \in I} \mathbf{A}_i$ to $\prod_{i \in I} \mathbf{A}_i/U$. Then let

$$\beta : \mathbf{B} \to v(\mathbf{B})$$

be the restriction of $\nu$ to **B**. As

$$\begin{aligned}\ker\beta &= \theta_U\restriction_B\\ &\subseteq \theta\end{aligned}$$

we have

$$\begin{aligned}\mathbf{A} &\cong \mathbf{B}/\theta\\ &\cong (\mathbf{B}/\ker\beta)/(\theta/\ker\beta).\end{aligned}$$

Now

$$\mathbf{B}/\ker\beta \cong \nu(\mathbf{B}) \leq \prod_{i\in I}\mathbf{A}_i/U$$

so

$$\mathbf{B}/\ker\beta \in ISP_U(K);$$

hence

$$\mathbf{A} \in HSP_U(K).$$

As every algebra in $V(K)$ is isomorphic to a subdirect product of subdirectly irreducible algebras, we have

$$V(K) = IP_SHSP_U(K).$$ □

One part of the previous proof has found so many applications that we isolate it in the following.

**Corollary 6.9** (Jónsson's Lemma). *If $V$ is a congruence-distributive variety and $\mathbf{A}_i \in V$, $i \in I$, if $\mathbf{B} \leq \prod_{i\in I}\mathbf{A}_i$, and $\theta \in \operatorname{Con}\mathbf{B}$ is such that $\mathbf{B}/\theta$ is a nontrivial subdirectly irreducible algebra, then there is an ultrafilter $U$ over $I$ such that*

$$\theta_U\restriction_B \subseteq \theta$$

*where $\theta_U$ is the congruence on $\prod_{i\in I}\mathbf{A}_i$ defined by*

$$\langle a,b\rangle \in \theta_U \quad \text{iff} \quad [\![a = b]\!] \in U.$$

**Corollary 6.10** (Jónsson). *If $K$ is a finite set of finite algebras and $V(K)$ is congruence-distributive, then the subdirectly irreducible algebras of $V(K)$ are in*

$$HS(K),$$

*and*

$$V(K) = IP_S(HS(K)).$$

Proof. By 6.5, $P_U(K) \subseteq I(K)$, so just apply 6.8. □

References

1. B. Jónsson [1967]
2. J. Łoś [1955]

EXERCISES §6

1. An ultrafilter $U$ over a set $I$ is *free* iff $\bigcap U = \varnothing$. Show that an ultrafilter $U$ over $I$ is free iff $I$ is infinite and the cofinite subsets of $I$ belong to $U$.

2. An ultrafilter $U$ over $I$ is *principal* if $\bigcap U \neq \varnothing$. Show that an ultrafilter $U$ is principal iff $U = \{J \subseteq I : i \in J\}$ for some $i \in I$.

3. If $U$ is a principal ultrafilter over $I$ and $\mathbf{A}_i$, $i \in I$, is a collection of algebras show that $\prod_{i \in I} \mathbf{A}_i/U \cong \mathbf{A}_j$ where $\bigcap U = \{j\}$.

4. Show that a finitely generated congruence distributive variety has only finitely many subvarieties. Show that the variety generated by the lattice $\mathbf{N}_5$ has exactly three subvarieties.

5. (Jónsson) If $\mathbf{A}_1,\mathbf{A}_2$ are two finite subdirectly irreducible algebras in a congruence-distributive variety and $\mathbf{A}_1 \not\cong \mathbf{A}_2$ show that there is an identity $p \approx q$ satisfied by one and not the other.

6. Given an uncountable set $I$ show that there is an ultrafilter $U$ over $I$ such that all members of $U$ are uncountable.

7. Show that for $I$ countably infinite there is a subset $S$ of the set of functions from $I$ to 2 which has cardinality equal to that of the continuum such that for $f \neq g$ with $f,g \in S$, $\{i \in I : f(i) = g(i)\}$ is finite. Conclude that $|A^I/U| \geq 2^\omega$ if $U$ is a nonprincipal ultrafilter over $I$ and $|A|$ is infinite.

# §7. Primal Algebras

When Rosenbloom presented his study of the variety of $n$-valued Post algebras in 1942 he proved that all finite members were isomorphic to direct powers of $\mathbf{P}_n$ (see II§1), just as in the case of Boolean algebras. However he thought that an analysis of the infinite members would prove to be far more complex than the corresponding study of infinite Boolean algebras. Then in 1953 Foster proved that every $n$-valued Post algebra was just a Boolean power of $\mathbf{P}_n$.

**Definition 7.1.** If $\mathbf{A}$ is an algebra and

$$f : A^n \to A$$

is an $n$-ary function on $A$, then $f$ is *representable by a term* if there is a term $p$ such that

$$f(a_1, \ldots, a_n) = p^{\mathbf{A}}(a_1, \ldots, a_n)$$

for $a_1, \ldots, a_n \in A$.

**Definition 7.2.** A finite algebra $\mathbf{A}$ is *primal* if every $n$-ary function on $A$, for every $n \geq 1$, is representable by a term.

In §10 we will give an easy test for primality, and show that the Post algebras $\mathbf{P}_n$ are primal. However, one can give a direct proof. A key tool here and in later sections is the switching function.

**Definition 7.3.** The function

$$s: A^4 \to A$$

on a set $A$ defined by

$$s(a,b,c,d) = \begin{cases} c & \text{if } a = b \\ d & \text{if } a \neq b \end{cases}$$

is called the *switching function* on $A$. A term $s(x,y,u,v)$ representing the switching function on an algebra $\mathbf{A}$ is called a *switching term* for $\mathbf{A}$.

**Theorem 7.4** (Foster). *Let* $\mathbf{P}$ *be a primal algebra. Then*

$$V(\mathbf{P}) = I\{\mathbf{P}[\mathbf{B}]^* : \mathbf{B} \textit{ is a Boolean algebra}\}.$$

PROOF. We only need to consider nontrivial $\mathbf{P}$. If $E$ is an equivalence relation on $P$ and

$$\langle a,b\rangle \notin E,$$
$$\langle c,d\rangle \in E$$

with $c \neq d$, then choose a term $p(x)$ such that

$$p(c) = a,$$
$$p(d) = b.$$

Thus

$$E \notin \operatorname{Con}(\mathbf{P});$$

hence $\mathbf{P}$ is simple. Also the only subalgebra of $\mathbf{P}$ is itself (as $P$ is the only subset of $P$ closed under all functions on $P$). As $\mathbf{P}$ has a majority term it follows that $V(\mathbf{P})$ is congruence–distributive, so by 6.8 and the above remarks

$$\begin{aligned} V(\mathbf{P}) &= IP_sHSP_U(\mathbf{P}) \\ &= IP_s(\mathbf{P}) \cup \{\text{trivial algebras}\}. \end{aligned}$$

Thus we only need to show every subdirect power of $\mathbf{P}$ is isomorphic to a Boolean power of $\mathbf{P}$. Let

$$\mathbf{A} \leq \mathbf{P}^I$$

be a nontrivial subdirect power of $\mathbf{P}$. Recall that for $p_1,p_2 \in P^I$ we let

$$[\![p_1 = p_2]\!] = \{i \in I : p_1(i) = p_2(i)\}.$$

In the following we will let $s(x,y,u,v)$ be a term which represents the switching function on $\mathbf{P}$.

**Claim i.** *The constant functions of* $P^I$ *are in* $A$.

This follows from noting that every constant function on P is represented by a term.

**Claim ii.** *The subsets* $[\![a_1 = a_2]\!]$, *for* $a_1,a_2 \in A$, *of* $I$ *form a subuniverse of the Boolean algebra* $\mathbf{Su}(I)$.

Let $c_1,c_2$ be two elements of $A$ with

$$[\![c_1 = c_2]\!] = \varnothing$$

(such must exist as we have assumed **P** is nontrivial). Then for $a_1,a_2,b_1,b_2 \in A$ the following observations suffice:

$$I = [\![c_1 = c_1]\!]$$
$$[\![a_1 = a_2]\!] \cup [\![b_1 = b_2]\!] = [\![s(a_1,a_2,b_1,b_2) = b_1]\!]$$
$$[\![a_1 = a_2]\!] \cap [\![b_1 = b_2]\!] = [\![s(a_1,a_2,b_1,a_1) = s(a_1,a_2,b_2,a_2)]\!].$$

Let **B** be the subalgebra of $\mathbf{Su}(I)$ with the universe

$$\{[\![a_1 = a_2]\!] : a_1,a_2 \in A\},$$

and let

$$X = \mathbf{B}^*.$$

**Claim iii.** *For* $a \in A$ *and* $U \in X$ *there is exactly one* $p \in P$ *such that* $a^{-1}(p) \in U$.

Since $P$ is finite this is an easy consequence of the facts

$$\bigcup_{p \in P} a^{-1}(p) = I \in U,$$

$U$ is an ultrafilter, and the $a^{-1}(p)$'s are pairwise disjoint.

So let us define $\sigma: A \times X \to P$ by

$$\sigma(a,U) = p \quad \text{iff} \quad a^{-1}(p) \in U.$$

Then let us define $\alpha: A \to P^X$ by

$$(\alpha a)(U) = \sigma(a,U).$$

Clearly all the constant functions of $P^X$ are in $\alpha A$ (just look at the images of the constant functions in $A$).

**Claim iv.** *For* $a,b \in A$, $[\![\alpha a = \alpha b]\!] = \{U \in X : [\![a = b]\!] \in U\}$.

To see this we have

$$\begin{aligned} [\![\alpha a = \alpha b]\!] &= \{U \in X : (\alpha a)(U) = (\alpha b)(U)\} \\ &= \{U \in X : \sigma(a, U) = \sigma(b, U)\} \\ &= \{U \in X : a^{-1}(p) \in U, b^{-1}(p) \in U \text{ for some } p \in P\} \\ &= \{U \in X : [\![a = b]\!] \in U\} \qquad \text{(why?).} \end{aligned}$$

Thus a typical clopen subset of $X$ is of the form $[\![\alpha a = \alpha b]\!]$.

Next for $a_1,a_2 \in A$ and $N$ a clopen subset of $X$ choose $b_1,b_2 \in A$ with

$$N = \{U \in X : [\![b_1 = b_2]\!] \in U\}.$$

Let

$$a = a_1\restriction_M \cup a_2\restriction_{I-M}$$

where

$$M = [\![b_1 = b_2]\!].$$

Then

$$a \in A$$

as

$$a = s(b_1,b_2,a_1,a_2).$$

Now

$$\begin{aligned}[\![\alpha a = \alpha a_1]\!] &= \{U \in X : [\![a = a_1]\!] \in U\}\\ &\supseteq \{U \in X : M \in U\}\\ &= N,\end{aligned}$$

and

$$\begin{aligned}[\![\alpha a = \alpha a_2]\!] &= \{U \in X : [\![a = a_2]\!] \in U\}\\ &\supseteq \{U \in X : I - M \in U\}\\ &= X - N,\end{aligned}$$

hence

$$\alpha a = \alpha a_1\restriction_N \cup \alpha a_2\restriction_{X-N}.$$

Then by 5.6 we see that

$$\alpha(A) = P[\mathbf{B}]^*.$$

The map $\alpha$ is actually a bijection, for if $a_1,a_2 \in A$ with

$$a_1 \neq a_2$$

then choosing, by 3.15(b), $U \in X$ with

$$[\![a_1 = a_2]\!] \notin U,$$

we have

$$(\alpha a_1)(U) \neq (\alpha a_2)(U).$$

Finally, to see that $\alpha$ is an isomorphism let $a_1, \ldots, a_n \in A$, and suppose $f$ is an $n$-ary function symbol. Then for $U \in X$ and $p$ such that

$$\sigma(f(a_1, \ldots, a_n), U) = p$$

we can use

$$f(a_1, \ldots, a_n)^{-1}(p) = \bigcup_{\substack{p_i \in P\\ f(p_1, \ldots, p_n) = p}} a_1^{-1}(p_1) \cap \cdots \cap a_n^{-1}(p_n)$$

and

$$f(a_1, \ldots, a_n)^{-1}(p) \in U$$

to show that, for some choice of $p_1, \ldots, p_n$ with $f(p_1, \ldots, p_n) = p$,

$$a_1^{-1}(p_1) \cap \cdots \cap a_n^{-1}(p_n) \in U.$$

Hence

$$a_i^{-1}(p_i) \in U, \qquad 1 \leq i \leq n,$$

and thus

$$\sigma(a_i,U) = p_i, \qquad 1 \le i \le n.$$

Consequently,

$$\begin{aligned}\alpha(f(a_1,\ldots,a_n))(U) &= \sigma(f(a_1,\ldots,a_n),U)\\ &= p\\ &= f(p_1,\ldots,p_n)\\ &= f(\sigma(a_1,U),\ldots,\sigma(a_n,U))\\ &= f((\alpha a_1)(U),\ldots,(\alpha a_n)(U))\\ &= f(\alpha a_1,\ldots,\alpha a_n)(U),\end{aligned}$$

so

$$\alpha f(a_1,\ldots,a_n) = f(\alpha a_1,\ldots,\alpha a_n).$$ □

REFERENCES

1. A. L. Foster [1953b]
2. P. C. Rosenbloom [1942]

EXERCISES §7

1. Show that a primal lattice is trivial.
2. Show that if **B** is a primal Boolean algebra then $|B| \le 2$.
3. Prove that for $p$ a prime number, $\langle Z/(p),+,\cdot,-,0,1\rangle$ is a primal algebra.
4. Prove that the Post algebras $\mathbf{P}_n$ are primal.
5. If $\mathbf{B}_1,\mathbf{B}_2$ are Boolean algebras and $\alpha:\mathbf{B}_1 \to \mathbf{B}_2$ is a homomorphism, let $\bar{\alpha}:\mathbf{B}_1^{**} \to \mathbf{B}_2^{**}$ be the corresponding homomorphism defined by $\bar{\alpha}(N_b) = N_{\alpha(b)}$. Then, given any algebra **A**, define $\alpha^*: A[\mathbf{B}_1]^* \to A[\mathbf{B}_2]^*$ by

   $$[\![\alpha^* c = c_a]\!] = \bar{\alpha}[\![c = c_a]\!], \quad \text{for } a \in A.$$

   Show that $\alpha^*$ is a homomorphism from $\mathbf{A}[\mathbf{B}_1]^*$ to $\mathbf{A}[\mathbf{B}_2]^*$.
6. If **P** is a primal algebra show that the only homomorphisms from $\mathbf{P}[\mathbf{B}_1]^*$ to $\mathbf{P}[\mathbf{B}_2]^*$ are of the form $\alpha^*$ described in Exercise 5.
7. If **P** is a nontrivial primal algebra show that $\mathbf{P}[\mathbf{B}_1]^* \cong \mathbf{P}[\mathbf{B}_2]^*$ iff $\mathbf{B}_1 \cong \mathbf{B}_2$.
8. (Sierpiński). Show that any finitary operation on a finite set $A$ is expressible as a composition of binary operations.

## §8. Boolean Products

Boolean products provide an effective generalization of the notion of Boolean power. Actually the construction that we call "Boolean product" has been known for several years as "the algebras of global sections of sheaves of algebras over Boolean spaces"; however the definition of the latter was unnecessarily involved.

**Definition 8.1.** A *Boolean product* of an indexed family $(\mathbf{A}_x)_{x \in X}$, $X \neq \varnothing$, of algebras is a subdirect product $\mathbf{A} \leq \prod_{x \in X} \mathbf{A}_x$, where $X$ can be endowed with a Boolean space topology so that

(i) $[\![a = b]\!]$ is clopen for $a, b \in A$, and
(ii) if $a,b \in A$ and $N$ is a clopen subset of $X$, then

$$a\restriction_N \cup b\restriction_{X-N} \in A.$$

We refer to condition (i) as "equalizers are clopen", and to condition (ii) as "the patchwork property" (draw a picture!). For a class of algebras $K$, let $\Gamma^a(K)$ denote the class of Boolean products which can be formed from subsets of $K$. Thus $\Gamma^a(K) \subseteq P_S(K)$.

Our definition of Boolean product is indeed very close to the description of Boolean powers given in 5.6. In this section we will develop a technique for establishing the existence of Boolean product representations, and apply it to biregular rings. But first we need to develop some lattice-theoretic notions and results.

**Definition 8.2.** Let $\mathbf{L}$ be a lattice. An *ideal $I$ of* $\mathbf{L}$ is a nonempty subset of $L$ such that

(i) $a \in I$, $b \in L$, *and* $b \leq a \Rightarrow b \in I$,
(ii) $a,b \in I \Rightarrow a \vee b \in I$.

$I$ is *proper* if $I \neq L$, and $I$ is *maximal* if $I$ is maximal among the proper ideals of $\mathbf{L}$. Similarly we define *filters*, *proper filters*, and *maximal filters* of $\mathbf{L}$.

Parallel to 3.7, 3.8, and 3.9 we have (using the same proofs) the following.

**Lemma 8.3.** *The set of ideals and the set of filters of a lattice are closed under finite intersection, and arbitrary intersection provided the intersection is not empty.*

**Definition 8.4.** Given a lattice $\mathbf{L}$ and a nonempty set $X \subseteq L$ let $I(X)$ denote the least ideal of $\mathbf{L}$ containing $X$, called the *ideal generated by* $X$, and let $F(X)$ denote the least filter of $\mathbf{L}$ containing $X$, called the *filter generated by* $X$.

**Lemma 8.5.** *For a lattice* $\mathbf{L}$ *and* $X \subseteq L$ *we have*

$$I(X) = \{a \in L : a \leq a_1 \vee \cdots \vee a_n \text{ for some } a_1, \ldots, a_n \in X\}$$
$$F(X) = \{a \in L : a \geq a_1 \wedge \cdots \wedge a_n \text{ for some } a_1, \ldots, a_n \in X\}.$$

*In particular if $J$ is an ideal of* $\mathbf{L}$ *and* $b \in L$, *then*

$$I(J \cup \{b\}) = \{a \in L : a \leq j \vee b \text{ for some } j \in J\}.$$

**Definition 8.6.** A lattice **L** is said to be *relatively complemented* if for

$$a \le b \le c$$

in **L** there exists $d \in L$ with

$$b \wedge d = a,$$
$$b \vee d = c.$$

$d$ is called a *relative complement* of $b$ in the interval $[a,c]$.

**Lemma 8.7.** *Suppose* **L** *is a relatively complemented distributive lattice with $I$ an ideal of* **L** *and $a \in L - I$. Then there is a maximal ideal $M$ of* **L** *with*

$$I \subseteq M, \qquad a \notin M.$$

*Furthermore, $L - M$ is a maximal filter of* **L**. *The same results hold interchanging the words ideal and filter.*

PROOF. Use Zorn's lemma to extend $I$ to an ideal $M$ which is maximal among the ideals of **L** containing $I$, but to which $a$ does not belong. It only remains to show that $M$ is actually a maximal ideal of **L** . For $b_1,b_2 \notin M$ we have

$$a \in I(M \cup \{b_i\}), \qquad i = 1,2;$$

hence for some $c_i \in M$, $i = 1,2$,

$$a \le b_1 \vee c_1,$$
$$a \le b_2 \vee c_2.$$

Hence

$$\begin{aligned} a &\le (b_1 \vee c_1) \wedge (b_2 \vee c_2) \\ &= (b_1 \wedge b_2) \vee [(b_1 \wedge c_2) \vee (c_1 \wedge b_2) \vee (c_1 \wedge c_2)]. \end{aligned}$$

As the element in brackets is in $M$, we must have

$$b_1 \wedge b_2 \notin M$$

as $a \notin M$. Thus it is easily seen that $L - M$ is a filter. Now given $b_1,b_2 \notin M$, choose $c \in M$ with

$$c \le b_1.$$

Then let $d_1 \in L$ be such that

$$b_1 \vee d_1 = b_1 \vee b_2,$$
$$b_1 \wedge d_1 = c,$$

i.e., $d_1$ is a relative complement of $b_1$ in the interval $[c,b_1 \vee b_2]$. As $L - M$ is a filter and $c \notin L - M$ it follows that $d_1 \in M$. But then

$$b_2 \le b_1 \vee d_1$$

says

$$b_2 \in I(M \cup \{b_1\});$$

hence

$$L = I(M \cup \{b_1\}).$$

Consequently $M$ is a maximal ideal. □

**Lemma 8.8.** *In a distributive lattice relative complements are unique if they exist.*

Proof. Suppose **L** is a distributive lattice and

$$a \leq b \leq c$$

in **L**. If $d_1$ and $d_2$ are relative complements of $b$ in the interval $[a,c]$, then

$$\begin{aligned} d_1 &= d_1 \wedge c \\ &= d_1 \wedge (b \vee d_2) \\ &= (d_1 \wedge b) \vee (d_1 \wedge d_2) \\ &= d_1 \wedge d_2. \end{aligned}$$

Likewise

$$d_2 = d_1 \wedge d_2,$$

so

$$d_1 = d_2.$$ □

**Definition 8.9.** If **L** is a relatively complemented distributive lattice with a least element 0 and $a,b \in L$, then $a\backslash b$ denotes the relative complement of $b$ in the interval $[0, a \vee b]$.

**Lemma 8.10.** *If* **L** *is a distributive lattice with a least element* 0 *such that for* $a,b \in L$ *the relative complement (denoted* $a\backslash b$*) of* $b$ *in the interval* $[0,a \vee b]$ *exists, then* **L** *is relatively complemented.*

Proof. Let

$$a \leq b \leq c$$

hold in **L**. Let

$$d = a \vee (c\backslash b).$$

Then

$$b \vee d = b \vee (c\backslash b) = c,$$

and

$$\begin{aligned} b \wedge d &= b \wedge [a \vee (c\backslash b)] \\ &= a \vee [b \wedge (c\backslash b)] \\ &= a, \end{aligned}$$

so $d$ is a relative complement of $b$ in $[a,c]$. □

Now we have all the facts we need about relatively complemented distributive lattices, so let us apply them to the study of Boolean products.

**Definition 8.11.** If **A** is an algebra then an embedding

$$\alpha : \mathbf{A} \to \prod_{x \in X} \mathbf{A}_x$$

gives a *Boolean product representation* of **A** if $\alpha(\mathbf{A})$ is a Boolean product of the $\mathbf{A}_x$'s.

**Theorem 8.12.** *Let* **A** *be an algebra. Suppose* **L** *is a sublattice of* **Con A** *such that*

(i) $\Delta \in L$,
(ii) *the congruences in* $L$ *permute,*
(iii) **L** *is a relatively complemented distributive lattice, and*
(iv) *for each* $a,b \in A$ *there is a smallest member* $\theta_{ab}$ *of* $L$ *with* $\langle a,b\rangle \in \theta_{ab}$.

*Let*

$$X = \{M : M \text{ is a maximal ideal of } \mathbf{L}\} \cup \{L\},$$

*and introduce a topology on* $X$ *with a subbasis*

$$\{N_\theta : \theta \in L\} \cup \{D_\theta : \theta \in L\}$$

*where*

$$N_\theta = \{M \in X : \theta \in M\},$$

*and*

$$D_\theta = \{M \in X : \theta \notin M\}.$$

*Then* $X$ *is a Boolean space,* $\bigcup M$ *is a congruence for each* $M \in X$*, and the map*

$$\alpha : A \to \prod_{M \in X} (A/\bigcup M)$$

*defined by*

$$(\alpha a)(M) = a/\bigcup M$$

*gives a Boolean product representation of* **A** *such that*

$$[\![\alpha a = \alpha b]\!] = N_{\theta_{ab}}.$$

*Consequently,*

$$\mathbf{A} \in I\Gamma^a(\{\mathbf{A}/\bigcup M : M \in X\}).$$

PROOF.

**Claim i.** *The subbasis*

$$\{N_\theta : \theta \in L\} \cup \{D_\theta : \theta \in L\}$$

*is a field of subsets of* $X$*, hence a basis for the topology. In particular,*

(a) $X = N_\Delta$, $\varnothing = D_\Delta$,
*and for* $\theta,\phi \in L$,
(b) $N_\theta \cup N_\phi = N_{\theta \cap \phi}$,
(c) $N_\theta \cap N_\phi = N_{\theta \vee \phi}$,
(d) $D_\theta \cup D_\phi = D_{\theta \vee \phi}$,
(e) $D_\theta \cap D_\phi = D_{\theta \cap \phi}$,

(f) $N_\theta \cup D_\phi = N_{\theta\backslash\phi}$,
(g) $N_\theta \cap D_\phi = D_{\phi\backslash\theta}$,
*and*
(h) $X = N_\theta \cup D_\theta$, $\varnothing = N_\theta \cap D_\theta$.

Proof. (a) Clearly

$$X = N_\Delta, \qquad \varnothing = D_\Delta.$$

The proofs below make frequent use of the fact that $L - M$ is a filter of $\mathbf{L}$ if $M \in X - \{L\}$.

(b) $M \in N_\theta \cup N_\phi$ iff $\theta \in M$ or $\phi \in M$
iff $\theta \cap \phi \in M$
iff $M \in N_{\theta \cap \phi}$.

One handles (c) similarly.

(d) $M \in D_\theta \cup D_\phi$ iff $\theta \notin M$ or $\phi \notin M$
iff $\theta \vee \phi \notin M$
iff $M \in D_{\theta \vee \phi}$.

One handles (e) similarly.

(f) From the statements

$$\phi \cap (\theta\backslash\phi) = \Delta$$
$$\theta\backslash\phi \subseteq \theta$$
$$\theta \subseteq \theta \vee \phi = \phi \vee (\theta\backslash\phi)$$

it follows, for $M \in X$, that

$$\phi \in M \quad \text{or} \quad \theta\backslash\phi \in M$$
$$\theta \notin M \quad \text{or} \quad \theta\backslash\phi \in M$$
$$\phi \notin M \quad \text{or} \quad \theta\backslash\phi \notin M \quad \text{or } \theta \in M.$$

The first two give

$$\theta\backslash\phi \notin M \Rightarrow \theta \notin M \text{ and } \phi \in M$$

and from the third

$$\theta\backslash\phi \in M \Rightarrow \theta \in M \text{ or } \phi \notin M.$$

Thus

$$\theta\backslash\phi \in M \Leftrightarrow \theta \in M \text{ or } \phi \notin M.$$

(g) This is an immediate consequence of (f).

(h) (These assertions are obvious.)

Thus we have a field of subsets of $X$. □

**Claim ii.** *$X$ is a Boolean space.*

Proof. If $M_1, M_2 \in X$ and $M_1 \neq M_2$ then without loss of generality let $\theta \in M_1 - M_2$. Then

$$M_1 \in N_\theta,$$
$$M_2 \in D_\theta,$$

so $X$ is Hausdorff. From claim (i) we have a basis of clopen subsets. So we only need to show $X$ is compact. Suppose

$$X = \bigcup_{i \in I} N_{\theta_i} \cup \bigcup_{j \in J} D_{\phi_j}.$$

As $L \in X$ it follows that $I \neq \varnothing$, say $i_0 \in I$. Let

$$D_{\theta'_i} = N_{\theta_i} \cap D_{\theta_{i_0}}$$

and

$$D_{\phi'_j} = D_{\phi_j} \cap D_{\theta_{i_0}}.$$

Then

$$\begin{aligned} D_{\theta_{i_0}} &= X \cap D_{\theta_{i_0}} \\ &= \bigcup_{i \in I} D_{\theta'_i} \cup \bigcup_{j \in J} D_{\phi'_j}. \end{aligned}$$

If the ideal of $\mathbf{L}$ generated by

$$\{\theta'_i : i \in I\} \cup \{\phi'_j : j \in J\}$$

does not contain $\theta_{i_0}$ then it can be extended to a maximal ideal $M$ of $\mathbf{L}$ such that $\theta_{i_0} \notin M$. But then

$$M \in D_{\theta_{i_0}} - \left(\bigcup_{i \in I} D_{\theta'_i} \cup \bigcup_{j \in J} D_{\phi'_j}\right),$$

which is impossible. Thus by 8.5 for some finite subsets $I_0$ (of $I$) and $J_0$ (of $J$) we have

$$\theta_{i_0} \leq \bigvee_{i \in I_0} \theta'_i \vee \bigvee_{j \in J_0} \phi'_j;$$

hence, by claim (i),

$$D_{\theta_{i_0}} \subseteq \bigcup_{i \in I_0} D_{\theta'_i} \cup \bigcup_{j \in J_0} D_{\phi'_j}.$$

As

$$\begin{aligned} D_{\theta'_i} &\subseteq N_{\theta_i}, \\ D_{\phi'_j} &\subseteq D_{\phi_j} \end{aligned}$$

we have

$$\begin{aligned} X &= N_{\theta_{i_0}} \cup D_{\theta_{i_0}} \\ &= N_{\theta_{i_0}} \cup \bigcup_{i \in I_0} N_{\theta_i} \cup \bigcup_{j \in J_0} D_{\phi_j}, \end{aligned}$$

so $X$ is compact. □

**Claim iii.** *$\alpha$ gives a Boolean product representation of* $\mathbf{A}$.

Proof. Certainly $\alpha$ is a homomorphism. If $a \neq b$ in $A$ then

$$\{\theta \in L : \langle a,b \rangle \in \theta\}$$

is a proper filter of **L**. Extend this to a maximal filter $F$ of **L**, and let

$$M = L - F,$$

a maximal ideal of **L**. Thus

$$\langle a,b\rangle \notin \bigcup M$$

as

$$\langle a,b\rangle \notin \theta$$

for $\theta \in M$. From this follows

$$\bigcap_{M \in X} (\bigcup M) = \Delta,$$

so $\alpha\mathbf{A}$ is a subdirect product of the $\mathbf{A}/\bigcup M$ by II§8.2.

For $a,b \in A$ we have

$$\begin{aligned}[\![\alpha a = \alpha b]\!] &= \{M \in X : \langle a,b\rangle \in \bigcup M\}\\ &= \{M \in X : \theta_{ab} \in M\}\\ &= N_{\theta_{ab}},\end{aligned}$$

so equalizers are clopen.

Next given $a,b \in A$ and $\theta \in L$ we want to show

$$(\alpha a)\restriction_{N_\theta} \cup (\alpha b)\restriction_{X - N_\theta} \in \alpha A.$$

Choose $\phi \in L$ such that

$$\langle a,b\rangle \in \phi.$$

Then

$$\langle a,b\rangle \in \theta \vee \phi = \theta \vee (\phi\backslash\theta),$$

so by the permutability of members of $L$ there is a $c \in A$ with

$$\begin{aligned}&\langle a,c\rangle \in \theta,\\ &\langle c,b\rangle \in \phi\backslash\theta.\end{aligned}$$

As

$$\begin{aligned}[\![\alpha a = \alpha c]\!] &= N_{\theta_{ac}}\\ &\supseteq N_\theta\end{aligned}$$

and

$$\begin{aligned}[\![\alpha c = \alpha b]\!] &= N_{\theta_{cb}}\\ &\supseteq N_{\phi\backslash\theta}\\ &= N_\phi \cup D_\theta\\ &\supseteq D_\theta\end{aligned}$$

we have

$$\alpha c = \alpha a\restriction_{N_\theta} \cup \alpha b\restriction_{D_\theta},$$

so $\alpha A$ has the patchwork property. $\square$

**Definition 8.13.** Given $\mathbf{A}$ let

$$\operatorname{Spec} \mathbf{A} = \{\phi \in \operatorname{Con} \mathbf{A} : \phi \text{ is a maximal congruence on } \mathbf{A}\} \cup \{\nabla\},$$

and let the topology on Spec $\mathbf{A}$ be generated by

$$\{E(a,b) \mid a,b \in A\} \cup \{D(a,b) \mid a,b \in A\},$$

where

$$E(a,b) = \{\phi \in \operatorname{Spec} \mathbf{A} : \langle a,b\rangle \in \phi\},$$
$$D(a,b) = \{\phi \in \operatorname{Spec} \mathbf{A} : \langle a,b\rangle \notin \phi\}.$$

**Corollary 8.14.** *Let* $\mathbf{A}$ *be an algebra such that the finitely generated congruences permute and form a sublattice* $\mathbf{L}$ *of* **Con A** *which is distributive and relatively complemented. Then the natural map*

$$\beta : \mathbf{A} \to \prod_{\theta \in \operatorname{Spec} \mathbf{A}} \mathbf{A}/\theta$$

*gives a Boolean product representation of* $\mathbf{A}$, *and for* $a,b \in A$,

$$[\![\beta a = \beta b]\!] = E(a,b).$$

Proof. Let $M \in X$, $X$ as defined in 8.12. If

$$M = L$$

then

$$\bigcup M = \nabla \in \operatorname{Spec} \mathbf{A}.$$

If

$$M \neq L,$$

then for some $a,b \in A$,

$$\Theta(a,b) \notin M,$$

so

$$\langle a,b\rangle \notin \bigcup M.$$

If $\bigcup M$ is not maximal then, for some $\theta \in \operatorname{Con} \mathbf{A}$,

$$\bigcup M \subseteq \theta \neq \nabla$$

and

$$\bigcup M \neq \theta.$$

But

$$\theta = \bigcup \{\phi \in L : \phi \subseteq \theta\},$$

so

$$I = \{\phi \in L : \phi \subseteq \theta\}$$

is a proper ideal of $L$ such that $M \subseteq I$ but $M \neq I$. This contradicts the maximality of $M$. Hence $M \in X$ implies

$$\bigcup M \in \operatorname{Spec} \mathbf{A}.$$

If $M_1, M_2 \in X$ with

$$M_1 \neq M_2$$

then it is readily verifiable that

$$\bigcup M_1 \neq \bigcup M_2.$$

And for $\theta \in \operatorname{Spec} \mathbf{A}$, clearly

$$\{\phi \in L : \phi \subseteq \theta\}$$

is in $X$. Thus the map

$$\sigma : X \to \operatorname{Spec} \mathbf{A}$$

defined by

$$\sigma M = \bigcup M$$

is a bijection. For $a,b \in A$ note that

$$\begin{aligned}\sigma(N_{\Theta(a,b)}) &= \sigma\{M \in X : \Theta(a,b) \in M\}\\ &= \{\textstyle\bigcup M : M \in X, \langle a,b\rangle \in \bigcup M\}\\ &= \{\theta \in \operatorname{Spec} \mathbf{A} : \langle a,b\rangle \in \theta\}\\ &= E(a,b);\end{aligned}$$

hence $\sigma$ is a homeomorphism from $X$ to $\operatorname{Spec} \mathbf{A}$. Thus

$$\beta : \mathbf{A} \to \prod_{\theta \in \operatorname{Spec} \mathbf{A}} \mathbf{A}/\theta$$

gives a Boolean product representation of $\mathbf{A}$ where

$$[\![\beta a = \beta b]\!] = E(a,b). \qquad \square$$

EXAMPLE (Dauns and Hofmann). A ring $\mathbf{R}$ is *biregular* if every principal ideal is generated by a central idempotent (we only consider two-sided ideals). For $r \in R$ let $I(r)$ denote the ideal of $\mathbf{R}$ generated by $r$. If $a$ and $b$ are central idempotents of $\mathbf{R}$ it is a simple exercise to verify

$$I(a) \vee I(b) = I(a + b - ab)$$

and

$$I(a) \wedge I(b) = I(ab).$$

Thus all finitely generated ideals are principal, and they form a sublattice of the lattice of all ideals of $\mathbf{R}$. From the above equalities one can readily check the distributive laws, and finally

$$I(b) \backslash I(a) = I(b - ab),$$

i.e., the finitely generated ideals of $\mathbf{R}$ form a relatively complemented distributive sublattice of the lattice of ideals of $\mathbf{R}$; and of course all rings have permutable congruences. Thus by 8.14, $\mathbf{R}$ is isomorphic to a Boolean product of simple rings and a trivial ring. (A lemma of Arens and Kaplansky shows that the simple rings have a unit element.)

REFERENCES

1. S. Burris and H. Werner [1979], [1980]
2. J. Dauns and K. Hofmann [1966]

EXERCISES §8

1. If **L** is a distributive lattice, $I$ is an ideal of **L**, and $a \in L - I$, show that there is an ideal $J$ which contains $I$ but $a \notin J$, and $L - J$ is a filter of **L**. However, show that $J$ cannot be assumed to be a maximal ideal of **L**.

2. (Birkhoff). Show that if **L** is a subdirectly irreducible distributive lattice then $|L| \leq 2$.

3. Verify the details of the example (due to Dauns and Hofmann) at the end of §8.

4. Let **A** be an algebra with subalgebra $\mathbf{A}_0$. Given a Boolean algebra **B** and a closed subset $Y$ of $\mathbf{B}^*$, let

$$C = \{c \in A[\mathbf{B}]^* : c(Y) \subseteq A_0\}.$$

Show that $C$ is a subuniverse of $\mathbf{A}[\mathbf{B}]^*$, and $\mathbf{C} \in \Gamma^a(\{A, A_0\})$.

5. If **A** is a Boolean product of $(\mathbf{A}_x)_{x \in X}$ and $Y$ is a subset of $X$, let $A\restriction_Y = \{a\restriction_Y : a \in A\}$, a subuniverse of $\prod_{x \in Y} \mathbf{A}_x$. Let the corresponding subalgebra be $\mathbf{A}\restriction_Y$. If $N$ is a clopen subset of $X$, $\varnothing \neq N \neq X$, show

$$\mathbf{A} \cong \mathbf{A}\restriction_N \times \mathbf{A}\restriction_{X-N}.$$

Hence conclude that if a variety $V$ can be expressed as $V = I\Gamma^a(K)$ then all the directly indecomposable members of $V$ are in $I(K)$.

# §9. Discriminator Varieties

In this section we look at the most successful generalization of Boolean algebras to date, successful because we obtain Boolean product representations (which can be used to provide a deep insight into algebraic and logical properties).

**Definition 9.1.** The *discriminator function* on a set $A$ is the function $t : A^3 \to A$ defined by

$$t(a,b,c) = \begin{cases} a & \text{if } a \neq b \\ c & \text{if } a = b. \end{cases}$$

A ternary term $t(x,y,z)$ representing the discriminator function on **A** is called a *discriminator term for* **A**.

**Lemma 9.2.** (a) *An algebra* **A** *has a discriminator term iff it has a switching term (see* §7).

(b) *An algebra* **A** *with a discriminator term is simple.*

PROOF. (a) ($\Rightarrow$) If $t(x,y,z)$ is a discriminator term for $\mathbf{A}$, let $s(x,y,u,v) = t(t(x,y,u), t(x,y,v),v)$.

($\Leftarrow$) If $s(x,y,u,v)$ is a switching term for $\mathbf{A}$, then let $t(x,y,z) = s(x,y,z,x)$.

(b) Let $s(x,y,u,v)$ be a switching term for $\mathbf{A}$. If $a,b,c,d \in A$ with $a \neq b$, we have

$$\langle c,d \rangle = \langle s(a,a,c,d), s(a,b,c,d) \rangle \in \Theta(a,b);$$

hence

$$a \neq b \Rightarrow \Theta(a,b) = \nabla.$$

Thus $A$ is simple. □

**Definition 9.3.** Let $K$ be a class of algebras with a common discriminator term $t(x,y,z)$. Then $V(K)$ is called a *discriminator variety*.

EXAMPLES. (1) If $\mathbf{P}$ is a *primal algebra* then $V(\mathbf{P})$ is a discriminator variety.

(2) The *cylindric algebras of dimension n* form a discriminator variety. To see this let $c(x) = c_0(c_1(\ldots(c_{n-1}(x)\ldots)$. From §3 Ex. 7 we know that a cylindric algebra $\mathbf{A}$ of dimension $n$ is subdirectly irreducible iff for $a \in A$,

$$a \neq 0 \Rightarrow c(a) = 1.$$

Thus the term $t(x,y,z)$ given by

$$[c(x + y) \wedge x] \vee [c(x + y)' \wedge z]$$

is a discriminator term on the subdirectly irreducible members. This ensures that the variety is a discriminator variety.

**Theorem 9.4** (Bulman–Fleming, Keimel, Werner). *Let $t(x,y,z)$ be a discriminator term for all algebras in $K$. Then*

(a) *$V(K)$ is an arithmetical variety.*
(b) *The indecomposable members of $V(K)$ are simple algebras, and*
(c) *The simple algebras are precisely the members of $ISP_U(K_+)$, where $K_+$ is $K$, augmented by a trivial algebra.*
(d) *Furthermore, every member of $V(K)$ is isomorphic to a Boolean product of simple algebras, i.e.,*

$$V(K) = I\Gamma^a SP_U(K_+).$$

PROOF. As $t(x,y,z)$ is a 2/3-minority term for $K$, we have an arithmetical variety by II§12.5. Hence the subdirectly irreducible members of $V(K)$ are in $HSP_U(K)$ by 6.8. For $A_i \in K$, $i \in I$, $U \in \mathbf{Su}(I)^*$, and $a,b,c \in \prod_{i \in I} A_i$, if

$$a/U = b/U$$

then

$$\begin{aligned} t(a/U,b/U,c/U) &= t(a,b,c)/U \\ &= c/U \end{aligned}$$

as

$$[\![t(a,b,c) = c]\!] \in U$$

since

$$[\![a = b]\!] \in U$$

and

$$[\![t(a,b,c) = c]\!] \supseteq [\![a = b]\!].$$

Likewise,

$$\begin{aligned} & [\![a = b]\!] \notin U \\ \Rightarrow\ & I - [\![a = b]\!] \in U \\ \Rightarrow\ & [\![t(a,b,c) = a]\!] \in U; \end{aligned}$$

hence

$$a/U \neq b/U \Rightarrow t(a/U,b/U,c/U) = a/U;$$

thus $t$ is a discriminator term for $\prod_{i \in I} \mathbf{A}_i/U$. If now

$$\mathbf{B} \leq \prod_{i \in I} \mathbf{A}_i/U$$

then $t$ is also a discriminator term for $\mathbf{B}$. Consequently, all members of $SP_U(K)$ are simple by 9.2. It follows by 6.8 that the subdirectly irreducible members of $V(K)$ are up to isomorphism precisely the members of $SP_U(K_+)$, and all subdirectly irreducible algebras are simple algebras with $t(x,y,z)$ as a discriminator term.

To see that we have Boolean product representations let

$$\mathbf{A} \in P_S SP_U(K_+),$$

say $\mathbf{A} \leq \prod_{i \in I} \mathbf{S}_i, \mathbf{S}_i \in SP_U(K_+)$. Let $s(x,y,u,v)$ be a switching term for $SP_U(K_+)$ (which must exist by 9.2). If $a,b,c,d \in A$ and

$$[\![a = b]\!] \subseteq [\![c = d]\!]$$

then

$$\langle c,d \rangle = \langle s(a,a,c,d), s(a,b,c,d) \rangle \in \Theta(a,b).$$

Thus

$$\langle a,b \rangle \in \{\langle c,d \rangle : [\![a = b]\!] \subseteq [\![c = d]\!]\} \subseteq \Theta(a,b).$$

The set

$$\{\langle c,d \rangle : [\![a = b]\!] \subseteq [\![c = d]\!]\}$$

is readily seen to be a congruence on $\mathbf{A}$; hence

$$\Theta(a,b) = \{\langle c,d \rangle : [\![a = b]\!] \subseteq [\![c = d]\!]\}.$$

From this it follows that

$$\begin{aligned} \Theta(a,b) \vee \Theta(c,d) &= \Theta(t(a,b,c), t(b,a,d)) \\ \Theta(a,b) \wedge \Theta(c,d) &= \Theta(s(a,b,c,d), c). \end{aligned}$$

Let us verify these two equalities. For $i \in I$,

$$t(a,b,c)(i) = t(b,a,d)(i)$$

holds iff

$$a(i) = b(i) \quad \text{and} \quad c(i) = d(i);$$

hence

$$[\![t(a,b,c) = t(b,a,d)]\!] = [\![a = b]\!] \cap [\![c = d]\!],$$

so

$$\langle a,b\rangle, \langle c,d\rangle \in \Theta(t(a,b,c), t(b,a,d)),$$

so

$$\Theta(a,b), \Theta(c,d) \subseteq \Theta(t(a,b,c), t(b,a,d)).$$

This gives

$$\Theta(a,b) \vee \Theta(c,d) \subseteq \Theta(t(a,b,c), t(b,a,d)).$$

Now clearly

$$\langle t(a,b,c), t(b,a,d)\rangle \in \Theta(a,b) \vee \Theta(c,d)$$

as

$$t(a,b,c)\Theta(a,b)t(a,a,c)\Theta(c,d)t(a,a,d)\Theta(a,b)t(b,a,d).$$

Thus

$$\langle t(a,b,c), t(b,a,d)\rangle \in \Theta(a,b) \vee \Theta(c,d),$$

so

$$\Theta(a,b) \vee \Theta(c,d) = \Theta(t(a,b,c), t(b,a,d)).$$

Next, note that

$$s(a,b,c,d)(i) = c(i) \quad \text{iff} \quad a(i) = b(i) \text{ or } c(i) = d(i);$$

hence

$$[\![s(a,b,c,d) = c]\!] = [\![a = b]\!] \cup [\![c = d]\!].$$

This immediately gives

$$\Theta(s(a,b,c,d), c) \subseteq \Theta(a,b), \Theta(c,d),$$

so

$$\Theta(s(a,b,c,d), c) \subseteq \Theta(a,b) \cap \Theta(c,d).$$

Conversely, if

$$\langle e_1,e_2\rangle \in \Theta(a,b) \cap \Theta(c,d)$$

then

$$[\![a = b]\!], [\![c = d]\!] \subseteq [\![e_1 = e_2]\!],$$

so

$$\begin{aligned}[\![s(a,b,c,d) = c]\!] &= [\![a = b]\!] \cup [\![c = d]\!] \\ &\subseteq [\![e_1 = e_2]\!],\end{aligned}$$

thus

$$\langle e_1,e_2\rangle \in \Theta(s(a,b,c,d), c).$$

This shows

$$\Theta(a,b) \cap \Theta(c,d) = \Theta(s(a,b,c,d), c).$$

The above equalities show that the finitely generated congruences on **A** form a sublattice **L** of **Con A**, and indeed they are all principal. As $V(K)$ is arithmetical **L** is a distributive lattice of permuting congruences. Next we show the existence of relative complements. For $a,b,c,d \in A$ note that

$$\begin{aligned}\Theta(c,d) \wedge \Theta(s(c,d,a,b),b) &= \Theta(s(c,d,s(c,d,a,b),b), s(c,d,a,b)) \\ &= \Delta\end{aligned}$$

as one can easily verify

$$s(c,d,s(c,d,a,b),b) = s(c,d,a,b);$$

and

$$\begin{aligned}\Theta(c,d) \vee \Theta(s(c,d,a,b), b) &= \Theta(t(c, d, s(c,d,a,b)), t(d,c,b)) \\ &= \Theta(t(c,d,a), t(d,c,b))\end{aligned}$$

(just verify that both of the corresponding equalizers are equal to $[\![c = d]\!] \cap [\![a = b]\!]$); hence

$$= \Theta(a,b) \vee \Theta(c,d).$$

Thus

$$\Theta(a,b)\backslash\Theta(c,d) = \Theta(s(c,d,a,b),b),$$

so $\mathbf{L}$ is relatively complemented.

Applying 8.14, we see that $\mathbf{A} \in I\Gamma^a SP_U(K_+)$.

Note that if a variety $V$ is such that $V = I\Gamma^a(K)$ then $V_{DI} \subseteq I(K)$, where $V_{DI}$ is the class of directly indecomposable members of $V$. □

## References

1. S. Burris and H. Werner [1979]
2. H. Werner [1978]

## Exercises §9

1. (a) Show that the variety of rings with identity generated by finitely many finite fields is a discriminator variety. (b) Show that the variety of rings generated by finitely many finite fields is a discriminator variety.

2. If $\mathbf{A}$ is a Boolean product of an indexed family $\mathbf{A}_x$, $x \in X$, of algebras with a common discriminator term, show that for each congruence $\theta$ on $\mathbf{A}$ there is a closed subset $Y$ of $X$ such that

$$\theta = \{\langle a,b\rangle \in A \times A : Y \subseteq [\![a = b]\!]\},$$

and hence for $\theta$ a maximal congruence on $\mathbf{A}$ there is an $x \in X$ such that

$$\theta = \{\langle a,b\rangle \in A \times A : a(x) = b(x)\}.$$

3. If $\mathbf{A}_1,\mathbf{A}_2$ are two nonisomorphic algebras with $\mathbf{A}_1 \leq \mathbf{A}_2$, and with a common ternary discriminator term, show that there is an algebra in $\Gamma^a(\{\mathbf{A}_1,\mathbf{A}_2\})$ which is not isomorphic to an algebra of the form $\mathbf{A}_1[\mathbf{B}_1]^* \times \mathbf{A}_2[\mathbf{B}_2]^*$.

The *spectrum* of a variety $V$, Spec $(V)$, is $\{|A| : \mathbf{A} \in V, \mathbf{A} \text{ is finite}\}$.

4. (Grätzer). For $S$ a subset of the natural numbers show that $S$ is the spectrum of some variety iff $1 \in S$ and $m,n \in S \Rightarrow m \cdot n \in S$. [Hint: Find a suitable discriminator variety.]

5. (Werner). Let **R** be a biregular ring, and for $a \in R$ let $a^*$ be the central idempotent which generates the same ideal as $a$. Show that the class of algebras $\langle R,+,\cdot,-,0,^*\rangle$ generates a discriminator variety, and hence deduce from 9.4 the Dauns–Hofmann theorem in the example at the end of §8.

## §10. Quasiprimal Algebras

Perhaps the most successful generalization of the two-element Boolean algebra was introduced by Pixley in 1970. But before looking at this, we want to consider two remarkable results which will facilitate the study of these algebras.

**Lemma 10.1** (Fleischer). *Let* **C** *be a subalgebra of* $\mathbf{A} \times \mathbf{B}$, *where* **A**,**B** *are in a congruence-permutable variety* $V$. *Let* $\mathbf{A}'$ *be the image of* **C** *under the first projection map* $\alpha$, *and let* $\mathbf{B}'$ *be the image of* **C** *under the second projection map* $\beta$. *Then*

$$C = \{\langle a,b\rangle \in A' \times B' : \alpha'(a) = \beta'(b)\}$$

*for some surjective homomorphisms* $\alpha' : \mathbf{A}' \to \mathbf{D}$, $\beta' : \mathbf{B}' \to \mathbf{D}$.

Proof. Let $\theta = \ker \alpha\restriction_C \vee \ker \beta\restriction_C$, and let $v$ be the natural map from **C** to $\mathbf{C}/\theta$. Next, define

$$\alpha' : \mathbf{A}' \to \mathbf{C}/\theta$$

to be the homomorphism such that

$$v = \alpha' \circ \alpha\restriction_C$$

and

$$\beta' : \mathbf{B}' \to \mathbf{C}/\theta$$

to be such that

$$v = \beta' \circ \beta\restriction_C.$$

(See Figure 29.) Suppose $c \in C$. Then

$$c = \langle \alpha c, \beta c\rangle \in A' \times B'$$

and

$$\alpha'(\alpha c) = vc$$
$$= \beta'(\beta c),$$

so

$$c \in \{\langle a,b\rangle \in A' \times B' : \alpha'(a) = \beta'(b)\}.$$

Conversely, if

$$\langle a,b\rangle \in A' \times B' \quad \text{and} \quad \alpha'(a) = \beta'(b)$$

let $c_1, c_2 \in C$ with

$$\alpha(c_1) = a, \qquad \beta(c_2) = b.$$

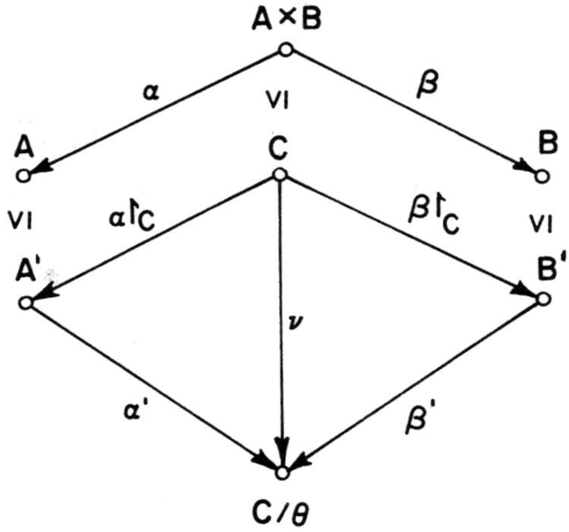


Figure 29

Then

$$\begin{aligned} \nu(c_1) &= \alpha'\alpha(c_1) \\ &= \alpha'(a) \\ &= \beta'(b) \\ &= \beta'(\beta c_2) \\ &= \nu(c_2), \end{aligned}$$

so

$$\langle c_1,c_2\rangle \in \theta;$$

hence $\langle c_1,c_2\rangle \in \ker \alpha \circ \ker \beta$ as **C** has permutable congruences. Choose $c \in C$ such that

$$c_1(\ker \alpha)c(\ker \beta)c_2.$$

Then

$$\alpha(c) = \alpha(c_1) = a,$$
$$\beta(c) = \beta(c_2) = b,$$

so

$$c = \langle a,b\rangle;$$

hence

$$\langle a,b\rangle \in C.$$

This proves

$$C = \{\langle a,b\rangle \in A' \times B' : \alpha'(a) = \beta'(b)\}. \qquad \square$$

**Corollary 10.2** (Foster-Pixley). *Let* $\mathbf{S}_1, \ldots, \mathbf{S}_n$ *be simple algebras in a congruence-permutable variety* $V$. *If*

$$\mathbf{C} \leq \mathbf{S}_1 \times \cdots \times \mathbf{S}_n$$

*is a subdirect product then*

$$\mathbf{C} \cong \mathbf{S}_{i_1} \times \cdots \times \mathbf{S}_{i_k}$$

for some $\{i_1, \ldots, i_k\} \subseteq \{1, \ldots, n\}$.

PROOF. Certainly the result is true if $n = 1$. So suppose $m > 1$ and the result is true for all $n < m$. Then $\mathbf{C}$ is isomorphic in an obvious way to a subalgebra $\mathbf{C}^*$ of $(\mathbf{S}_1 \times \cdots \times \mathbf{S}_{m-1}) \times \mathbf{S}_m$. Let

$$\mathbf{A} = \mathbf{S}_1 \times \cdots \times \mathbf{S}_{m-1},$$
$$\mathbf{B} = \mathbf{S}_m.$$

Let

$$\alpha' : \mathbf{A}' \to \mathbf{D},$$
$$\beta' : \mathbf{B}' \to \mathbf{D}$$

be as in 10.1 (Of course $\mathbf{B}' = \mathbf{B}$.) As $\beta'$ is surjective and $\mathbf{B}'$ is simple it follows that $\mathbf{D}$ is simple.

If $\mathbf{D}$ is nontrivial, then $\beta'$ is an isomorphism. In this case

$$C^* = \{\langle a,b\rangle \in A' \times B' : \alpha' a = \beta' b\}$$

implies

$$C^* = \{\langle a, \beta'^{-1}\alpha' a\rangle : a \in A'\},$$

so

$$\mathbf{A}' \cong \mathbf{C}^*$$

under the map

$$a \mapsto \langle a, \beta'^{-1}\alpha' a\rangle$$

(just use the fact that $\beta'^{-1}\alpha'$ is a homomorphism from $\mathbf{A}'$ to $\mathbf{B}'$), and hence $\mathbf{C} \cong \mathbf{A}'$. As

$$\mathbf{A}' \leq \mathbf{S}_1 \times \cdots \times \mathbf{S}_{m-1}$$

is a subdirect product then the induction hypothesis implies $\mathbf{C}$ is isomorphic to a product of some of the $\mathbf{S}_i$, $1 \leq i \leq m$.

The other case to consider is that in which $\mathbf{D}$ is trivial. But then

$$\begin{aligned} \mathbf{C}^* &= \{\langle a,b\rangle \in A' \times B' : \alpha' a = \beta' b\} \\ &= A' \times B' \end{aligned}$$

so

$$\mathbf{C} \cong \mathbf{A}' \times \mathbf{B}'.$$

As $\mathbf{A}'$ is isomorphic to some product of the $\mathbf{S}_i$ and $\mathbf{B}'$ is isomorphic to $\mathbf{S}_m$, we have $\mathbf{C}$ isomorphic to a product of suitable $\mathbf{S}_i$'s. □

**Definition 10.3.** Let $f$ be a function from $A^n \to A$. Define $f$ on $A^2$ by

$$f(\langle a_1,b_1\rangle, \ldots, \langle a_n,b_n\rangle) = \langle f(a_1, \ldots, a_n), f(b_1, \ldots, b_n)\rangle.$$

For an algebra $\mathbf{A}$ we say $f$ *preserves subalgebras of* $\mathbf{A}^2$ if, for any $\mathbf{B} \leq \mathbf{A}^2$,

$$f(B^n) \subseteq B,$$

i.e., $B$ is closed under $f$.

**Lemma 10.4** (Baker–Pixley). *Let* $\mathbf{A}$ *be a finite algebra of type* $\mathscr{F}$ *with a majority term* $M(x,y,z)$. *Then for any function*

$$f: A^n \to A, \qquad n \geq 1,$$

*which preserves subalgebras of* $\mathbf{A}^2$ *there is a term* $p(x_1, \ldots, x_n)$ *of type* $\mathscr{F}$ *representing* $f$ *on* $\mathbf{A}$.

PROOF. First note that for $\mathbf{B} \leq \mathbf{A}$ we have

$$f(B^n) \subseteq B,$$

as

$$C = \{\langle b,b \rangle : b \in B\}$$

is a subuniverse of $A^2$; hence

$$f(C^n) \subseteq C,$$

i.e., if we are given $b_1, \ldots, b_n \in B$ there is a $b \in B$ such that

$$f(\langle b_1,b_1 \rangle, \ldots, \langle b_n,b_n \rangle) = \langle b,b \rangle.$$

But then

$$f(b_1, \ldots, b_n) = b.$$

Thus given any $n$-tuple $\langle a_1, \ldots, a_n \rangle \in A^n$ we can find a term $p$ with

$$p(a_1, \ldots, a_n) = f(a_1, \ldots, a_n)$$

as

$$f(a_1, \ldots, a_n) \in \mathrm{Sg}(\{a_1, \ldots, a_n\})$$

(see II§10.3). Also given any two elements

$$\langle a_1, \ldots, a_n \rangle, \langle b_1, \ldots, b_n \rangle \in A^n,$$

we have

$$f(\langle a_1,b_1 \rangle, \ldots, \langle a_n,b_n \rangle) \in \mathrm{Sg}(\{\langle a_1,b_1 \rangle, \ldots, \langle a_n,b_n \rangle\});$$

hence there is a term $q$ with

$$q(\langle a_1,b_1 \rangle, \ldots, \langle a_n,b_n \rangle) = f(\langle a_1,b_1 \rangle, \ldots, \langle a_n,b_n \rangle),$$

so

$$q(a_1, \ldots, a_n) = f(a_1, \ldots, a_n)$$

and

$$q(b_1, \ldots, b_n) = f(b_1, \ldots, b_n).$$

Now suppose that for every $k$ elements of $A^n$, $k \geq 2$, we can find a term function $p$ which agrees with $f$ on those $k$ elements. If $k \neq |A|^n$, let $S$ be a set of $k + 1$ elements of $A^n$. Choose three distinct members $\langle a_1, \ldots, a_n \rangle$,

$\langle b_1,\ldots,b_n\rangle$, $\langle c_1,\ldots,c_n\rangle$ of $S$, and then choose terms $p_1,p_2,p_3$ such that $p_1$ agrees with $f$ on the set $S - \{\langle a_1,\ldots,a_n\rangle\}$, etc. Let

$$p(x_1,\ldots,x_n) = M(p_1(x_1,\ldots,x_n), p_2(x_1,\ldots,x_n), p_3(x_1,\ldots,x_n)).$$

Since for any member of $S$ at least two of $p_1,p_2,p_3$ agree with $f$, it follows that $p$ agrees with $f$ on $S$. By iterating this procedure we are able to construct a term which agrees with $f$ everywhere. □

**Definition 10.5.** An algebra **S** is *hereditarily simple* if every subalgebra is simple.

**Definition 10.6.** A finite algebra **A** with a discriminator term is said to be *quasiprimal.*

**Theorem 10.7.** (Pixley). *A finite algebra* **A** *is quasiprimal iff* $V(\mathbf{A})$ *is arithmetical and* **A** *is hereditarily simple.*

Proof. (⇒) In §9 we verified that if **A** has a discriminator term then **A** is hereditarily simple and $V(\mathbf{A})$ is arithmetical.

(⇐) Let $t: A^3 \to A$ be the discriminator function on $A$. Since $V(\mathbf{A})$ is arithmetical it suffices by 10.4 and II§12.5 to show that $t$ preserves subalgebras of $\mathbf{A}^2$. So let **C** be a subalgebra of $\mathbf{A}^2$. Let $\mathbf{A}'$ be the image of **C** under the first projection map, and $\mathbf{A}''$ the image of **C** under the second projection map. By 10.1 there is an algebra **D** and surjective homomorphisms

$$\alpha': \mathbf{A}' \to \mathbf{D},$$
$$\beta': \mathbf{A}'' \to \mathbf{D}$$

such that

$$C = \{\langle a',a''\rangle \in A' \times A'' : \alpha' a' = \beta' a''\}.$$

As **A** is hereditarily simple, it follows that either $\alpha'$ and $\beta'$ are both isomorphisms, or **D** is trivial. In the first case

$$C = \{\langle a',\beta'^{-1}\alpha' a'\rangle : a' \in A'\},$$

and in the second case

$$C = A' \times A''.$$

Now let

$$\langle a',a''\rangle, \langle b',b''\rangle, \langle c',c''\rangle \in A^2,$$

and let $C$ be the subuniverse of $\mathbf{A}^2$ generated by these three elements. If $C$ is of the form

$$\{\langle a',\gamma a'\rangle : a' \in A'\}$$

for some isomorphism

$$\gamma: A' \to A''$$

($\gamma$ was $\beta'^{-1}\alpha'$ above) then

$$\langle a',a''\rangle = \langle b',b''\rangle \quad \text{iff} \quad a' = b';$$

hence

$$t(\langle a',a''\rangle,\langle b',b''\rangle,\langle c',c''\rangle) = \langle t(a',b',c'), t(a'',b'',c'')\rangle = \begin{cases} \langle c',c''\rangle & \text{if } a' = b' \\ \langle a',a''\rangle & \text{if } a' \neq b', \end{cases}$$

and in either case it belongs to $C$. If $C$ is

$$A' \times A'',$$

then as

$$t(\langle a',a''\rangle,\langle b',b''\rangle,\langle c',c''\rangle) \in \{\langle a',a''\rangle,\langle a',c''\rangle,\langle c',c''\rangle,\langle c',a''\rangle\} \subseteq C$$

we see that this, combined with the previous sentence, shows $t$ preserves subalgebras of $\mathbf{A}^2$. □

**Corollary 10.8** (Foster–Pixley). *For a finite algebra* $\mathbf{A}$ *the following are equivalent*:

(a) $\mathbf{A}$ *is primal*,
(b) $V(\mathbf{A})$ *is arithmetical and* $\mathbf{A}$ *is simple with no subalgebras except itself, and the only automorphism of* $\mathbf{A}$ *is the identity map, and*
(c) $\mathbf{A}$ *is quasiprimal and* $\mathbf{A}$ *has only one subalgebra* (*itself*) *and only one automorphism* (*the identity map*).

Proof. (a ⇒ b) If $\mathbf{A}$ is primal then there is a discriminator term for $\mathbf{A}$ so $V(\mathbf{A})$ is arithmetical and $\mathbf{A}$ is simple by §9. As all unary functions on $A$ are represented by terms, $\mathbf{A}$ has no subalgebras except $\mathbf{A}$, and only one automorphism.

(b ⇒ c) This is immediate from 10.7.

(c ⇒ a) $\mathbf{A}^2$ can have only $A^2$ and $\{\langle a,a\rangle : a \in A\}$ as subuniverses in view of the details of the proof of 10.7. Thus for $f: A^n \to A$, $n \geq 1$, it is clear that $f$ preserves subalgebras of $\mathbf{A}^2$. By 10.4, $f$ is representable by a term $p$, so $\mathbf{A}$ is primal. □

Examples. (1) The *ring* $\mathbf{Z}/(p) = \langle Z/(p), +, \cdot, -, 0, 1\rangle$ is primal for $p$ a prime number as $Z/(p) = \{1, 1+1, \ldots\}$; hence $\mathbf{Z}/(p)$ has no subalgebras except itself, and only one automorphism. A discriminator term is given by

$$t(x,y,z) = (x-y)^{p-1} \cdot x + [1 - (x-y)^{p-1}] \cdot z.$$

(2) The *Post algebra* $\mathbf{P}_n = \langle\{0,1,\ldots,n-1\}, \vee, \wedge, ', 0, 1\rangle$ is primal as $P_n = \{0, 0', \ldots, 0^{(n-1)}\}$, where $a^{(k)}$ means $k$ applications of $'$ to $a$; hence $\mathbf{P}_n$ has no subalgebras except $\mathbf{P}_n$, and no automorphisms except the identity map. For the discriminator term we can proceed as follows. For $a,b \in P_n$,

$$\bigwedge_{1 \leq k \leq n} a^{(k)} \vee b^{(k)} = 0 \quad \text{iff} \quad a = b.$$

$$\left(\bigwedge_{1 \leq j \leq n-1} a^{(j)}\right)' = \begin{cases} 0 & \text{if } a = 0 \\ 1 & \text{if } a \neq 0. \end{cases}$$

Thus let

$$g(a,b) = \left[ \bigwedge_{1 \le j \le n-1} \left( \bigwedge_{1 \le k \le n} a^{(k)} \vee b^{(k)} \right)^{(j)} \right]'.$$

Then

$$g(a,b) = \begin{cases} 0 & \text{if } a = b \\ 1 & \text{if } a \ne b \end{cases}.$$

Now we can let

$$t(x,y,z) = [g(x,y) \wedge z] \vee [g(g(x,y), 1) \wedge z].$$

It is fairly safe to wager that the reader will think that quasiprimal algebras are highly specialized and rare—however Murskiĭ proved in (6) below that almost all finite algebras are quasiprimal.

## References


1. K. Baker and A. F. Pixley [1975]
2. I. Fleischer [1955]
3. A. L. Foster and A. F. Pixley [1964a], [1964b]
4. A. F. Pixley [1971]
5. H. Werner [1978]
6. V. L. Murskiĭ [1975]


## Exercises §10

1. Show that one cannot replace the "congruence-permutable" hypothesis of 10.1 by "congruence-distributive". [It suffices to choose **C** to be a three-element lattice.]

2. Show that every finite subdirect power of the alternating group $\mathbf{A}_5$ is isomorphic to a direct power of $\mathbf{A}_5$.

3. If $V$ is a congruence-permutable variety such that every subdirectly irreducible algebra is simple show that every finite algebra in $V$ is isomorphic to a direct product of simple algebras.

4. (Pixley). Show that a finite algebra **A** is quasiprimal iff every $n$-ary function, $n \ge 1$, on **A** which preserves the subuniverses of $\mathbf{A}^2$ consisting of the isomorphisms between subalgebras of **A** can be represented by a term.

5. (Quackenbush). An algebra **A** is *demi-semi-primal* if it is quasiprimal and each isomorphism between nontrivial subalgebras of **A** can be extended to an automorphism of **A**. Show that a finite algebra **A** is demi-semi-primal iff every $n$-ary function, $n \ge 1$, on **A** which preserves the subalgebras of **A** and the subuniverses of $\mathbf{A}^2$ consisting of the automorphisms of **A** can be represented by a term.

6. (Foster-Pixley). An algebra **A** is *semiprimal* if it is quasiprimal with distinct nontrivial subalgebras being nonisomorphic, and no subalgebra of **A** has a proper automorphism. Show that a finite algebra **A** is semiprimal iff every $n$-ary function, $n \ge 1$, on **A** which preserves the subalgebras of **A** can be represented by a term.

# §11. Functionally Complete Algebras and Skew-free Algebras

A natural generalization of primal algebras would be to consider those finite algebras **A** such that every finitary function on $A$ could be represented by a polynomial (see II§13.3). Given an algebra **A** of type $\mathscr{F}$, recall the definition of $\mathscr{F}_A$ and $\mathbf{A}_A$ given in II§13.3.

**Definition 11.1.** A finite algebra **A** is *functionally complete* if $\mathbf{A}_A$ is primal, i.e., if every finitary function on **A** is representable by a polynomial.

In this section we will prove Werner's remarkable characterization of functionally complete algebras **A**, given that $V(\mathbf{A})$ is congruence-permutable.

**Definition 11.2.** Let $\mathbf{2}_L$ denote the two-element distributive lattice $\langle 2, \vee, \wedge \rangle$ where $2 = \{0,1\}$ and $0 < 1$.

**Lemma 11.3.** *Let* **S** *be a finite simple algebra such that* $V(\mathbf{S})$ *is congruence-permutable and*

$$\mathbf{Con}(\mathbf{S}_S^n) \cong \mathbf{2}_L^n$$

*for* $n < \omega$. *Then* **S** *is functionally complete.*

Proof. For brevity let **F** denote $\mathbf{F}_{V(\mathbf{S}_S)}(\bar{x},\bar{y},\bar{z})$. From II§11.10 it follows that $\mathbf{F} \in ISP(\mathbf{S}_S)$. As $\mathbf{S}_S$ has no proper subalgebras, **F** is subdirectly embeddable in $\mathbf{S}_S^k$ for some $k$. Then from 10.2, we have

$$\mathbf{F} \cong \mathbf{S}_S^n$$

for some $n$, so by hypothesis

$$\mathbf{Con}(\mathbf{F}) \cong \mathbf{2}_L^n.$$

Thus **Con F** is distributive, so by II§12.7, $V(\mathbf{S}_S)$ is congruence-distributive. Since $V(\mathbf{S})$ is congruence-permutable so is $V(\mathbf{S}_S)$ (just use the same Mal'cev term for permutability); hence $V(\mathbf{S}_S)$ is arithmetical. As $\mathbf{S}_S$ has only one automorphism we see from 10.8 that $\mathbf{S}_S$ is primal, so **S** is functionally complete. □

The rest of this section is devoted to improving the formulation of 11.3.

**Definition 11.4.** Let $\theta_i \in \operatorname{Con} \mathbf{A}_i$, $1 \le i \le n$. The *product congruence*

$$\theta_1 \times \cdots \times \theta_n$$

on $\mathbf{A}_1 \times \cdots \times \mathbf{A}_n$ is defined by

$$\langle\langle a_1, \ldots, a_n\rangle, \langle b_1, \ldots, b_n\rangle\rangle \in \theta_1 \times \cdots \times \theta_n$$

iff

$$\langle a_i,b_i\rangle \in \theta_i \quad \text{for } 1 \le i \le n.$$

(We leave the verification that $\theta_1 \times \cdots \times \theta_n$ is a congruence on $\mathbf{A}_1 \times \cdots \times \mathbf{A}_n$ to the reader.)

**Definition 11.5.** A subdirect product

$$\mathbf{B} \le \mathbf{B}_1 \times \cdots \times \mathbf{B}_k$$

of finitely many algebras is *skew-free* if all the congruences on $\mathbf{B}$ are of the form

$$(\theta_1 \times \cdots \times \theta_k) \cap B^2,$$

where $\theta_i \in \operatorname{Con} \mathbf{B}_i$, i.e., the congruences on $\mathbf{B}$ are precisely the restrictions of the product congruences on $\mathbf{B}_1 \times \cdots \times \mathbf{B}_k$ to $\mathbf{B}$. A finite set of algebras $\{\mathbf{A}_1, \ldots, \mathbf{A}_n\}$ is *totally skew-free* if every subdirect product

$$\mathbf{B} \le \mathbf{B}_1 \times \cdots \times \mathbf{B}_k$$

is skew-free, where $\mathbf{B}_i \in \{\mathbf{A}_1, \ldots, \mathbf{A}_n\}$.

**Lemma 11.6.** *The subdirect product*

$$\mathbf{B} \le \mathbf{B}_1 \times \cdots \times \mathbf{B}_k$$

*is skew-free iff*

$$\theta = (\theta \vee \rho_1) \cap \cdots \cap (\theta \vee \rho_k)$$

*for* $\theta \in \operatorname{Con} \mathbf{B}$, *where*

$$\rho_i = (\ker \pi_i) \cap B^2$$

*and* $\pi_i$ *is the* $i$th *projection map on* $B_1 \times \cdots \times B_k$.

Proof. ($\Rightarrow$) Given $\mathbf{B}$ skew-free let $\theta \in \operatorname{Con} \mathbf{B}$. Then

$$\theta = (\theta_1 \times \cdots \times \theta_k) \cap B^2$$

for suitable $\theta_i \in \operatorname{Con} \mathbf{B}_i$. Let

$$\nu_i : \mathbf{B} \to \mathbf{B}/(\theta \vee \rho_i)$$

be the canonical homomorphism, and let

$$\hat{\pi}_i : \mathbf{B} \to \mathbf{B}_i$$

be the $i$th projection of $\mathbf{B}_1 \times \cdots \times \mathbf{B}_k$ restricted to $\mathbf{B}$. Then as

$$\ker \hat{\pi}_i = \rho_i \subseteq \theta \vee \rho_i = \ker \nu_i$$

there is a homomorphism

$$\alpha_i : \mathbf{B}_i \to \mathbf{B}/(\theta \vee \rho_i)$$

such that

$$\nu_i = \alpha_i \hat{\pi}_i.$$

Now for $a,b \in B$ we have

$$\begin{aligned}\langle a,b\rangle \in \theta \vee \rho_i \quad &\text{iff} \quad v_i(a) = v_i(b)\\ &\text{iff} \quad \alpha_i\pi_i(a) = \alpha_i\pi_i(b)\\ &\text{iff} \quad \alpha_i a_i = \alpha_i b_i\\ &\text{iff} \quad \langle a_i,b_i\rangle \in \ker \alpha_i;\end{aligned}$$

hence

$$\theta \vee \rho_i = (\nabla \times \cdots \times \ker \alpha_i \times \cdots \times \nabla) \cap B^2.$$

Also since

$$\langle a,b\rangle \in \rho_i \Rightarrow a_i = b_i$$

it is clear that

$$\langle a,b\rangle \in \theta \vee \rho_i \Rightarrow \langle a_i,b_i\rangle \in \theta_i;$$

hence

$$\ker \alpha_i \subseteq \theta_i.$$

Thus

$$\theta \vee \rho_i \subseteq (\nabla \times \cdots \times \theta_i \times \cdots \times \nabla) \cap B^2,$$

and then

$$\begin{aligned}\theta &\subseteq (\theta \vee \rho_1) \cap \cdots \cap (\theta \vee \rho_k)\\ &\subseteq (\theta_1 \times \nabla \times \cdots \times \nabla) \cap \cdots \cap (\nabla \times \cdots \times \nabla \times \theta_k) \cap B^2\\ &= (\theta_1 \times \cdots \times \theta_k) \cap B^2\\ &= \theta,\end{aligned}$$

so the first half of the theorem is proved.

($\Leftarrow$) For this direction just note that the above assertion

$$\theta \vee \rho_i = (\nabla \times \cdots \times \ker \alpha_i \times \cdots \times \nabla) \cap B^2,$$

for $\theta \in \operatorname{Con} \mathbf{B}$, does not depend on the skew-free property. Thus

$$\begin{aligned}\theta &= (\theta \vee \rho_1) \cap \cdots \cap (\theta \vee \rho_k)\\ &= (\ker \alpha_1 \times \nabla \times \cdots \times \nabla) \cap \cdots \cap (\nabla \times \cdots \times \nabla \times \ker \alpha_k) \cap B^2\\ &= (\ker \alpha_1 \times \cdots \times \ker \alpha_k) \cap B^2,\end{aligned}$$

so $\theta$ is the restriction of a product congruence. □

Now we can finish off the technical lemmas concerning the congruences in the abstract setting of lattice theory.

**Lemma 11.7.** *Suppose* $\mathbf{L}$ *is a modular lattice with a largest element* 1. *Also suppose that* $a_1,a_2 \in L$ *have the property*:

$$c \in [a_1 \wedge a_2, 1] \Rightarrow c = (c \vee a_1) \wedge (c \vee a_2).$$

*Then for any* $b \in L$,

$$c \in [a_1 \wedge a_2 \wedge b, b] \Rightarrow c = (c \vee (a_1 \wedge b)) \wedge (c \vee (a_2 \wedge b)).$$

PROOF. Let $c \in [a_1 \wedge a_2 \wedge b, b]$. Then

$$\begin{aligned} c &= c \vee (b \wedge a_1 \wedge a_2) \\ &= b \wedge (c \vee (a_1 \wedge a_2)) \\ &= b \wedge (c \vee a_1) \wedge (c \vee a_2) \\ &= [c \vee (a_1 \wedge b)] \wedge [c \vee (a_2 \wedge b)] \end{aligned}$$

follows from the modular law and our hypotheses. □

**Lemma 11.8.** *Let* **L** *be a modular lattice with a largest element* 1. *Then if* $a_1, \ldots, a_n \in L$ *have the property*

$$c \in [a_i \wedge a_j, 1] \Rightarrow c = (c \vee a_i) \wedge (c \vee a_j),$$

$1 \leq i, j \leq n$, *then*

$$c \in [a_1 \wedge \cdots \wedge a_n, 1] \Rightarrow c = (c \vee a_1) \wedge \cdots \wedge (c \vee a_n).$$

PROOF. Clearly the lemma holds if $n \leq 2$. So let us suppose it holds for all $n < m$, where $m \geq 3$. Then for $c \in [a_1 \wedge \cdots \wedge a_m, 1]$,

$$\begin{aligned} c &= c \vee (a_1 \wedge c) \\ &= c \vee \{[(a_1 \wedge c) \vee (a_1 \wedge a_2)] \wedge \cdots \wedge [(a_1 \wedge c) \vee (a_1 \wedge a_m)]\}. \qquad (*) \end{aligned}$$

The last equation follows by replacing **L** by the sublattice of elements $x$ of $L$ such that $x \leq a_1$, and noting that $a_1 \wedge a_2, \ldots, a_1 \wedge a_m$ satisfy the hypothesis of 11.8 in view of 11.7. By the induction hypothesis we have for this sublattice

$$a_1 \wedge c = [(a_1 \wedge c) \vee (a_1 \wedge a_2)] \wedge \cdots \wedge [(a_1 \wedge c) \vee (a_1 \wedge a_m)].$$

Now applying the modular law and the hypotheses to (*) we have

$$\begin{aligned} c &= c \vee \{a_1 \wedge [c \vee (a_1 \wedge a_2)] \wedge \cdots \wedge [c \vee (a_1 \wedge a_m)]\} \\ &= c \vee \{a_1 \wedge [(c \vee a_1) \wedge (c \vee a_2)] \wedge \cdots \wedge [(c \vee a_1) \wedge (c \vee a_m)]\} \\ &= (c \vee a_1) \wedge \cdots \wedge (c \vee a_m). \end{aligned}$$

This finishes the induction step. □

**Lemma 11.9.** *Let* $\{\mathbf{A}_1, \ldots, \mathbf{A}_n\}$ *be a set of algebras in a congruence-modular variety such that for any subdirect product* **D** *of any two (not necessarily distinct) members, say* $\mathbf{D} \leq \mathbf{A}_i \times \mathbf{A}_j$, *the only congruences on* **D** *are restrictions of product congruences. Then* $\{\mathbf{A}_1, \ldots, \mathbf{A}_n\}$ *is totally skew-free.*

PROOF. Let

$$\mathbf{B} \leq \mathbf{B}_1 \times \cdots \times \mathbf{B}_k$$

be a subdirect product of members of $\{\mathbf{A}_1, \ldots, \mathbf{A}_n\}$, and let

$$\rho_i = (\ker \pi_i) \cap B^2$$

as before. For $1 \leq i \leq j \leq k$, $\mathbf{B}/(\rho_i \cap \rho_j)$ is isomorphic to a subalgebra of $\mathbf{B}_i \times \mathbf{B}_j$, which is a subdirect product of $\mathbf{B}_i \times \mathbf{B}_j$ obtained by using a projection map on $\mathbf{B}$. From this and the correspondence theorem it follows that if $\theta \in \operatorname{Con} \mathbf{B}$, then

$$\rho_i \cap \rho_j \subseteq \theta$$

implies

$$\theta = (\theta \vee \rho_i) \cap (\theta \vee \rho_j)$$

by our assumption on $\mathbf{D}$ above and 11.6. Now we can invoke 11.8, noting that $B \times B$ is the largest element of $\mathbf{Con\ B}$, to show that, for $\theta \in \operatorname{Con} \mathbf{B}$,

$$\theta = (\theta \vee \rho_1) \cap \cdots \cap (\theta \vee \rho_k)$$

because $\rho_1 \cap \cdots \cap \rho_k$ is the smallest congruence of $\mathbf{B}$. By 11.6, $\{\mathbf{A}_1, \ldots, \mathbf{A}_n\}$ must be totally skew-free. □

**Lemma 11.10.** *Suppose $\mathbf{A}_1, \ldots, \mathbf{A}_n$ belong to a congruence–distributive variety. Then $\{\mathbf{A}_1, \ldots, \mathbf{A}_n\}$ is totally skew-free.*

PROOF. For any subdirect product

$$\mathbf{B} \leq \mathbf{B}_1 \times \cdots \times \mathbf{B}_k,$$

where $\mathbf{B}_1, \ldots, \mathbf{B}_k$ belong to a congruence–distributive variety, let $\rho_i$ be as defined in 11.6. Then for $\theta \in \operatorname{Con} \mathbf{B}$,

$$\begin{aligned} \theta = \theta \vee \Delta &= \theta \vee (\rho_1 \wedge \cdots \wedge \rho_k) \\ &= (\theta \vee \rho_1) \wedge \cdots \wedge (\theta \vee \rho_k), \end{aligned}$$

so $\mathbf{B}$ is skew-free by 11.6. Hence $\{\mathbf{A}_1, \ldots, \mathbf{A}_n\}$ is totally skew-free. □

**Lemma 11.11.** *Let $\mathbf{P}$ be a nontrivial primal algebra. Then*

$$\mathbf{Con\ P}^2 \cong \mathbf{2}_L^2.$$

PROOF. As $V(\mathbf{P})$ is congruence–distributive the congruences of $\mathbf{P}^2$ are precisely the product congruences $\theta_1 \times \theta_2$ by 11.10. As $\mathbf{P}$ is simple, $\mathbf{Con\ P}^2$ is isomorphic to $\mathbf{2}_L^2$. □

**Theorem 11.12** (Werner). *Let $\mathbf{A}$ be a nontrivial finite algebra such that $V(\mathbf{A})$ is congruence-permutable. Then $\mathbf{A}$ is functionally complete iff $\mathbf{Con\ A}^2 \cong \mathbf{2}_L^2$.*

**Proof.** (⇒) Suppose **A** is functionally complete. Note that

$$\mathbf{Con}\ \mathbf{A}^n = \mathbf{Con}\ \mathbf{A}^n_{A^n}$$

(adding constants does not affect the congruences). As $\mathbf{A}_A$ is primal, we have by 11.11,

$$\mathbf{Con}\ \mathbf{A}^2 \cong \mathbf{2}^2_L.$$

(⇐) As

$$\mathbf{Con}\ \mathbf{A}^2 \cong \mathbf{2}^2_L$$

again

$$\mathbf{Con}\ \mathbf{A}^2_A \cong \mathbf{2}^2_L.$$

Thus $\mathbf{A}_A$ must be simple (otherwise there would be other product congruences on $\mathbf{A}^2_A$), and having the constants of $A$ ensures $\mathbf{A}_A$ has no subalgebras and no automorphisms. A (now familiar) application of Fleischer's lemma shows that the only subdirect powers contained in $\mathbf{A}_A \times \mathbf{A}_A$ are $\mathbf{A}^2_A$ and **D**, where

$$D = \{\langle a,a\rangle : a \in A\}.$$

The congruences on $\mathbf{A}^2_A$ are product congruences since there are at least four product congruences $\Delta \times \Delta$, $\Delta \times \nabla$, $\nabla \times \Delta$, $\nabla \times \nabla$, and from above

$$\mathbf{Con}\ \mathbf{A}^n_A \cong \mathbf{2}^n_L,$$

The congruences on **D** are $(\nabla \times \nabla) \cap D^2$ and $(\Delta \times \Delta) \cap D^2$ as $\mathbf{D} \cong \mathbf{A}_A$. Thus by 11.9, $\{\mathbf{A}_A\}$ is totally skew-free. Consequently,

$$\mathbf{Con}\ \mathbf{A}^n_A \cong \mathbf{2}^n_L,$$

so **A** is functionally complete by 11.3. □

**Corollary 11.13** (Maurer–Rhodes). *A finite group* **G** *is functionally complete iff* **G** *is nonabelian and simple or* **G** *is trivial.*

**Proof.** The variety of groups is congruence-permutable, hence congruence-modular. If

$$\mathbf{Con}\ \mathbf{G}^2 \cong \mathbf{2}^2_L$$

then **G** is simple.

The nontrivial simple abelian groups are of the form $\mathbf{Z}/(p)$; and

$$|\mathrm{Con}(\mathbf{Z}/(p) \times \mathbf{Z}/(p))| > 4$$

as

$$\{\langle a,a\rangle : a \in Z/(p)\}$$

is a normal subgroup of $\mathbf{Z}/(p) \times \mathbf{Z}/(p)$, so $\mathbf{Z}/(p)$ cannot be functionally complete. Hence **G** is nonabelian and simple.

If **G** is nonabelian simple and **N** is a normal subgroup of $\mathbf{G}^2$, suppose $\langle a,b\rangle \in N$ with $a \neq 1$. Choose $c \in G$ such that

$$cac^{-1} \neq a.$$

Then

$$\langle cac^{-1},b\rangle = \langle c,b\rangle\langle a,b\rangle\langle c^{-1},b^{-1}\rangle \in N;$$

hence

$$\langle cac^{-1}a^{-1},1\rangle = \langle cac^{-1},b\rangle\langle a^{-1},b^{-1}\rangle \in N.$$

As **G** is simple, it follows that

$$\langle cac^{-1}a^{-1},1\rangle$$

generates the normal subgroup ker $\pi_2$ since

$$cac^{-1}a^{-1} \neq 1,$$

so

$$\ker \pi_2 \subseteq N.$$

Similarly,

$$b \neq 1 \Rightarrow \ker \pi_1 \subseteq N.$$

If both $a,b \neq 1$, then

$$\ker \pi_1, \ker \pi_2 \subseteq N$$

implies

$$G^2 = N.$$

Thus $\mathbf{G}^2$ has only four normal subgroups, so

$$\mathbf{Con}\ \mathbf{G}^2 \cong \mathbf{2}_L^2,$$

This finishes the proof that **G** is functionally complete. □

## References

1. S. Burris [1975a]
2. H. Werner [1974]

## Exercises §11

1. If **A** is a finite algebra belonging to an arithmetical variety, show that **A** is functionally complete iff **A** is simple.
2. If $\mathbf{R}_1,\mathbf{R}_2$ are rings with identity show that $\mathbf{R}_1 \times \mathbf{R}_2$ is skew-free. Does this hold if we do not require an identity?
3. Describe all functionally complete rings with identity.
4. Describe all functionally complete lattices.
5. Describe all functionally complete Heyting algebras.
6. Describe all functionally complete semilattices.
7. Show the seven-element Steiner quasigroup is functionally complete.

8. (Day) Show that a finitely generated congruence-distributive variety has the CEP iff each subdirectly irreducible member has the CEP.

## §12. Semisimple Varieties

Every nontrivial Boolean algebra is isomorphic to a subdirect power of the simple two-element algebra, and in 9.4 we proved that every algebra in a discriminator variety is isomorphic to a subdirect product of simple algebras. We can generalize this in the following manner.

**Definition 12.1.** An algebra is *semisimple* if it is isomorphic to a subdirect product of simple algebras. A variety $V$ is semisimple if every member of $V$ is semisimple.

**Lemma 12.2.** *A variety $V$ is semisimple iff every subdirectly irreducible member of $V$ is simple.*

PROOF. (⇒) Let $\mathbf{A}$ be a subdirectly irreducible member of $V$. Then $\mathbf{A}$ can be subdirectly embedded in a product of simple algebras, say by

$$\alpha : \mathbf{A} \to \prod_{i \in I} \mathbf{S}_i.$$

As $\mathbf{A}$ is subdirectly irreducible, there is a projection map

$$\pi_i : \prod_{i \in I} \mathbf{S}_i \to \mathbf{S}_i$$

such that $\pi_i \circ \alpha$ is an isomorphism. Thus

$$\mathbf{A} \cong \mathbf{S}_i,$$

so $\mathbf{A}$ is simple.

(⇐) For this direction use the fact that every algebra is isomorphic to a subdirect product of subdirectly irreducible algebras. □

**Definition 12.3.** Let $\mathbf{A}$ be an algebra and let $\theta \in \operatorname{Con} \mathbf{A}$. In the proof of II§5.5 we showed that $\theta$ is a subuniverse of $\mathbf{A} \times \mathbf{A}$. Let $\boldsymbol{\theta}$ denote the subalgebra of $\mathbf{A} \times \mathbf{A}$ with universe $\theta$.

**Lemma 12.4** (Burris). *Let $\mathbf{A}$ be a nonsimple directly indecomposable algebra in a congruence-distributive variety. If $\theta \in \operatorname{Con} \mathbf{A}$ is maximal or the smallest congruence above $\Delta$, then $\boldsymbol{\theta}$ is directly indecomposable.*

PROOF. We have

$$\boldsymbol{\theta} \leq \mathbf{A} \times \mathbf{A}.$$

By 11.10, $\boldsymbol{\theta}$ is skew-free. Thus suppose

$$(\phi_1 \times \phi_2) \cap \theta^2,$$

and

$$(\phi_1^* \times \phi_2^*) \cap \theta^2$$

are a pair of factor congruences on $\boldsymbol{\theta}$, where $\phi_i, \phi_i^* \in \operatorname{Con} \mathbf{A}$, $i = 1,2$. From

$$[(\phi_1 \times \phi_2) \cap \theta^2] \circ [(\phi_1^* \times \phi_2^*) \cap \theta^2] = \nabla_\theta$$

it follows that

$$\phi_i \circ \phi_i^* = \nabla_A,$$

$i = 1,2$. To see this let $a,b \in A$. Then

$$\langle\langle a,a\rangle, \langle b,b\rangle\rangle \in \theta^2,$$

so for some $c,d \in A$,

$$\langle a,a\rangle[(\phi_1 \times \phi_2) \cap \theta^2]\langle c,d\rangle[(\phi_1^* \times \phi_2^*) \cap \theta^2]\langle b,b\rangle.$$

Thus

$$a\phi_1 c\phi_1^* b,$$
$$a\phi_2 d\phi_2^* b.$$

Next, from

$$[(\phi_1 \times \phi_2) \cap \theta^2] \cap [(\phi_1^* \times \phi_2^*) \cap \theta^2] = \Delta_\theta$$

it follows that

$$\phi_i \cap \phi_i^* \cap \theta = \Delta_A$$

for $i = 1,2$. To see this, suppose

$$\langle a,b\rangle \in \phi_1 \cap \phi_1^* \cap \theta,$$

with $a \neq b$. Then

$$\langle\langle a,b\rangle, \langle b,b\rangle\rangle \in [(\phi_1 \times \phi_2) \cap \theta^2] \cap [(\phi_1^* \times \phi_2^*) \cap \theta^2],$$

which is impossible as

$$\langle a,b\rangle \neq \langle b,b\rangle.$$

Likewise, we show

$$\phi_2 \cap \phi_2^* \cap \theta = \Delta_A.$$

Suppose $\theta$ is a maximal congruence on $\mathbf{A}$. If

$$\phi_i \cap \phi_i^* \neq \Delta_A$$

for $i = 1,2$, then

$$\theta \vee (\phi_i \cap \phi_i^*) = \nabla_A$$

as

$$\phi_i \cap \phi_i^* \nsubseteq \theta;$$

and

$$\theta \cap (\phi_i \cap \phi_i^*) = \Delta_A,$$

so $\phi_i \cap \phi_i^*$ is the complement of $\theta$ in Con **A**, $i = 1,2$. In distributive lattices complements are unique, so

$$\phi_1 \cap \phi_1^* = \phi_2 \cap \phi_2^*.$$

Then choose $\langle a,b \rangle \in \phi_1 \cap \phi_1^*$ with $a \neq b$. This leads to

$$\langle \langle a,a \rangle, \langle b,b \rangle \rangle \in [(\phi_1 \times \phi_2) \cap \theta^2] \cap [(\phi_1^* \times \phi_2^*) \cap \theta^2],$$

which is impossible as

$$\langle a,a \rangle \neq \langle b,b \rangle.$$

Now we can assume without loss of generality that

$$\phi_1 \cap \phi_1^* = \Delta_A.$$

Thus, by the above, $\phi_1,\phi_1^*$ is a pair of factor congruences on **A**. As **A** is directly indecomposable we must have

$$\{\phi_1,\phi_1^*\} = \{\Delta_A,\nabla_A\},$$

say

$$\phi_1 = \nabla_A, \qquad \phi_1^* = \Delta_A.$$

Then

$$\begin{aligned}(\phi_1^* \times \phi_2^*) \cap \theta^2 &= (\Delta_A \times \phi_2^*) \cap \theta^2 \\ &= [\Delta_A \times (\phi_2^* \cap \theta)] \cap \theta^2;\end{aligned}$$

hence

$$\phi_2 \circ (\phi_2^* \cap \theta) = \nabla_A.$$

As

$$\phi_2 \cap (\phi_2^* \cap \theta) = \Delta_A$$

and **A** is directly indecomposable we must have

$$\phi_2^* \cap \theta = \Delta_A,$$

so

$$\begin{aligned}(\phi_1^* \times \phi_2^*) \cap \theta^2 &= (\Delta_A \times \Delta_A) \cap \theta^2 \\ &= \Delta_\theta.\end{aligned}$$

This shows that $\boldsymbol{\theta}$ has only one pair of factor congruences, namely

$$\{\Delta_\theta,\nabla_\theta\},$$

hence $\boldsymbol{\theta}$ is directly indecomposable.

Next suppose $\theta$ is the smallest congruence in Con $\mathbf{A} - \{\Delta_A\}$. Then

$$\theta \cap (\phi_i \cap \phi_i^*) = \Delta_A$$

immediately gives

$$\phi_i \cap \phi_i^* = \Delta_A,$$

so we must have

$$\{\phi_i,\phi_i^*\} = \{\Delta_A,\nabla_A\}$$

as

$$\phi_i \circ \phi_i^* = \nabla_A,$$

$i = 1,2$. If

$$\phi_1 \neq \phi_2,$$

say

$$\phi_1 = \nabla_A, \phi_2 = \Delta_A,$$

then

$$(\phi_1 \times \phi_2) \cap \theta^2 = (\theta \times \Delta_A) \cap \theta^2,$$

which implies

$$(\phi_1^* \times \phi_2^*) \cap \theta^2 = (\Delta_A \times \theta) \cap \theta^2.$$

But if $\langle a,b \rangle \notin \theta$ then

$$\langle \langle a,a \rangle, \langle b,b \rangle \rangle \notin (\theta \times \Delta_A) \cap \theta^2 \circ (\Delta_A \times \theta) \cap \theta^2,$$

so we do not have factor congruences. Hence necessarily

$$\phi_1 = \phi_2,$$
$$\phi_1^* = \phi_2^*,$$

and this leads to the factor congruences

$$\{\nabla_\theta, \Delta_\theta\},$$

so $\boldsymbol{\theta}$ is directly indecomposable. □

**Theorem 12.5** (Burris). *If $V$ is a congruence-distributive variety such that every directly indecomposable member is subdirectly irreducible then $V$ is semisimple.*

Proof. Suppose $\mathbf{A} \in V$ where $\mathbf{A}$ is a nonsimple subdirectly irreducible algebra. Let $\theta$ be the least congruence in Con $\mathbf{A} - \{\Delta\}$. Note that $\theta \neq \nabla_A$. Then $\boldsymbol{\theta}$ is a directly indecomposable member of $V$ which is not subdirectly irreducible (as

$$\rho_1 \cap \rho_2 = \Delta_\theta$$

where

$$\rho_i = (\ker \pi_i) \cap \theta^2,$$

$$\pi_i : A \times A \to A,$$

$i = 1,2$). □

Reference

1. S. Burris [a]

Exercises §12

1. Let $V$ be a finitely generated congruence-distributive variety such that every directly indecomposable is subdirectly irreducible. Prove that $V$ is semisimple arithmetical.
2. Give an example of a finitely generated semisimple congruence-distributive variety which is not arithmetical.

3. Given **A** as in 12.4 can one conclude for any congruence $\theta$ such that $\Delta < \theta < \nabla$ that $\boldsymbol{\theta}$ is directly indecomposable?

4. Given **A**,$\theta$ as in 12.4 and $B$ a subuniverse of **Su**$(I)$ let $\mathbf{A}[B,\theta]^*$ be the subalgebra of $\mathbf{A}^I$ with universe $\{a \in A^I : a^{-1}(i) \in B,\ a^{-1}(i)/\theta \in \{\varnothing, I\}, \text{ for } i \in I\}$. Show that $\mathbf{A}[B,\theta]^*$ is directly indecomposable.

# §13. Directly Representable Varieties

One of the most striking features of the variety of Boolean algebras is the fact that, up to isomorphism, there is only one nontrivial directly indecomposable member, namely **2** (see Corollary 1.9). From this we have a detailed classification of the finite Boolean algebras. A natural generalization is the following.

**Definition 13.1.** A variety $V$ is *directly representable* if it is finitely generated and has (up to isomorphism) only finitely many finite directly indecomposable members.

After special cases of directly representable varieties had been investigated by Taylor, Quackenbush, Clark and Krauss, and McKenzie in the mid-1970's, a remarkable analysis was made by McKenzie in late 1979. Most of this section is based on his work.

**Lemma 13.2** (Pólya). *Let $c_1, \ldots, c_t$ be a finite sequence of natural numbers such that not all are equal to the same number. Then the sequence*

$$s_n = c_1^n + \cdots + c_t^n, \qquad n \geq 1,$$

*has the property that the set of prime numbers $p$ for which one can find an $n$ such that $p$ divides $s_n$ is infinite.*

Proof. Suppose that $c_1, \ldots, c_t$ is such a sequence and that the only primes $p$ such that $p$ divides at least one of $\{s_n : n \geq 1\}$ are $p_1, \ldots, p_r$. Without loss of generality we can assume that the greatest common divisor of $c_1, \ldots, c_t$ is 1.

**Claim.** *For $p$ a prime and for $n \geq 1$, $k \geq 1$, $t < p^{k+1}$,*

$$p^{k+1} \nmid c_1^{(p-1)p^k \cdot n} + \cdots + c_t^{(p-1)p^k \cdot n}.$$

To see this, note that from Euler's Theorem we have

$$p \nmid c_i \Rightarrow c_i^{\phi(p^{k+1})} \equiv 1 (\text{mod } p^{k+1});$$

and furthermore

$$\begin{aligned} p \mid c_i &\Rightarrow c_i \equiv 0 \ (\text{mod } p) \\ &\Rightarrow c_i^{k+1} \equiv 0 \ (\text{mod } p^{k+1}) \\ &\Rightarrow c_i^{\phi(p^{k+1})} \equiv 0 \ (\text{mod } p^{k+1}). \end{aligned}$$

Let $u$ be the number of integers $i \in [1,t]$ such that $p \nmid c_i$, i.e., $c_i^{\phi(p^{k+1})} \equiv 1 \pmod{p^{k+1}}$. Then $u \geq 1$ as g.c.d.$(c_1, \ldots, c_t) = 1$. Furthermore, for $n \geq 1$,

$$c_1^{\phi(p^{k+1}) \cdot n} + \cdots + c_t^{\phi(p^{k+1}) \cdot n} \equiv u (\text{mod } p^{k+1}).$$

Since $1 \leq u \leq t < p^{k+1}$, $p^{k+1} \nmid u$, and hence the claim is proved.

Now if we set

$$m = \phi(p_1^{k+1}) \cdots \phi(p_r^{k+1})$$

then for $n \geq 1$, $1 \leq j \leq r$, $t < p^{k+1}$, the claim implies

$$p_j^{k+1} \nmid c_1^{mn} + \cdots + c_t^{mn},$$

so

$$s_{mn} \leq p_1^{k+1} \cdots p_r^{k+1}$$

as $p_1, \ldots, p_r$ are the only possible prime divisors of $s_{mn}$. Thus the sequence $(s_{mn})_{n \geq 1}$ is bounded. But this can happen only if $a_1 = \cdots = a_t = 1$, which is a contradiction. □

**Definition 13.3.** A congruence $\theta$ on $\mathbf{A}$ is *uniform* if for every $a,b \in A$,

$$|a/\theta| = |b/\theta|.$$

An algebra $\mathbf{A}$ is *congruence-uniform* if every congruence on $\mathbf{A}$ is uniform.

**Theorem 13.4** (McKenzie). *If $V$ is a directly representable variety then every finite member of $V$ is congruence-uniform.*

Proof. If $V$ is directly representable then there exist (up to isomorphism) finitely many finite algebras $\mathbf{D}_1, \ldots, \mathbf{D}_k$ of $V$ which are directly indecomposable; hence every finite member of $V$ is isomorphic to some $\mathbf{D}_1^{m_1} \times \cdots \times \mathbf{D}_k^{m_k}$. Thus there are only finitely many prime numbers $p$ such that $p \,|\, |A|$ for some finite $\mathbf{A} \in V$.

Now if $\mathbf{A}$ is a finite member of $V$ which is not congruence-uniform choose $\theta \in \text{Con}\,\mathbf{A}$ such that for some $a,b \in A$, $|a/\theta| \neq |b/\theta|$. For $n \geq 1$, let $\mathbf{B}_n$ be the subalgebra of $\mathbf{A}^n$ whose universe is given by

$$B_n = \{a \in A^n : a(i)\theta a(j) \text{ for } 0 \leq i, j < n\}.$$

Let the cosets of $\theta$ be $S_1, \ldots, S_t$ and have sizes $c_1, \ldots, c_t$ respectively. Then

$$|B_n| = c_1^n + \cdots + c_t^n;$$

hence by Pólya's lemma there are infinitely many primes $p$ such that for some $\mathbf{B}_n$, $p \,|\, |B_n|$. As $\mathbf{B}_n \in SP(\mathbf{A}) \subseteq V$ this is impossible. Thus every finite member of $V$ is congruence-uniform. □

**Lemma 13.5** (McKenzie). *If $\mathbf{A}$ is a finite algebra such that each member of $S(\mathbf{A} \times \mathbf{A})$ is congruence-uniform then the congruences on $\mathbf{A}$ permute.*

PROOF. Given $\theta_1,\theta_2 \in \operatorname{Con} \mathbf{A}$ let $\mathbf{B}$ be the subalgebra of $\mathbf{A} \times \mathbf{A}$ whose universe is given by

$$B = \theta_1 \circ \theta_2.$$

Let

$$\phi = \theta_2 \times \theta_2 \restriction_B,$$

a congruence on $\mathbf{B}$. For $a \in A$,

$$a/\theta_2 \times a/\theta_2 \subseteq \theta_2 \subseteq B;$$

hence

$$\langle a,a \rangle/\phi = a/\theta_2 \times a/\theta_2.$$

Since $\mathbf{A} \in IS(\mathbf{A} \times \mathbf{A})$, both $\theta_2$ on $\mathbf{A}$ and $\phi$ on $\mathbf{B}$ are uniform congruences. If $r$ is the size of cosets of $\theta_2$ and $s$ is the size of cosets of $\phi$ it follows that $s = r^2$. Now for $\langle a,b \rangle \in B$, we have

$$\langle a,b \rangle/\phi \subseteq a/\theta_2 \times b/\theta_2,$$
$$|\langle a,b \rangle/\phi| = s,$$
$$|a/\theta_2| = |b/\theta_2| = r,$$

and $s = r^2$; hence

$$\langle a,b \rangle/\phi = a/\theta_2 \times b/\theta_2.$$

Now for $c,d \in A$,

$$\langle c,d \rangle \in \theta_2 \circ \theta_1 \circ \theta_2 \circ \theta_2$$

iff

$$\langle c,d \rangle \in a/\theta_2 \times b/\theta_2 \quad \text{for some } \langle a,b \rangle \in B,$$

so

$$\theta_2 \circ \theta_1 \circ \theta_2 \circ \theta_2 \subseteq B = \theta_1 \circ \theta_2,$$

hence

$$\theta_2 \circ \theta_1 \subseteq \theta_1 \circ \theta_2,$$

so the congruences on $\mathbf{A}$ permute. □

**Theorem 13.6** (Clark-Krauss). *If $V$ is a locally finite variety all of whose finite algebras are congruence-uniform then $V$ is congruence-permutable.*

PROOF. As $\mathbf{F}_V(\overline{x},\overline{y},\overline{z})$ is finite, by 13.5 it has permutable congruences; hence $V$ is congruence-permutable. □

**Corollary 13.7** (McKenzie). *If $V$ is a directly representable variety then $V$ is congruence-permutable.*

PROOF. Just combine 13.4 and 13.6. □

**Theorem 13.8** (Burris). *Let $V$ be a finitely generated congruence-distributive variety. Then $V$ is directly representable iff $V$ is semisimple arithmetical.*

PROOF. ($\Rightarrow$) From 12.4 and 12.5 $V$ is semisimple, and by 13.7 $V$ is congruence-permutable. Hence $V$ is semisimple arithmetical.

($\Leftarrow$) If $V$ is semisimple arithmetical then every finite subdirectly irreducible member of $V$ is a simple algebra, hence every finite member of $V$ is isomorphic to a subdirect product of finitely many simples. Then by 10.2 every finite member of $V$ is isomorphic to a direct product of simple algebras. By 6.10 there are only finitely many simple members of $V$, so $V$ is directly representable. □

**Theorem 13.9** (McKenzie). *If $V = I\Gamma^a(K)$, where $K$ is a finite set of finite algebras, then $V$ is congruence-permutable.*

PROOF. As every finite Boolean space is discrete it follows that every finite member of $V$ is in $IP(K_+)$; hence $V$ is directly representable, so 13.7 applies. □

A definitive treatment of directly representable varieties is given in (1) below.

REFERENCE

1. R. McKenzie [c]

EXERCISES §13

1. Which finitely generated varieties of Heyting algebras are directly representable?
2. Which finitely generated varieties of lattices are directly representable?
3. If **G** is a finite Abelian group show that $V(\mathbf{G})$ is directly representable.
4. If **R** is a finite ring with identity show that $V(\mathbf{R})$ is directly representable if **R** is a product of fields.

# CHAPTER V
# Connections with Model Theory

Since the 1950's, a branch of logic called model theory has developed under the leadership of Tarski. Much of what is considered universal algebra can be regarded as an extensively developed fragment of model theory, just as field theory is part of ring theory. In this chapter we will look at several results in universal algebra which require some familiarity with model theory. The chapter is self-contained, so the reader need not have had any previous exposure to a basic course in logic.

## §1. First-order Languages, First-order Structures, and Satisfaction

Model theory has been primarily concerned with connections between first-order properties and first-order structures. First-order languages are very restrictive (when compared to English), and many interesting questions cannot be discussed using them. On the other hand, they have a precise grammar and there are beautiful results (such as the compactness theorem) connecting first-order properties and the structures which satisfy these properties.

**Definition 1.1.** A (*first-order*) *language* $\mathscr{L}$ consists of a set $\mathscr{R}$ of *relation symbols* and a set $\mathscr{F}$ of *function symbols*, and associated to each member of $\mathscr{R}$ [of $\mathscr{F}$] is a natural number [a nonnegative integer] called the *arity* of the symbol. $\mathscr{F}_n$ denotes the set of function symbols in $\mathscr{F}$ of arity $n$, and $\mathscr{R}_n$ denotes the set of relation symbols in $\mathscr{R}$ of arity $n$. $\mathscr{L}$ is a *language of algebras* if $\mathscr{R} = \varnothing$, and it is a *language of relational structures* if $\mathscr{F} = \varnothing$.

**Definition 1.2.** If we are given a nonempty set $A$ and a positive integer $n$ we say that $r$ is an *$n$-ary relation* on $A$ if $r \subseteq A^n$. $r$ is *unary* if $n = 1$, *binary* if $n = 2$, and *ternary* if $n = 3$. A relation is *finitary* if it is $n$-ary for some $n$, $1 \leq n < \omega$. When $r$ is a binary relation we frequently write $arb$ for $\langle a,b \rangle \in r$.

**Definition 1.3.** If $\mathscr{L}$ is a first-order language then a (*first-order*) *structure of type* $\mathscr{L}$ (or $\mathscr{L}$*-structure*) is an ordered pair $\mathbf{A} = \langle A,L \rangle$ with $A \neq \varnothing$, where $L$ consists of a family $R$ of *fundamental relations* $r^{\mathbf{A}}$ on $A$ indexed by $\mathscr{R}$ (with the arity of $r^{\mathbf{A}}$ equal to the arity of $r$, for $r \in \mathscr{R}$) and a family $F$ of *fundamental operations* $f^{\mathbf{A}}$ on $A$ indexed by $\mathscr{F}$ (with the arity of $f^{\mathbf{A}}$ equal to the arity of $f$, for $f \in \mathscr{F}$). $A$ is called the *universe of* $\mathbf{A}$, and in practice we usually write just $r$ for $r^{\mathbf{A}}$ and $f$ for $f^{\mathbf{A}}$. If $\mathscr{R} = \varnothing$ then $\mathbf{A}$ is an *algebra*; if $\mathscr{F} = \varnothing$ then $\mathbf{A}$ is a *relational structure*. If $\mathscr{L}$ is finite, say $\mathscr{F} = \{f_1, \ldots, f_m\}$, $\mathscr{R} = \{r_1, \ldots, r_n\}$, then we often write

$$\langle A, f_1, \ldots, f_m, r_1, \ldots, r_n \rangle$$

instead of $\langle A,L \rangle$.

EXAMPLES. (1) If $\mathscr{L} = \{+,\cdot,\leq\}$, then the *linearly ordered field of rationals* $\langle Q,+,\cdot,\leq \rangle$ is a structure of type $\mathscr{L}$.

(2) If $\mathscr{L} = \{\leq\}$, then a *partially ordered set* $\langle P,\leq \rangle$ is a relational structure of type $\mathscr{L}$.

**Definition 1.4.** If $\mathscr{L}$ is a first-order language and $X$ is a set (members of $X$ are called *variables*) we define the *terms of type* $\mathscr{L}$ *over* $X$ to be the terms of type $\mathscr{F}$ over $X$ (see II§11). The *atomic formulas of type* $\mathscr{L}$ *over* $X$ are expressions of the form

$p \approx q$ where $p,q$ are terms of type $\mathscr{L}$ over $X$

$r(p_1, \ldots, p_n)$ where $r \in \mathscr{R}_n$ and $p_1, \ldots, p_n$ are terms of type $\mathscr{L}$ over $X$.

EXAMPLE. For the language $\mathscr{L} = \{+,\cdot,\leq\}$ we see that

$$(x \cdot y) \cdot z \approx x \cdot y, \qquad (x \cdot y) \cdot z \leq x \cdot z$$

are examples of atomic formulas, where of course we are writing binary functions and binary relations in the everyday manner, namely we write $u \cdot v$ for $\cdot(u,v)$, and $u \leq v$ for $\leq(u,v)$. If we were to rewrite the above atomic formulas using only the original definition of terms we would have the expressions

$$\cdot(\cdot(x,y),z) \approx \cdot(x,y), \qquad \leq(\cdot(\cdot(x,y),z), \cdot(x,z)).$$

**Definition 1.5.** Let $\mathscr{L}$ be a first-order language and $X$ a set of variables. The set of (*first-order*) *formulas of type* $\mathscr{L}$ (or $\mathscr{L}$*-formulas*) *over* $X$, written $\mathscr{L}(X)$, is the smallest collection of strings of symbols from $\mathscr{L} \cup X \cup \{(,)\} \cup \{\&, \vee, \neg, \rightarrow, \leftrightarrow, \forall, \exists, \approx\} \cup \{,\}$ containing the atomic formulas of type $\mathscr{L}$

over $X$, and such that if $\Phi, \Phi_1, \Phi_2 \in \mathscr{L}(X)$ then

$$\begin{aligned}
(\Phi_1) \& (\Phi_2) &\in \mathscr{L}(X), \\
(\Phi_1) \vee (\Phi_2) &\in \mathscr{L}(X), \\
\neg(\Phi) &\in \mathscr{L}(X), \\
(\Phi_1) \rightarrow (\Phi_2) &\in \mathscr{L}(X), \\
(\Phi_1) \leftrightarrow (\Phi_2) &\in \mathscr{L}(X), \\
\forall x(\Phi) &\in \mathscr{L}(X), \\
\exists x(\Phi) &\in \mathscr{L}(X).
\end{aligned}$$

The symbols & (and), $\vee$ (or), $\neg$ (not), $\rightarrow$ (implies), and $\leftrightarrow$ (iff) are called the *propositional connectives.* $\forall$ is the *universal quantifier*, and $\exists$ is the *existential quantifier*; we refer to them simply as *quantifiers.* $p \not\approx q$ denotes $\neg(p \approx q)$.

Example. With $\mathscr{L} = \{+,\cdot,\leq\}$ we see that

$$(\forall x(x \cdot y \approx y + u)) \rightarrow (\exists y(x \cdot y \leq y + u))$$

is in $\mathscr{L}(\{x,y,u\})$, but

$$\forall x(x \& y \approx u)$$

does not belong to $\mathscr{L}(\{x,y,u\})$.

**Definition 1.6.** A formula $\Phi_1$ is a *subformula* of a formula $\Phi$ if there is a consecutive string of symbols in the formula $\Phi$ which is precisely the formula $\Phi_1$.

Example. The subformulas of

$$(\forall x(x \cdot y \approx y + u)) \rightarrow (\exists y(x \cdot y \leq y + u))$$

are itself,

$$\begin{aligned}
&\forall x(x \cdot y \approx y + u), \\
&x \cdot y \approx y + u, \\
&\exists y(x \cdot y \leq y + u),
\end{aligned}$$

and

$$x \cdot y \leq y + u.$$

*Remark.* Note that the definition of subformula does not apply to the string of symbols

$$(\forall x(x \cdot y \approx y + u)) \rightarrow (\exists y(x \cdot y \leq y + u));$$

for clearly $y \approx y$ is a consecutive string of symbols in this expression which gives a formula, but we would not want this to be a subformula. However if one translates the above into the formula

$$(\forall x(\cdot(x,y) \approx +(y,u)) \rightarrow (\exists y(\leq(\cdot(x,y),+(y,u)),$$

then the subformulas, retranslated, are just those listed in the example above.

**Definition 1.7.** A particular variable $x$ may appear several times in the string of symbols which constitute a formula $\Phi$; each of these is called an *occurrence* of $x$. Similarly we may speak of occurrences of subformulas. Since strings are written linearly we can speak of the first occurrence, etc., reading from left to right.

EXAMPLE. There are three occurrences of $x$ in the formula

$$(\forall x(x \cdot y \approx y + u)) \rightarrow (\exists y(x \cdot y \leq y + u)).$$

**Definition 1.8.** A particular occurrence of a variable $x$ in a formula $\Phi$ is said to *belong to* an occurrence of a subformula $\Phi_1$ of $\Phi$ if the occurrence of $x$ is a component of the string of symbols which form the occurrence of $\Phi_1$. An occurrence of $x$ in $\Phi$ is *free* if $x$ does not belong to any occurrence of a subformula of the form $\forall x(\Psi)$ or $\exists x(\Psi)$. Otherwise, an occurrence of $x$ is *bound* in $\Phi$. A variable $x$ is free in $\Phi$ if some occurrence of $x$ is free in $\Phi$. To say that $x$ is not free in $\Phi$ we write simply $x \notin \Phi$. A *sentence* is a formula with no free variables. When we write $\Phi(x_1, \ldots, x_n)$ we will mean a formula all of whose free variables are *among* $\{x_1, \ldots, x_n\}$. We find it convenient to express $\Phi(x_1, \ldots, x_m, y_1, \ldots, y_n, \ldots)$ by $\Phi(\vec{x}, \vec{y}, \ldots)$. If $x_i$ is free in $\Phi(x_1, \ldots, x_n)$ then this notation is assumed to refer to *all* the free occurrences of $x_i$. Thus, given a formula $\Phi(x_1, \ldots, x_n)$, when we write

$$\Phi(x_1, \ldots, x_{i-1}, y, x_{i+1}, \ldots, x_n)$$

we mean the formula obtained by replacing *all* free occurrences of $x_i$ by $y$.

EXAMPLE. Let $\Phi(x,y,u)$ be the formula

$$(\forall x(x \cdot y \approx y + u)) \rightarrow (\exists y(x \cdot y \leq y + u)).$$

The first two occurrences of $x$ in $\Phi(x,y,u)$ are bound, the third is free. $\Phi(x,x,u)$ is the formula

$$(\forall x(x \cdot x \approx x + u)) \rightarrow (\exists y(x \cdot y \leq y + u)).$$

**Definition 1.9.** If **A** is a structure of type $\mathscr{L}$ we let $\mathscr{L}_A$ denote the language obtained by adding a nullary function symbol $a$ to $\mathscr{L}$ for each $a \in A$. Given $\Phi(x_1, \ldots, x_n)$ of type $\mathscr{L}_A$ and $a \in A$ the formula

$$\Phi(x_1, \ldots, x_{i-1}, a, x_{i+1}, \ldots, x_n)$$

is the formula obtained by replacing every free occurrence of $x_i$ by $a$. We sometimes refer to formulas of type $\mathscr{L}_A$ as formulas of type $\mathscr{L}$ with *parameters* from A.

When desirable we give ourselves the option of inserting or removing parentheses to improve readability, and sometimes we use brackets [, ] and braces {, } instead of parentheses.

Next we want to capture the intuitive understanding of what it means for a first-order formula to be true in a first-order structure. A precise definition of truth (i.e., definition of satisfaction) will allow us to do proofs by induction later on. From now on we will frequently drop parentheses. For example we will write $\Phi_1 \,\&\, \Phi_2$ instead of $(\Phi_1)\,\&\,(\Phi_2)$, and $\forall x \exists y \Phi$ instead of $\forall x(\exists y(\Phi))$; but we would not write $\Phi_1 \,\&\, \Phi_2 \vee \Phi_3$ for $(\Phi_1)\,\&\,(\Phi_2 \vee \Phi_3)$.

**Definition 1.10.** Let $\mathbf{A}$ be a structure of type $\mathscr{L}$. For *sentences* $\Phi$ in $\mathscr{L}_A(X)$ we define the notion $\mathbf{A} \models \Phi$ (read: "$\mathbf{A}$ satisfies $\Phi$" or "$\Phi$ is true in $\mathbf{A}$" or "$\Phi$ holds in $\mathbf{A}$") recursively as follows:

(i) if $\Phi$ is atomic:
 (a) $\mathbf{A} \models p(a_1, \ldots, a_n) \approx q(a_1, \ldots, a_n)$ iff $p^{\mathbf{A}}(a_1, \ldots, a_n) = q^{\mathbf{A}}(a_1, \ldots, a_n)$
 (b) $\mathbf{A} \models r(a_1, \ldots, a_n)$ iff $r^{\mathbf{A}}(a_1, \ldots, a_n)$ holds in $\mathbf{A}$
(ii) $\mathbf{A} \models \Phi_1 \,\&\, \Phi_2$ iff $\mathbf{A} \models \Phi_1$ and $\mathbf{A} \models \Phi_2$
(iii) $\mathbf{A} \models \Phi_1 \vee \Phi_2$ iff $\mathbf{A} \models \Phi_1$ or $\mathbf{A} \models \Phi_2$
(iv) $\mathbf{A} \models \neg\Phi$ iff it is not the case that $\mathbf{A} \models \Phi$ (which we abbreviate to: $\mathbf{A} \not\models \Phi$)
(v) $\mathbf{A} \models \Phi_1 \rightarrow \Phi_2$ iff $\mathbf{A} \not\models \Phi_1$ or $\mathbf{A} \models \Phi_2$
(vi) $\mathbf{A} \models \Phi_1 \leftrightarrow \Phi_2$ iff ($\mathbf{A} \not\models \Phi_1$ and $\mathbf{A} \not\models \Phi_2$) or ($\mathbf{A} \models \Phi_1$ and $\mathbf{A} \models \Phi_2$)
(vii) $\mathbf{A} \models \forall x \Phi(x)$ iff $\mathbf{A} \models \Phi(a)$ for *every* $a \in A$
(viii) $\mathbf{A} \models \exists x \Phi(x)$ iff $\mathbf{A} \models \Phi(a)$ for *some* $a \in A$.

For a *formula* $\Phi \in \mathscr{L}_A(X)$ we say

$$\mathbf{A} \models \Phi$$

iff

$$\mathbf{A} \models \forall x_1 \ldots \forall x_n \Phi,$$

where $x_1, \ldots, x_n$ are the free variables of $\Phi$. For a class $K$ of $\mathscr{L}$-structures and $\Phi \in \mathscr{L}(X)$ we say

$$\mathrm{K} \models \Phi \quad \text{iff} \quad \mathbf{A} \models \Phi \quad \text{for every } \mathbf{A} \in K,$$

and for $\Sigma$ a set of $\mathscr{L}$-formulas

$$\mathbf{A} \models \Sigma \quad \text{iff} \quad A \models \Phi \quad \text{for every } \Phi \in \Sigma$$
$$\mathrm{K} \models \Sigma \quad \text{iff} \quad K \models \Phi \quad \text{for every } \Phi \in \Sigma.$$

(If $\mathbf{A} \models \Sigma$ we also say $\mathbf{A}$ is a *model of* $\Sigma$.) Then we say

$$\Sigma \models \Phi \quad \text{iff} \quad \mathbf{A} \models \Sigma \quad \text{implies} \quad \mathbf{A} \models \Phi, \quad \text{for every } \mathbf{A},$$

(read: "$\Sigma$ yields $\Phi$"), and

$$\Sigma \models \Sigma_1 \quad \text{iff} \quad \Sigma \models \Phi \quad \text{for every } \Phi \in \Sigma_1.$$

EXAMPLE. A *graph* is a structure $\langle A,r \rangle$ where $r$ is a binary relation which is irreflexive and symmetric, i.e., for $a,b \in A$ we do not have $r(a,a)$, and if $r(a,b)$ holds so does $r(b,a)$. Graphs are particularly nice to work with because of the possibility of drawing numerous examples. Let $\mathbf{A} = \langle A,r \rangle$ be the graph in Figure 30, where an edge between two points means they are

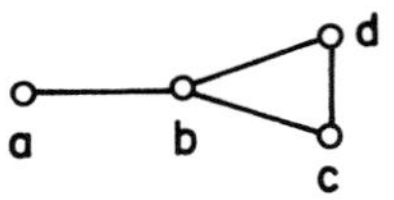

Figure 30

related by $r$. Let us find out if

$$\mathbf{A} \models \forall x \exists y \forall z (r(x,z) \vee r(y,z)).$$

This sentence will be true in $\mathbf{A}$ iff the following four assertions hold:

(i) $\mathbf{A} \models \exists y \forall z (r(a,z) \vee r(y,z))$
(ii) $\mathbf{A} \models \exists y \forall z (r(b,z) \vee r(y,z))$
(iii) $\mathbf{A} \models \exists y \forall z (r(c,z) \vee r(y,z))$
(iv) $\mathbf{A} \models \exists y \forall z (r(d,z) \vee r(y,z))$.

Let us examine (i). It will hold iff one of the following hold:

($\mathrm{i}_a$) $\mathbf{A} \models \forall z (r(a,z) \vee r(a,z))$
($\mathrm{i}_b$) $\mathbf{A} \models \forall z (r(a,z) \vee r(b,z))$
($\mathrm{i}_c$) $\mathbf{A} \models \forall z (r(a,z) \vee r(c,z))$
($\mathrm{i}_d$) $\mathbf{A} \models \forall z (r(a,z) \vee r(d,z))$.

The validity of ($\mathrm{i}_b$) depends upon all of the following holding:

($\mathrm{i}_{ba}$) $\mathbf{A} \models r(a,a) \vee r(b,a)$
($\mathrm{i}_{bb}$) $\mathbf{A} \models r(a,b) \vee r(b,b)$
($\mathrm{i}_{bc}$) $\mathbf{A} \models r(a,c) \vee r(b,c)$
($\mathrm{i}_{bd}$) $\mathbf{A} \models r(a,d) \vee r(b,d)$.

($\mathrm{i}_b$) is true, hence (i) holds. Likewise, the reader can verify that (ii), (iii), and (iv) hold. But this means the graph $\mathbf{A}$ satisfies the original sentence.

It is useful to be able to work with sentences in some sort of normal form.

**Definition 1.11.** Let $\Phi_1(x_1, \ldots, x_n)$ and $\Phi_2(x_1, \ldots, x_n)$ be two formulas in $\mathscr{L}(X)$. We say that $\Phi_1$ and $\Phi_2$ are *logically equivalent*, written $\Phi_1 \sim \Phi_2$, if for every structure $\mathbf{A}$ of type $\mathscr{L}$ and every $a_1, \ldots, a_n \in A$ we have

$$\mathbf{A} \models \Phi_1(a_1, \ldots, a_n) \quad \text{iff} \quad \mathbf{A} \models \Phi_2(a_1, \ldots, a_n).$$

If for all $\mathscr{L}$-structures $\mathbf{A}$, $\mathbf{A} \models \Phi$, where $\Phi$ is an $\mathscr{L}$-formula, we write

$$\models \Phi.$$

The reader should readily recognize the logical equivalence of the following pairs of formulas.

**Lemma 1.12.** *Suppose $\Phi$, $\Phi_1$, $\Phi_2$ and $\Phi_3$ are formulas in some $\mathscr{L}(X)$. Then the following pairs of formulas are logically equivalent:*

| | | |
|---|---|---|
| $\Phi \,\&\, \Phi$ | $\Phi$ | idempotent laws |
| $\Phi \vee \Phi$ | $\Phi$ | |
| $\Phi_1 \,\&\, \Phi_2$ | $\Phi_2 \,\&\, \Phi_1$ | commutative laws |
| $\Phi_1 \vee \Phi_2$ | $\Phi_2 \vee \Phi_1$ | |
| $\Phi_1 \,\&\, (\Phi_2 \,\&\, \Phi_3)$ | $(\Phi_1 \,\&\, \Phi_2) \,\&\, \Phi_3$ | associative laws |
| $\Phi_1 \vee (\Phi_2 \vee \Phi_3)$ | $(\Phi_1 \vee \Phi_2) \vee \Phi_3$ | |
| $\Phi_1 \,\&\, (\Phi_2 \vee \Phi_3)$ | $(\Phi_1 \,\&\, \Phi_2) \vee (\Phi_1 \,\&\, \Phi_3)$ | distributive laws |
| $\Phi_1 \vee (\Phi_2 \,\&\, \Phi_3)$ | $(\Phi_1 \vee \Phi_2) \,\&\, (\Phi_1 \vee \Phi_3)$ | |
| $\neg(\Phi_1 \,\&\, \Phi_2)$ | $(\neg\Phi_1) \vee (\neg\Phi_2)$ | de Morgan laws |
| $\neg(\Phi_1 \vee \Phi_2)$ | $(\neg\Phi_1) \,\&\, (\neg\Phi_2)$ | |
| $\Phi_1 \leftrightarrow \Phi_2$ | $(\Phi_1 \to \Phi_2) \,\&\, (\Phi_2 \to \Phi_1)$ | |
| $\Phi_1 \to \Phi_2$ | $(\neg\Phi_1) \vee \Phi_2$ | |
| $\neg\neg\Phi$ | $\Phi$ | |

Proof. (Exercise.) □

The next list of equivalent formulas, involving quantifiers, may not be so familiar to the reader.

**Lemma 1.13.** *If $\Phi$, $\Phi_1$ and $\Phi_2$ are formulas in some $\mathscr{L}(X)$ then the following pairs of formulas are logically equivalent:*

| | | |
|---|---|---|
| $\forall x(\Phi_1 \,\&\, \Phi_2)$ | $(\forall x\Phi_1) \,\&\, (\forall x\Phi_2)$ | |
| $\exists x(\Phi_1 \vee \Phi_2)$ | $(\exists x\Phi_1) \vee (\exists x\Phi_2)$ | |
| $\forall x\Phi$ | $\Phi$ | *if* $x \notin \Phi$ |
| $\exists x\Phi$ | $\Phi$ | *if* $x \notin \Phi$ |
| $\forall x(\Phi_1 \vee \Phi_2)$ | $(\forall x\Phi_1) \vee \Phi_2$ | *if* $x \notin \Phi_2$ |
| $\exists x(\Phi_1 \,\&\, \Phi_2)$ | $(\exists x\Phi_1) \,\&\, \Phi_2$ | *if* $x \notin \Phi_2$ |
| $\neg\forall x\Phi(x)$ | $\exists x\neg\Phi(x)$ | |
| $\neg\exists x\Phi(x)$ | $\forall x\neg\Phi(x)$ | |
| $\forall x(\Phi_1 \to \Phi_2)$ | $\Phi_1 \to (\forall x\Phi_2)$ | *if* $x \notin \Phi_1$ |
| $\exists x(\Phi_1 \to \Phi_2)$ | $\Phi_1 \to (\exists x\Phi_2)$ | *if* $x \notin \Phi_1$ |
| $\forall x(\Phi_1 \to \Phi_2)$ | $(\exists x\Phi_1) \to \Phi_2$ | *if* $x \notin \Phi_2$ |
| $\exists x(\Phi_1 \to \Phi_2)$ | $(\forall x\Phi_1) \to \Phi_2$ | *if* $x \notin \Phi_2$ |
| $\forall x\Phi(x)$ | $\forall y\Phi(y)$ | *provided replacing all free occurrences of x in $\Phi(x)$ by y does not lead to any new bound occurences of y.* |
| $\exists x\Phi(x)$ | $\exists y\Phi(y)$ | |

Proof. All of these are immediate consequences of the definition of satisfaction. In the last two cases let us point out what happens if one does not heed the "provided . . ." clause. Consider the formula $\Phi(x)$ given by $\exists y(x \not\approx y)$. Replacing $x$ by $y$ gives $\exists y(y \not\approx y)$. Now the sentence $\forall x\exists y(x \not\approx y)$ is true in any structure **A** with at least two elements, whereas $\forall x\exists y(y \not\approx y)$ is logically equivalent to $\exists y(y \not\approx y)$, which is never true. □

**Definition 1.14.** If $\Phi \in \mathscr{L}(X)$ we define the *length* $l(\Phi)$ of $\Phi$ to be the number of occurrences of the symbols $\&, \vee, \neg, \rightarrow, \leftrightarrow, \forall$, and $\exists$ in $\Phi$.

$$\text{Note that } l(\Phi) = 0 \quad \text{iff} \quad \Phi \text{ is atomic.}$$

**Lemma 1.15.** *If $\Phi_1$ is a subformula of $\Phi$ and $\Phi_1$ is logically equivalent to $\Phi_2$, then replacing an occurrence of $\Phi_1$ by $\Phi_2$ gives a formula $\Phi^*$ which is logically equivalent to $\Phi$.*

PROOF. We proceed by induction on $l(\Phi)$.

If $l(\Phi) = 0$ then $\Phi$ is atomic, so the only subformula of $\Phi$ is $\Phi$ itself, and the lemma is obvious in this case. So suppose $l(\Phi) \geq 1$ and for any $\Psi$ such that $l(\Psi) < l(\Phi)$ the replacement of an occurence of a subformula of $\Psi$ by a logically equivalent formula leads to a formula which is logically equivalent to $\Psi$. Let $\Phi_1$ be a subformula of $\Phi$ and suppose $\Phi_1$ is logically equivalent to $\Phi_2$. The case in which $\Phi_1 = \Phi$ is trivial, so we assume $l(\Phi_1) < l(\Phi)$. There are now seven cases to consider. Suppose $\Phi$ is $\Phi' \;\&\; \Phi''$. Then the occurence of $\Phi_1$ being considered is an occurrence in $\Phi'$ or $\Phi''$, say it is an occurrence in $\Phi'$. Let $\Phi'^*$ be the result of replacing $\Phi_1$ in $\Phi'$ by $\Phi_2$. By the induction assumption $\Phi'^*$ is logically equivalent to $\Phi'$. Let $\Phi^*$ be the result of replacing the occurrence of $\Phi_1$ in $\Phi$ by $\Phi_2$. Then $\Phi^*$ is $\Phi'^* \;\&\; \Phi''$, and this is easily argued to be logically equivalent to $\Phi' \;\&\; \Phi''$, i.e., to $\Phi$. Likewise one handles the four cases involving $\vee, \neg, \rightarrow, \leftrightarrow$. If $\Phi$ is $\forall x \Phi'(x,\vec{y})$ then let $\Phi'^*(x,\vec{y})$ be the result of replacing the occurrence of $\Phi_1$ in $\Phi'(x,\vec{y})$ by $\Phi_2$. Then by the induction hypothesis $\Phi'^*(x,\vec{y})$ is logically equivalent to $\Phi'(x,\vec{y})$, so given a structure $\mathbf{A}$ of type $\mathscr{L}$ we have

$$\mathbf{A} \models \Phi'^*(x,\vec{y}) \leftrightarrow \Phi'(x,\vec{y});$$

hence

$$\mathbf{A} \models \Phi'^*(a,\vec{y}) \leftrightarrow \Phi'(a,\vec{y})$$

for $a \in A$, so

$$\mathbf{A} \models \forall x \Phi'^*(x,\vec{y}) \quad \text{iff} \quad \mathbf{A} \models \forall x \Phi'(x,\vec{y});$$

thus $\Phi$ is logically equivalent to $\forall x \Phi'^*(x,\vec{y})$. Similarly, we can handle the case $\exists x \Phi'(x,\vec{y})$. □

**Definition 1.16.** An *open formula* is a formula in which there are no occurrences of quantifiers.

**Definition 1.17.** A formula $\Phi$ is in *prenex form* if it looks like

$$Q_1 x_1 \ldots Q_n x_n \Phi'(x_1, \ldots, x_n)$$

where each $Q_i$ is a quantifier and $\Phi'(x_1, \ldots, x_n)$ is an open formula. $\Phi'$ is called the *matrix* of $\Phi$.

Here, and in all future references to prenex form, we have the convention that no quantifiers need appear in the formula $\Phi$.

**Theorem 1.18.** *Every formula is logically equivalent to a formula in prenex form.*

PROOF. This follows from 1.12, 1.13, and 1.15. First, if necessary, change some of the bound variables to new variables so that for any variable $x$ there is at most one occurrence of $\forall x$ as well as $\exists x$ in the formula, both do not occur in the formula, and no variable occurs both as a bound variable and a free variable. Then one simply pulls the quantifiers out front using 1.13. □

EXAMPLE. The following shows how to put the formula $\forall x \neg (r(x,y) \rightarrow \exists x r(x, z))$ in prenex form.

$$\begin{aligned}\forall x \neg (r(x,y) \rightarrow \exists x r(x,z)) &\sim \forall x \neg(r(x,y) \rightarrow \exists w r(w,z))\\ &\sim \forall x \neg \exists w(r(x,y) \rightarrow r(w,z))\\ &\sim \forall x \forall w \neg(r(x,y) \rightarrow r(w,z)).\end{aligned}$$

In view of the associative law for & and $\vee$, we will make it a practice of dropping parentheses in formulas when the ambiguity is only "up to logical equivalence". Thus $\Phi_1$ & $\Phi_2$ & $\Phi_3$ replaces $(\Phi_1$ & $\Phi_2)$ & $\Phi_3$ and $\Phi_1$ & $(\Phi_2$ & $\Phi_3)$, etc. Also, we find it convenient to replace $\Phi_1$ & $\cdots$ & $\Phi_n$ by $\&_{1 \le i \le n} \Phi_i$ (called the *conjunction* of the $\Phi_i$), and $\Phi_1 \vee \cdots \vee \Phi_n$ by $\bigvee_{1 \le i \le n} \Phi_i$ (called the *disjunction* of the $\Phi_i$).

**Definition 1.19.** An open formula is in *disjunctive form* if it is in the form

$$\bigvee_i \mathop{\&}_j \Phi_{ij}$$

where each $\Phi_{ij}$ is atomic or negated atomic (i.e., the negation of an atomic formula). An open formula is in *conjunctive form* if it is in the form

$$\mathop{\&}_i \bigvee_j \Phi_{ij}$$

where again each $\Phi_{ij}$ is atomic or negated atomic.

**Theorem 1.20.** *Every open formula is logically equivalent to an open formula in disjunctive form, as well as to one in conjunctive form.*

PROOF. This is easily proved by induction on the length of the formula by using the generalized distributive laws

$$\left(\bigvee_i \Phi_i\right) \& \left(\bigvee_j \Psi_j\right) \sim \bigvee_i \bigvee_j (\Phi_i \& \Psi_j),$$

$$\left(\mathop{\&}_i \Phi_i\right) \vee \left(\mathop{\&}_j \Psi_j\right) \sim \mathop{\&}_i \mathop{\&}_j (\Phi_i \vee \Psi_j),$$

the generalized De Morgan laws

$$\neg\left(\bigvee_i \Phi_i\right) \sim \mathop{\&}_i (\neg \Phi_i),$$

$$\neg\left(\mathop{\&}_i \Phi_i\right) \sim \bigvee_i (\neg \Phi_i),$$

and the elimination of $\rightarrow$, $\leftrightarrow$, and $\neg\neg$. □

EXAMPLE. Let $\Phi$ be the formula (with $\mathscr{L} = \{\cdot, <\}$)

$$(x \cdot y \approx z) \rightarrow \neg[(x < z) \vee (x \approx 0)].$$

Then

$$\begin{aligned} \Phi &\sim \neg(x \cdot y \approx z) \vee \neg[(x < z) \vee (x \approx 0)] \\ &\sim \neg(x \cdot y \approx z) \vee [\neg(x < z) \,\&\, \neg(x \approx 0)] \qquad \text{(in disjunctive form)} \\ &\sim [\neg(x \cdot y \approx z) \vee \neg(x < z)] \,\&\, [\neg(x \cdot y \approx z) \vee \neg(x \approx 0)] \\ &\qquad \text{(in conjunctive form).} \end{aligned}$$

The notions of subalgebra, isomorphism, and embedding can be easily generalized to first-order structures.

**Definition 1.21.** Let **A** and **B** be first-order structures of type $\mathscr{L}$. We say **A** is a *substructure* of **B**, written $\mathbf{A} \leq \mathbf{B}$, if $A \subseteq B$ and the fundamental operations and relations of **A** are precisely the restrictions of the corresponding fundamental operations and relations of **B** to $A$. If $X \subseteq B$ let $\text{Sg}(X)$ be the smallest subset of $B$ which is closed under the fundamental operations of **B**. The substructure $\text{Sg}(X)$ with universe $\text{Sg}(X)$ (assuming $\text{Sg}(X) \neq \varnothing$) is called the *substructure generated by X*. As in II§3 we have $|\text{Sg}(X)| \leq |X| + |\mathscr{F}| + \omega$. If $K$ is a class of structures of type $\mathscr{L}$, let $S(K)$ be *the class of all substructures of members of K*.

A very restrictive notion of substructure which we will encounter again in the next section is the following.

**Definition 1.22.** Let **A**,**B** be two first-order structures of type $\mathscr{L}$. **A** is an *elementary substructure* of **B** if $\mathbf{A} \leq \mathbf{B}$ and for any sentence $\Phi$ of type $\mathscr{L}_A$ (and hence of type $\mathscr{L}_B$),

$$\mathbf{A} \models \Phi \quad \text{iff} \quad \mathbf{B} \models \Phi.$$

In this case we write

$$\mathbf{A} \prec \mathbf{B}.$$

$S^{(\prec)}(K)$ denotes *the class of elementary substructures of members of K*.

EXAMPLE. Let us find the elementary substructures of the group of integers $\mathbf{Z} = \langle Z, +, -, 0 \rangle$. Suppose $\mathbf{A} \prec \mathbf{Z}$. As **Z** is a group it follows that **A** is a group.

$$\mathbf{Z} \models \exists x \exists y (x \not\approx y),$$

so

$$\mathbf{A} \models \exists x \exists y(x \not\approx y);$$

hence **A** is nontrivial. Thus for some $n > 0$, $n \in A$. As

$$\mathbf{Z} \models \exists x(x + x + \cdots + x \approx n),$$

where there are $n$ $x$'s added together, it follows that **A** satisfies the same; hence

$$1 \in A.$$

But then

$$\mathbf{A} = \mathbf{Z}.$$

**Definition 1.23.** Let **A** and **B** be first-order structures of type $\mathscr{L}$ and suppose $\alpha: A \to B$ is a bijection such that

$$\alpha f(a_1, \ldots, a_n) = f(\alpha a_1, \ldots, \alpha a_n)$$

for $f$ a fundamental operation, and that $r(a_1, \ldots, a_n)$ holds in **A** iff $r(\alpha a_1, \ldots, \alpha a_n)$ holds in **B**. Then $\alpha$ is an *isomorphism* from **A** to **B**, and **A** is *isomorphic* to **B** (written $\mathbf{A} \cong \mathbf{B}$). If $\alpha: A \to B$ is an isomorphism from **A** to a substructure of **B**, we say $\alpha$ is an *embedding* of **A** into **B**. Let $I(K)$ denote *the closure of* $K$ *under isomorphism.* An embedding $\alpha: \mathbf{A} \to \mathbf{B}$ such that $\alpha\mathbf{A} \prec \mathbf{B}$ is called an *elementary embedding.*

## Exercises §1

1. In the language of semigroups $\{\cdot\}$ find formulas expressing (a) "$x$ is of order dividing $n$", where $n$ is a positive integer (b) "$x$ is of order at most $n$" (c) "$x$ is of order at least $n$".

2. Find formulas which express the following properties of structures: (a) **A** "has size at most $n$" (b) **A** "has size at least $n$".

3. Given a finite structure **A** for a finite language show that there is a first-order formula $\Phi$ such that for any structure **B** of the same type, $\mathbf{B} \models \Phi$ iff $\mathbf{B} \cong \mathbf{A}$.

Given a graph $\langle G, r\rangle$ and $g \in G$, the *valence* or *degree* of $g$ is $|\{h \in G: hrg\}|$.

4. In the language of graphs $\{r\}$ find formulas to express (a) "$x$ has valence at most $n$" (b) "$x$ has valence at least $n$" (c) "$x$ and $y$ are connected by a path of length at most $n$".

5. Show that the following properties of groups can be expressed by first-order formulas: (a) **G** "is centerless" (b) **G** "is a group of exponent $n$" (c) **G** "is nilpotent of class $k$" (d) "$x$ and $y$ are conjugate elements".

A property $P$ of first-order structures is *first-order* (or *elementary*) *relative to* $K$, where $K$ is a class of first-order structures, if there is a set $\Sigma$ of first-order formulas such that for $\mathbf{A} \in K$, **A** has $P$ iff $\mathbf{A} \models \Sigma$. If we can choose $\Sigma$ to be finite we say that $P$ is *strictly* first-order (or *strictly* elementary). Similarly one can define properties of elements of first-order structures relative to $K$.

6. Show that "being of infinite size" is a first-order property (relative to any $K$).

7. Relative to the class of graphs show that the following properties are first-order: (a) "$x$ has infinite valence" (b) "$x$ and $y$ are not connected".

8. Prove that if $\mathbf{A} \cong \mathbf{B}$ then $\mathbf{A} \models \Phi$ iff $\mathbf{B} \models \Phi$, for any $\Phi$.

9. Let $K = \{\mathbf{N}\}$ where $\mathbf{N}$ is the natural numbers $\langle N,+,\cdot,1\rangle$. Show that relative to $K$ the following can be expressed by first-order formulas: (a) "$x < y$" (b) "$x \mid y$" (c) "$x$ is a prime number".

10. Put the following formula in prenex form with the matrix in conjunctive form:

$$\forall x[xry \to \exists y(xry \to \exists x(yrx \mathbin{\&} xry))].$$

11. Does the following binary structure (Figure 31) satisfy

$$\forall x[\exists y(xry \leftrightarrow \exists x(xry))]?$$

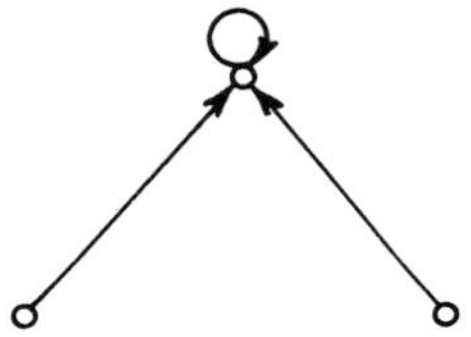

Figure 31

12. Express the following in the language $\{r\}$, where $r$ is a binary relation symbol:

(a) $\langle A,r\rangle$ "is a partially ordered set"
(b) $\langle A,r\rangle$ "is a linearly ordered set"
(c) $\langle A,r\rangle$ "is a dense linearly ordered set"
(d) $r$ "is an equivalence relation on $A$"
(e) $r$ "is a function on $A$"
(f) $r$ "is a surjective function on $A$"
(g) $r$ "is an injective function on $A$".

13. A sentence $\Phi$ is *universal* if $\Phi$ is in prenex form and looks like

$$\forall x_1 \ldots \forall x_n \Psi$$

where $\Psi$ is open, i.e., $\Phi$ contains no existential quantifier. Show that substructures preserve universal sentences, i.e., if $\mathbf{A} \leq \mathbf{B}$ and $\Phi$ is a universal sentence then

$$\mathbf{B} \models \Phi \Rightarrow \mathbf{A} \models \Phi.$$

14. Show that in the language $\{\cdot\}$ the property of being a reduct (see II§1 Ex. 1) of a group is first-order, but not definable by universal sentences.

15. Show that any two countable dense linearly ordered sets without endpoints are isomorphic. [Hint: Build the isomorphism step-by-step by selecting the elements alternately from the first and second sets.]

16. Can one embed:

(a) $\langle \omega,\leq,+,0\rangle$ in $\langle \omega,\leq,\cdot,1\rangle$?
(b) $\langle \omega,\leq,\cdot,1\rangle$ in $\langle \omega,\leq,+,0\rangle$?

17. Let **A** be a finite structure. Describe all possible elementary substructures of **A**.

18. Let **A** be a countable dense linearly ordered set without endpoints. If **B** is a substructure of **A** which is also dense in **A** show $\mathbf{B} \prec \mathbf{A}$.

19. Find all elementary substructures of the graph (called a rooted dyadic tree) pictured in Figure 32.

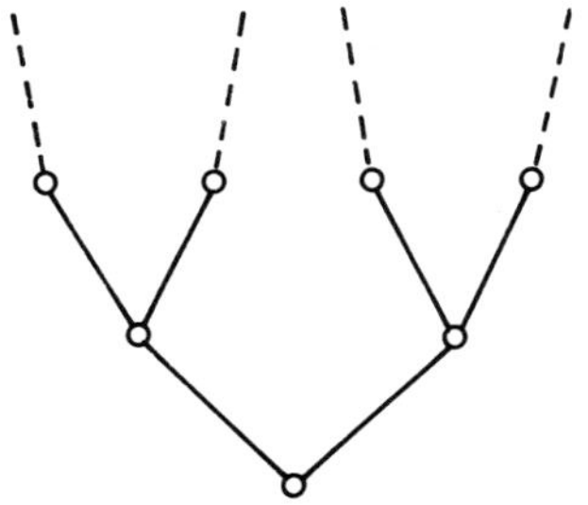

Figure 32

If we are given two structures **A** and **B** of type $\mathscr{L}$, then a mapping $\alpha: A \to B$ is a *homomorphism* if (i) $\alpha f(a_1, \ldots, a_n) = f(\alpha a_1, \ldots, \alpha a_n)$ for $f \in \mathscr{F}$ (ii) $r(a_1, \ldots, a_n) \Rightarrow r(\alpha a_1, \ldots, \alpha a_n)$ for $r \in \mathscr{R}$.

If $\alpha$ is a homomorphism we write, as before, $\alpha: \mathbf{A} \to \mathbf{B}$. The *image* of **A** under $\alpha$, denoted by $\alpha\mathbf{A}$, is the substructure of **B** with universe $\alpha A$. The homomorphism $\alpha$ is an *embedding* if the map $\alpha: \mathbf{A} \to \alpha\mathbf{A}$ is an isomorphism. A sentence $\Phi$ is *positive* if it is in prenex form and the matrix uses only the propositional connectives & and $\vee$.

20. Suppose $\alpha: \mathbf{A} \to \mathbf{B}$ is a homomorphism and $\Phi$ is a positive sentence with $\mathbf{A} \models \Phi$. Show $\alpha\mathbf{A} \models \Phi$; hence homomorphisms preserve positive sentences.

21. Let $\mathscr{L} = \{f\}$ where $f$ is a unary function symbol. Is the sentence $\forall x \forall y(fx \approx fy \to x \approx y)$ logically equivalent to a positive sentence?

22. Is (a) the class of 4-colorable graphs, (b) the class of cubic graphs, definable by positive sentences in the language $\{r\}$?

23. Show every poset $\langle P, \leq \rangle$ can be embedded in a distributive lattice $\langle D, \leq \rangle$.

A family $\mathscr{C}$ of structures is a *chain* if for each $\mathbf{A},\mathbf{B} \in \mathscr{C}$ either $\mathbf{A} \leq \mathbf{B}$ or $\mathbf{B} \leq \mathbf{A}$. If $\mathscr{C}$ is a chain of structures define the structure $\bigcup\mathscr{C}$ by letting its universe be $\bigcup\{A : \mathbf{A} \in \mathscr{C}\}$, and defining $f(a_1, \ldots, a_n)$ to agree with $f^{\mathbf{A}}(a_1, \ldots, a_n)$ for any $\mathbf{A} \in \mathscr{C}$ with $a_1, \ldots, a_n \in A$, and letting $r(a_1, \ldots, a_n)$ hold iff it holds for some $\mathbf{A} \in \mathscr{C}$.

A sentence $\Phi$ is an $\forall\exists$-sentence iff it is in prenex form and it looks like $\forall x_1 \ldots \forall x_m \exists y_1 \ldots \exists y_n \Psi$, where $\Psi$ is open.

24. If $\mathscr{C}$ is a chain of structures and $\Phi$ is an $\forall\exists$-*sentence* such that $\mathbf{A} \models \Phi$ for $\mathbf{A} \in \mathscr{C}$ show that $\bigcup\mathscr{C} \models \Phi$.

25. Show that the class of algebraically closed fields is definable by $\forall\exists$-sentences in the language $\{+, \cdot, -, 0, 1\}$.

26. The class of semigroups which are reducts of monoids can be axiomatized by

$$\forall x \forall y \forall z[(x \cdot y) \cdot z \approx x \cdot (y \cdot z)]$$
$$\exists x \forall y(y \cdot x \approx x \cdot y \ \& \ y \cdot x \approx y).$$

Can this class be axiomatized by $\forall\exists$-sentences?

Given a nonempty indexed family $(\mathbf{A}_i)_{i \in I}$ of structures of type $\mathscr{L}$, define the *direct product* $\prod_{i \in I} \mathbf{A}_i$ to be the structure whose universe is the set $\prod_{i \in I} A_i$, and where fundamental operations and relations are specified by

$$f(a_1, \ldots, a_n)(i) = f^{\mathbf{A}_i}(a_1(i), \ldots, a_n(i))$$
$$r(a_1, \ldots, a_n) \text{ holds iff for all } i \in I, r^{\mathbf{A}_i}(a_1(i), \ldots, a_n(i)) \text{ holds.}$$

27. Given homomorphisms $\alpha_i : \mathbf{A} \to \mathbf{B}_i$, $i \in I$, show that the natural map $\alpha : A \to \prod_{i \in I} B_i$ is a homomorphism from $\mathbf{A}$ to $\prod_{i \in I} \mathbf{B}_i$.

28. Show that a projection map on $\prod_{i \in I} \mathbf{A}_i$ is a surjective homomorphism.

A *Horn formula* $\Phi$ is a formula in prenex form which looks like

$$Q_1 x_1 \ldots Q_n x_n \left( \underset{i}{\&} \Phi_i \right)$$

where each $Q_i$ is a quantifier, and each $\Phi_i$ is a formula of the form

$$\Psi_1 \vee \cdots \vee \Psi_k,$$

in which each $\Psi_j$ is atomic or negated atomic, and at most one of the $\Psi_j$ is atomic.

29. Show that the following can be expressed by Horn formulas: (a) "the cancellation law" (for semigroups), (b) "of size at least $n$", (c) any atomic formula, (d) "inverses exist" (for monoids), (e) "being centerless" (for groups).

30. If $\Phi$ is a Horn formula and $\mathbf{A}_i \models \Phi$ for $i \in I$ show that

$$\prod_{i \in I} \mathbf{A}_i \models \Phi.$$

A substructure $\mathbf{A}$ of a direct product $\prod_{i \in I} \mathbf{A}_i$ is a *subdirect product* if $\pi_i(A) = A_i$ for all $i \in I$. An embedding $\alpha : \mathbf{A} \to \prod_{i \in I} \mathbf{A}_i$ is a *subdirect embedding* if $\alpha\mathbf{A}$ is a subdirect product.

A sentence $\Phi$ is a *special Horn sentence* if it is of the form

$$\underset{i}{\&} \forall \vec{x}(\Phi_i \to \Psi_i)$$

where each $\Phi_i$ is positive and each $\Psi_i$ is atomic.

31. Show that a special Horn sentence is logically equivalent to a Horn sentence.

32. Show that if $\mathbf{A}$ is a subdirect product of $\mathbf{A}_i$, $i \in I$, and $\Phi$ is a special Horn sentence such that $\mathbf{A}_i \models \Phi$ for all $i \in I$, then $\mathbf{A} \models \Phi$; hence subdirect products preserve special Horn sentences.

33. Can the class of cubic graphs be defined by special Horn sentences?

A *complete graph* $\langle G,r\rangle$ is one satisfying

$$\forall x \forall y(x \not\approx y \rightarrow xry).$$

A complete graph with one edge removed is *almost complete.*

34. Show that every graph is subdirectly embedded in a product of complete and/or almost complete graphs.

35. If **A** is an algebra of type $\mathscr{F}$ with a discriminator term $t(x,y,z)$ [and switching term $s(x,y,u,v)$] show that **A** satisfies (see IV§9)

$$(p \approx q \,\&\, \hat{p} \approx \hat{q}) \leftrightarrow t(p,q,\hat{p}) \approx t(q,p,\hat{q})$$
$$(p \approx q \vee \hat{p} \approx \hat{q}) \leftrightarrow s(p,q,\hat{p},\hat{q}) \approx \hat{p}$$
$$(p \approx q \vee \hat{p} \not\approx \hat{q}) \leftrightarrow s(\hat{p},\hat{q},p,q) \approx q$$

and if **A** is nontrivial,

$$(p \not\approx q) \leftrightarrow \forall x[t(p,q,x) \approx p].$$

Show that, consequently, if **A** is nontrivial then for every [universal] $\mathscr{F}$-formula $\phi$ there is an [universal] $\mathscr{F}$-formula $\phi^*$ whose matrix is an equation $p \approx q$ such that **A** satisfies

$$\phi \leftrightarrow \phi^*.$$

Define the *spectrum* of an $\mathscr{L}$-formula $\phi$, Spec $\phi$, to be $\{|A| : \mathbf{A}$ is an $\mathscr{L}$-structure, $\mathbf{A} \models \phi$, $\mathbf{A}$ is finite$\}$.

36. (McKenzie). If $\phi$ is an $\mathscr{F}$-formula, where $\mathscr{F}$ is a type of algebras, show that there is a (finitely axiomatizable) variety $V$ such that Spec $V$ (see IV§9 Ex. 4) is the closure of Spec $\phi$ under finite products.

## §2. Reduced Products and Ultraproducts

Reduced products result from a certain combination of the direct product and quotient constructions. They were introduced in the 1950's by Łoś, and the special case of ultraproducts has been a subject worthy of at least one book. In the following you will need to recall the definition of $[\![a = b]\!]$ from IV§5.5, and that of direct products of structures from p. 204.

**Definition 2.1.** Let $(\mathbf{A}_i)_{i \in I}$ be a nonempty indexed family of structures of type $\mathscr{L}$, and suppose $F$ is a filter over $I$. Define the binary relation $\theta_F$ on $\prod_{i \in I} A_i$ by

$$\langle a,b\rangle \in \theta_F \quad \text{iff} \quad [\![a = b]\!] \in F.$$

(When discussing reduced products we will always assume $\varnothing \notin F$, i.e., $F$ is *proper*.)

**Lemma 2.2.** *For* $(\mathbf{A}_i)_{i\in I}$ *and* $F$ *as above, the relation* $\theta_F$ *is an equivalence relation on* $\prod_{i\in I} A_i$. *For a fundamental n-ary operation* $f$ *of* $\prod_{i\in I} \mathbf{A}_i$ *and for*

$$\langle a_1,b_1\rangle, \ldots, \langle a_n,b_n\rangle \in \theta_F$$

*we have*

$$\langle f(a_1,\ldots,a_n), f(b_1,\ldots,b_n)\rangle \in \theta_F,$$

*i.e.,* $\theta_F$ *is a congruence for the "algebra part of* **A**".

PROOF. Clearly $\theta_F$ is reflexive and symmetric. If

$$\langle a,b\rangle,\langle b,c\rangle \in \theta_F$$

then

$$[\![a = b]\!], [\![b = c]\!] \in F,$$

hence

$$[\![a = b]\!] \cap [\![b = c]\!] \in F.$$

Now from

$$[\![a = c]\!] \supseteq [\![a = b]\!] \cap [\![b = c]\!]$$

it follows that

$$[\![a = c]\!] \in F,$$

so

$$\langle a,c\rangle \in \theta_F.$$

Consequently, $\theta_F$ is an equivalence relation. Next with $f$ and $\langle a_i,b_i\rangle$ as in the statement of the lemma note that

$$[\![f(a_1,\ldots,a_n) = f(b_1,\ldots,b_n)]\!] \supseteq [\![a_1 = b_1]\!] \cap \cdots \cap [\![a_n = b_n]\!];$$

hence

$$[\![f(a_1,\ldots,a_n) = f(b_1,\ldots,b_n)]\!] \in F,$$

so

$$\langle f(a_1,\ldots,a_n), f(b_1,\ldots,b_n)\rangle \in \theta_F.$$

□

**Definition 2.3.** Given a nonempty indexed family of structures $(\mathbf{A}_i)_{i\in I}$ of type $\mathscr{L}$ and a proper filter $F$ over $I$ define the *reduced product* $\prod_{i\in I} \mathbf{A}_i/F$ as follows. Let its universe $\prod_{i\in I} A_i/F$ be the set $\prod_{i\in I} A_i/\theta_F$, and let $a/F$ denote the element $a/\theta_F$. For $f$ an $n$-ary function symbol and for $a_1,\ldots,a_n \in \prod_{i\in I} A_i$ let

$$f(a_1/F,\ldots,a_n/F) = f(a_1,\ldots,a_n)/F,$$

and for $r$ an $n$-ary relation symbol let $r(a_1/F,\ldots,a_n/F)$ hold iff

$$\{i \in I: \mathbf{A}_i \models r(a_1(i),\ldots,a_n(i))\} \in F.$$

If $K$ is a nonempty class of structures of type $\mathscr{L}$ let $P_R(K)$ denote *the class of all reduced products* $\prod_{i\in I} \mathbf{A}_i/F$, where $\mathbf{A}_i \in K$.

In view of Definition 2.3, it is reasonable to extend our use of the $[\![\ ]\!]$ notation as follows.

**Definition 2.4.** If $(\mathbf{A}_i)_{i \in I}$ is a nonempty indexed family of structures of type $\mathscr{L}$ and if $\Phi(a_1, \ldots, a_n)$ is a sentence of type $\mathscr{L}_A$, where $\mathbf{A} = \prod_{i \in I} \mathbf{A}_i$, let

$$[\![\Phi(a_1, \ldots, a_n)]\!] = \{i \in I : \mathbf{A}_i \models \Phi(a_1(i), \ldots, a_n(i))\}.$$

Thus given a reduced product $\prod_{i \in I} \mathbf{A}_i/F$ and an atomic sentence $\Phi(a_1, \ldots, a_n)$ we see that

$$\prod_{i \in I} \mathbf{A}_i/F \models \Phi(a_1/F, \ldots, a_n/F) \quad \text{iff} \quad [\![\Phi(a_1, \ldots, a_n)]\!] \in F.$$

Determining precisely which sentences are preserved by reduced products has been one of the milestones in the history of model theory. Our next theorem is concerned with the easy half of this study.

**Definition 2.5.** A *Horn formula* is a formula in prenex form with a matrix consisting of conjunctions of formulas $\Phi_1 \vee \cdots \vee \Phi_n$ where each $\Phi_i$ is atomic or negated atomic, and at most one $\Phi_i$ is atomic in each such disjunction. Such disjunctions of atomic and negated atomic formulas are called *basic Horn formulas.*

The following property of direct products is useful in induction proofs on reduced products.

**Lemma 2.6** (The maximal property). *Let $\mathbf{A}_i$, $i \in I$, be a nonempty indexed family of structures of type $\mathscr{L}$. If we are given a formula $\exists x \Phi(x, y_1, \ldots, y_n)$ of type $\mathscr{L}$ and $a_1, \ldots, a_n \in \prod_{i \in I} A_i$ then there is an $a \in \prod_{i \in I} A_i$ such that*

$$[\![\exists x \Phi(x, a_1, \ldots, a_n)]\!] = [\![\Phi(a, a_1, \ldots, a_n)]\!].$$

Proof. For

$$i \in [\![\exists x \Phi(x, a_1, \ldots, a_n)]\!]$$

choose $a(i) \in A_i$ such that

$$\mathbf{A}_i \models \Phi(a(i), a_1(i), \ldots, a_n(i)),$$

and for other $i$'s in $I$ let $a(i)$ be arbitrary. Then it is readily verified that such an $a$ satisfies the lemma. □

**Theorem 2.7.** *Let $\prod_{i \in I} \mathbf{A}_i/F$ be a reduced product of structures of type $\mathscr{L}$, and suppose $\Phi(x_1, \ldots, x_n)$ is a Horn formula of type $\mathscr{L}$. If*

$$a_1, \ldots, a_n \in \prod_{i \in I} A_i$$

and

$$[\![\Phi(a_1, \ldots, a_n)]\!] \in F$$

then

$$\prod_{i \in I} \mathbf{A}_i/F \models \Phi(a_1/F, \ldots, a_n/F).$$

Proof. First let us suppose $\Phi$ is a basic Horn formula

$$\Phi_1(x_1, \ldots, x_n) \vee \cdots \vee \Phi_k(x_1, \ldots, x_n).$$

Our assumption

$$\left[\!\!\left[ \bigvee_{1 \le i \le k} \Phi_i(a_1, \ldots, a_n) \right]\!\!\right] \in F$$

is equivalent to

$$\bigcup_{1 \le i \le k} [\![\Phi_i(a_1, \ldots, a_n)]\!] \in F.$$

If for some $\Phi_i$ which is a negated atomic formula we have

$$I - [\![\Phi_i(a_1, \ldots, a_n)]\!] \notin F$$

then, by the definition of reduced product,

$$\prod_{i \in I} \mathbf{A}_i/F \vDash \Phi_i(a_1/F, \ldots, a_n/F);$$

hence

$$\prod_{i \in I} \mathbf{A}_i/F \vDash \Phi(a_1/F, \ldots, a_n/F).$$

If now for each negated atomic formula $\Phi_i$ we have

$$I - [\![\Phi_i(a_1, \ldots, a_n)]\!] \in F$$

then there must be one of the $\Phi_i$'s, say $\Phi_k$, which is atomic. (Otherwise

$$I - [\![\Phi(a_1, \ldots, a_n)]\!] = I - \bigcup_{1 \le i \le k} [\![\Phi_i(a_1, \ldots, a_n)]\!] \in F,$$

which is impossible as $F$ is closed under intersection and $\varnothing \notin F$.) Now in this case

$$[\![\neg \Phi_i(a_1, \ldots, a_n)]\!] \in F$$

for $1 \le i \le k - 1$, so

$$\left[\!\!\left[ \mathop{\&}_{1 \le i \le k-1} \neg \Phi_i(a_1, \ldots, a_n) \right]\!\!\right] \in F.$$

Since

$$[\![\Phi(a_1, \ldots, a_n)]\!] \in F,$$

taking the intersection we have

$$\left[\!\!\left[ \left( \mathop{\&}_{1 \le i \le k-1} \neg \Phi_i(a_1, \ldots, a_n) \right) \& \, \Phi_k(a_1, \ldots, a_n) \right]\!\!\right] \in F,$$

so

$$[\![\Phi_k(a_1, \ldots, a_n)]\!] \in F.$$

This says

$$\prod_{i \in I} \mathbf{A}_i/F \vDash \Phi_k(a_1/F, \ldots, a_n/F);$$

hence

$$\prod_{i\in I} \mathbf{A}_i/F \models \Phi(a_1/F, \ldots, a_n/F).$$

If $\Phi$ is a conjunction

$$\Psi_1 \,\&\, \cdots \,\&\, \Psi_k$$

of basic Horn formulas then

$$[\![\Psi_1(a_1, \ldots, a_n) \,\&\, \cdots \,\&\, \Psi_k(a_1, \ldots, a_n)]\!] \in F$$

leads to

$$[\![\Psi_i(a_1, \ldots, a_n)]\!] \in F$$

for $1 \leq i \leq k$, so

$$\prod_{i\in I} \mathbf{A}_i/F \models \Psi_i(a_1/F, \ldots, a_n/F),$$

$1 \leq i \leq k$, and thus

$$\prod_{i\in I} \mathbf{A}_i/F \models \Phi(a_1/F, \ldots, a_n/F).$$

Next we look at the general case in which $\Phi$ is in the form

$$Q_1 y_1 \ldots Q_m y_m \Psi(y_1, \ldots, y_m, x_1, \ldots, x_n)$$

with $\Psi$ being a conjunction of basic Horn formulas. We use induction on the number of occurrences of quantifiers in $\Phi$. If there are no quantifiers then we have finished this case in the last paragraph. So suppose that the theorem is true for any Horn formula with fewer than $m$ occurrences of quantifiers. In $\Phi$ above let us first suppose $Q_1$ is the universal quantifier, i.e.,

$$\Phi = \forall y_1 \Phi^*(y_1, x_1, \ldots, x_n).$$

If we are given $a \in \prod_{i\in I} A_i$, then from

$$[\![\Phi(a_1, \ldots, a_n)]\!] \in F$$

it follows that

$$[\![\Phi^*(a, a_1, \ldots, a_n)]\!] \in F$$

as

$$[\![\Phi(a_1, \ldots, a_n)]\!] \subseteq [\![\Phi^*(a, a_1, \ldots, a_n)]\!].$$

By the induction hypothesis

$$\prod_{i\in I} \mathbf{A}_i/F \models \Phi^*(a/F, a_1/F, \ldots, a_n/F);$$

hence

$$\prod_{i\in I} \mathbf{A}_i/F \models \Phi(a_1/F, \ldots, a_n/F).$$

Next suppose $Q_1$ is the existential quantifier, i.e.,

$$\Phi = \exists y_1 \Phi^*(y_1, x_1, \ldots, x_n).$$

Choose by 2.6 $a \in \prod_{i \in I} A_i$ such that

$$[\![\Phi(a_1, \ldots, a_n)]\!] = [\![\Phi^*(a, a_1, \ldots, a_n)]\!].$$

Then again by the induction hypothesis

$$\prod_{i \in I} \mathbf{A}_i/F \models \Phi^*(a/F, a_1/F, \ldots, a_n/F);$$

hence

$$\prod_{i \in I} \mathbf{A}_i/F \models \Phi(a_1/F, \ldots, a_n/F). \qquad \square$$

The following generalizes the definition of ultraproducts in IV§6 to arbitrary first-order structures.

**Definition 2.8.** A reduced product $\prod_{i \in I} \mathbf{A}_i/U$ is called an *ultraproduct* if $U$ is an ultrafilter over $I$. If all the $\mathbf{A}_i = \mathbf{A}$ then we write $\mathbf{A}^I/U$ and call it an *ultrapower* of $\mathbf{A}$. The *class of all ultraproducts of members of* $K$ is denoted by $P_U(K)$.

For the following recall the basic properties of ultrafilters from IV§3. We abbreviate $a_1, \ldots, a_n$ by $\vec{a}$, and $a_1/U, \ldots, a_n/U$ by $\vec{a}/U$.

**Theorem 2.9** (Łoś). *Given structures* $\mathbf{A}_i$, $i \in I$, *of type* $\mathscr{L}$, *if* $U$ *is an ultrafilter over* $I$ *and* $\Phi$ *is any first-order formula of type* $\mathscr{L}$ *then*

$$\prod_{i \in I} \mathbf{A}_i/U \models \Phi(a_1/U, \ldots, a_n/U)$$

*iff*

$$[\![\Phi(a_1, \ldots, a_n)]\!] \in U.$$

Proof. (By induction on $l(\Phi)$.) For $l(\Phi) = 0$ we have already observed that the theorem is true. So suppose $l(\Phi) > 0$ and the theorem holds for all $\Psi$ such that $l(\Psi) < l(\Phi)$. If

$$\Phi = \Phi_1 \,\&\, \Phi_2$$

then

$$\begin{aligned} [\![\Phi_1(\vec{a}) \,\&\, \Phi_2(\vec{a})]\!] \in U \quad &\text{iff} \quad [\![\Phi_1(\vec{a})]\!] \cap [\![\Phi_2(\vec{a})]\!] \in U \\ &\text{iff} \quad [\![\Phi_i(\vec{a})]\!] \in U \quad \text{for } i = 1,2 \\ &\text{iff} \quad \prod_{i \in I} \mathbf{A}_i/U \models \Phi_i(\vec{a}/U) \quad \text{for } i = 1,2 \\ &\text{iff} \quad \prod_{i \in I} \mathbf{A}_i/U \models \Phi_1(\vec{a}/U) \,\&\, \Phi_2(\vec{a}/U). \end{aligned}$$

One handles the logical connectives $\vee, \neg, \rightarrow, \leftrightarrow$ in a similar fashion. If

$$\Phi(\vec{a}) = \exists x \hat{\Phi}(x, \vec{a})$$

choose $a \in \prod_{i \in I} A_i$ such that

$$[\![\exists x \hat{\Phi}(x, \vec{a})]\!] = [\![\hat{\Phi}(a, \vec{a})]\!].$$

Then

$$\begin{aligned}
[\![\Phi(\vec{a})]\!] \in U \quad &\text{iff} \quad [\![\exists x\hat{\Phi}(x,\vec{a})]\!] \in U \\
&\text{iff} \quad [\![\hat{\Phi}(a,\vec{a})]\!] \in U \quad \text{for some } a \\
&\text{iff} \quad \prod_{i\in I} \mathbf{A}_i/U \models \hat{\Phi}(a/U,\vec{a}/U) \quad \text{for some } a \\
&\text{iff} \quad \prod_{i\in I} \mathbf{A}_i/U \models \exists x\hat{\Phi}(x,\vec{a}/U) \\
&\text{iff} \quad \prod_{i\in I} \mathbf{A}_i/U \models \Phi(\vec{a}/U).
\end{aligned}$$

Finally, if

$$\Phi(\vec{a}) = \forall x\hat{\Phi}(x,\vec{a})$$

then one can find a $\Psi(\vec{a})$ such that the quantifier $\forall$ does not appear in $\Psi$ and $\Phi \sim \Psi$ (by 1.13), hence from what we have just proved,

$$\begin{aligned}
[\![\Phi(\vec{a})]\!] \in U \quad &\text{iff} \quad [\![\Psi(\vec{a})]\!] \in U \\
&\text{iff} \quad \prod_{i\in I} \mathbf{A}_i/U \models \Psi(\vec{a}/U) \\
&\text{iff} \quad \prod_{i\in I} \mathbf{A}_i/U \models \Phi(\vec{a}/U).
\end{aligned}$$

□

**Lemma 2.10.** *Let* **A** *be a first-order structure, $I$ a nonempty index set and $F$ a proper filter over $I$. For $a \in A$ let $c_a$ denote the element of $A^I$ with*

$$c_a(i) = a, \qquad i \in I.$$

*The map*

$$\alpha : A \to A^I/F$$

*defined by*

$$\alpha a = c_a/F$$

*is an embedding of* **A** *into $\mathbf{A}^I/F$. The map $\alpha$ is called the natural embedding of* **A** *into $\mathbf{A}^I/F$.*

Proof. (Exercise.) □

**Theorem 2.11.** *If* **A** *is a first-order structure of type $\mathscr{L}$, $I$ is an index set, and $U$ is an ultrafilter over $I$, then the natural embedding $\alpha$ of* **A** *into $\mathbf{A}^I/U$ is an elementary embedding.*

Proof. Just note that for formulas $\Phi(x_1, \ldots, x_n)$ of type $\mathscr{L}$ we have

$$[\![\Phi(c_{a_1}, \ldots, c_{a_n})]\!] = I \quad \text{if} \quad \mathbf{A} \models \Phi(a_1, \ldots, a_n),$$

and

$$[\![\Phi(c_{a_1}, \ldots, c_{a_n})]\!] = \varnothing \quad \text{if} \quad \mathbf{A} \not\models \Phi(a_1, \ldots, a_n).$$

Thus

$$\alpha\mathbf{A} \models \Phi(\alpha a_1, \ldots, \alpha a_n) \quad \text{iff} \quad \mathbf{A}^I/U \models \Phi(\alpha a_1, \ldots, \alpha a_n).$$

□

Next we prove one of the most celebrated theorems of logic.

**Theorem 2.12** (The Compactness Theorem). *Let $\Sigma$ be a set of first-order sentences of type $\mathscr{L}$ such that for every finite subset $\Sigma_0$ of $\Sigma$ there is a structure satisfying $\Sigma_0$. Then $\mathbf{A} \models \Sigma$ for some $\mathbf{A}$ of type $\mathscr{L}$.*

PROOF. Let $I$ be the family of finite subsets of $\Sigma$, and for $i \in I$ let $\mathbf{A}_i$ be a structure satisfying the sentences in $i$. For $i \in I$ let

$$J_i = \{j \in I : i \subseteq j\}.$$

Then

$$J_{i_1} \cap J_{i_2} = J_{i_1 \cup i_2},$$

so the collection of $J_i$'s is closed under finite intersection. As no $J_i = \varnothing$ it follows that

$$F = \{J \subseteq I : J_i \subseteq J \text{ for some } i \in I\}$$

is a proper filter over $I$, so by IV§3.17 we can extend it to an ultrafilter $U$ over $I$; hence each $J_i$ belongs to $U$. Now for $\Phi \in \Sigma$ we have

$$\{\Phi\} \in I,$$

so

$$\mathbf{A}_j \models \Phi \quad \text{for } j \in J_{\{\Phi\}}$$

as $\Phi \in j$. Looking at $\prod_{i \in I} \mathbf{A}_i$ we see that

$$[\![\Phi]\!] \supseteq J_{\{\Phi\}}$$

so

$$\prod_{i \in I} \mathbf{A}_i/U \models \Phi;$$

hence

$$\prod_{i \in I} \mathbf{A}_i/U \models \Sigma.$$ □

**Corollary 2.13.** *If $\Sigma$ is a set of sentences of type $\mathscr{L}$ and $\Phi$ is a sentence of type $\mathscr{L}$ such that*

$$\Sigma \models \Phi$$

*then, for some finite subset $\Sigma_0$ of $\Sigma$,*

$$\Sigma_0 \models \Phi.$$

PROOF. If the above fails then for some $\Sigma$ and $\Phi$ and for every finite subset $\Sigma_0$ of $\Sigma$ there is a structure $\mathbf{A}$ which satisfies $\Sigma_0$ but not $\Phi$; hence $\Sigma_0 \cup \{\neg\Phi\}$ is satisfied by some $\mathbf{A}$. But then 2.12 says $\Sigma \cup \{\neg\Phi\}$ is satisfied by some $\mathbf{A}$, which is impossible as $\mathbf{A} \models \Sigma$ implies $\mathbf{A} \models \Phi$. □

A slight variation of the proof of the compactness theorem gives us the following.

**Theorem 2.14.** *Every first-order structure* $\mathbf{A}$ *can be embedded in an ultraproduct of its finitely generated substructures.*

PROOF. Let $I$ be the family of nonempty finite subsets of $A$, and for $i \in I$ let $\mathbf{A}_i$ be the substructure of $\mathbf{A}$ generated by $i$. Also, for $i \in I$ let

$$J_i = \{j \in I : i \subseteq j\}.$$

As in 2.12 extend the family of $J_i$'s to an ultrafilter $U$ over $I$. For $a \in A$ let $\lambda a$ be any element of $\prod_{i \in I} A_i$ such that

$$(\lambda a)(i) = a$$

if $a \in i$. Then let

$$\alpha : \mathbf{A} \to \prod_{i \in I} \mathbf{A}_i / U$$

be defined by

$$\alpha a = (\lambda a)/U.$$

For $\Phi(x_1, \ldots, x_n)$ an atomic or negated atomic formula and $a_1, \ldots, a_n \in A$ such that

$$\mathbf{A} \models \Phi(a_1, \ldots, a_n),$$

we have

$$[\![\Phi(\lambda a_1, \ldots, \lambda a_n)]\!] \supseteq J_{\{a_1, \ldots, a_n\}};$$

hence

$$\alpha(\mathbf{A}) \models \Phi(\lambda a_1/U, \ldots, \lambda a_n/U).$$

This is easily seen to guarantee that $\alpha$ is an embedding. □

For the remainder of this section we will assume that we are working with some convenient fixed countably infinite set of variables $X$, i.e. all formulas will be over this $X$.

**Definition 2.15.** A class $K$ of first-order $\mathscr{L}$-structures is an *elementary class* (or a *first-order class*) if there is a set $\Sigma$ of first-order formulas such that

$$\mathbf{A} \in K \quad \text{iff} \quad \mathbf{A} \models \Sigma.$$

$K$ is said to be *axiomatized* (or *defined*) by $\Sigma$ in this case, and $\Sigma$ is a set of *axioms* for $K$. Let $\mathrm{Th}(K)$ be the set of first-order sentences of type $\mathscr{L}$ satisfied by $K$, called the *theory* of $K$.

**Theorem 2.16.** *Let* $K$ *be a class of first-order structures of type* $\mathscr{L}$. *Then the following are equivalent*:

(a) $K$ *is an elementary class.*
(b) $K$ *is closed under* $I, S^{(\prec)}$, *and* $P_U$.
(c) $K = IS^{(\prec)}P_U(K^*)$, for some class $K^*$.

PROOF. For (a) ⇒ (b) use the fact that each of $I, S^{(\prec)}$ and $P_U$ preserve first-order properties. (b) ⇒ (c) is trivial, for let $K^* = K$. For (c) ⇒ (*a*) we claim that $K$ is axiomatizable by $\mathrm{Th}(K^*)$ where $K^*$ is as in (c). Note that $K \models \mathrm{Th}(K^*)$. Suppose

$$\mathbf{A} \models \mathrm{Th}(K^*).$$

Let $\mathrm{Th}^*(\mathbf{A})$ be the set of sentences $\Phi(a_1, \ldots, a_n)$ of type $\mathscr{L}_A$ satisfied by $\mathbf{A}$, and let $I$ be the collection of finite subsets of $\mathrm{Th}^*(\mathbf{A})$. If

$$\Phi(a_1, \ldots, a_n) \in \mathrm{Th}^*(\mathbf{A})$$

then for some $\mathbf{B} \in K^*$,

$$\mathbf{B} \models \exists x_1 \ldots \exists x_n \Phi(x_1, \ldots, x_n).$$

For otherwise

$$K^* \models \forall x_1 \ldots \forall x_n \neg \Phi(x_1, \ldots, x_n),$$

which is impossible as

$$\mathbf{A} \models \exists x_1 \ldots \exists x_n \Phi(x_1, \ldots, x_n)$$

and $\mathbf{A} \models \mathrm{Th}(K^*)$. Consequently, for $i \in I$ we can choose $\mathbf{A}_i \in K^*$ and elements $\hat{a}(i) \in A_i$ for $a \in A$ such that the formulas in $i$ become true of $\mathbf{A}_i$ when $a$ is interpreted as $\hat{a}(i)$, for $a \in A$. Let

$$J_i = \{j \in I : i \subseteq j\},$$

and, as before, let $U$ be an ultrafilter over $I$ such that $J_i \in U$ for $i \in I$. Let $\hat{a}$ be the element in $\prod_{i \in I} A_i$ whose $i$th coordinate is $\hat{a}(i)$. Then for

$$\Phi(a_1, \ldots, a_n) \in \mathrm{Th}^*(\mathbf{A})$$

we have

$$[\![\Phi(\hat{a}_1, \ldots, \hat{a}_n)]\!] \supseteq J_i \in U$$

where

$$i = \{\Phi(a_1, \ldots, a_n)\};$$

hence

$$[\![\Phi(\hat{a}_1, \ldots, \hat{a}_n)]\!] \in U.$$

Thus

$$\prod_{i \in I} \mathbf{A}_i/U \models \Phi(\hat{a}_1/U, \ldots, \hat{a}_n/U).$$

By considering the atomic and negated atomic sentences in $\mathrm{Th}^*(\mathbf{A})$ we see that the mapping

$$\alpha : A \to \prod_{i \in I} A_i/U$$

defined by

$$\alpha a = \hat{a}/U$$

gives an embedding of $\mathbf{A}$ into $\prod_{i \in I} \mathbf{A}_i/U$, and then again from the above it follows that the embedding is elementary. Thus $\mathbf{A} \in IS^{(\prec)}P_U(K^*)$. ☐

**Definition 2.17.** An elementary class $K$ is a *strictly first-order* (or *strictly elementary*) class if $K$ can be axiomatized by finitely many formulas, or equivalently by a single formula.

**Corollary 2.18.** *An elementary class K of first-order structures is a strictly elementary class iff the complement K′ of K is closed under ultraproducts.*

Proof. If $K$ is axiomatized by $\Phi$ then the complement of $K$ is axiomatized by $\neg\Phi$; hence $K'$ is an elementary class, so $K'$ is closed under $P_U$. Conversely suppose $K'$ is closed under $P_U$. Let $I$ be the collection of finite subsets of $\mathrm{Th}(K)$. If $K$ is not finitely axiomatizable, for each $i \in I$ there must be a structure $\mathbf{A}_i$ such that

$$\mathbf{A}_i \models i$$

but

$$\mathbf{A}_i \notin K.$$

Let

$$J_i = \{j \in i : i \subseteq j\},$$

and construct $U$ as before. Then

$$\prod_{i \in I} \mathbf{A}_i/U \models \Phi$$

for $\Phi \in \mathrm{Th}(K)$ as

$$[\![\Phi]\!] \supseteq J_{\{\Phi\}} \in U.$$

Thus

$$\prod_{i \in I} \mathbf{A}_i/U \models \mathrm{Th}(K),$$

so

$$\prod_{i \in I} \mathbf{A}_i/U \in K.$$

But this is impossible since by the assumption

$$\prod_{i \in I} \mathbf{A}_i/U \in K'$$

as each $\mathbf{A}_i \in K'$. Thus $K$ must be a strictly elementary class. □

**Definition 2.19.** A first-order formula $\Phi$ is a *universal formula* if it is in prenex form and all the quantifiers are universal. An elementary class is a *universal class* if it can be axiomatized by universal formulas.

**Theorem 2.20.** *Let K be a class of structures of type* $\mathscr{L}$. *Then the following are equivalent*:

(a) *K is a universal class,*
(b) *K is closed under I, S, and* $P_U$,
(c) $K = ISP_U(K^*)$, *for some* $K^*$.

Proof. (a) ⇒ (b) is easily checked and (b) ⇒ (c) is straightforward. For (c) ⇒ (a) let $\mathrm{Th}_\forall(K^*)$ be the set of universal sentences of type $\mathscr{L}$ which are satisfied by $K^*$, and suppose $\mathbf{A} \models \mathrm{Th}_\forall(K^*)$. Let $\mathrm{Th}_\forall^*(\mathbf{A})$ be the set of universal sentences of type $\mathscr{L}_A$ which are satisfied by $\mathbf{A}$. Now we just repeat the last part of the proof of 2.16, replacing $\mathrm{Th}^*$ by $\mathrm{Th}_\forall^*$. □

**Definition 2.21.** A first-order formula $\Phi$ is a *universal Horn formula* if it is both a universal and a Horn formula. A class $K$ of structures is a *universal Horn class* if it can be axiomatized by universal Horn formulas.

Before looking at classes defined by universal Horn formulas we need a technical lemma.

**Lemma 2.22.** *The following inequalities on class operators hold*:

(a) $P \leq IP_R$.
(b) $P_R P_R \leq IP_R$.
(c) $P_R \leq ISPP_U$.

PROOF. (a) Given $\prod_{i \in I} \mathbf{A}_i$ let $F = \{I\}$ be the smallest filter over $I$. Then one sees that

$$\prod_{i \in I} \mathbf{A}_i \cong \prod_{i \in I} \mathbf{A}_i/F$$

using the map $\alpha(a) = a/F$.

(b) Given a set $J$ and a family of pairwise disjoint sets $I_j$, $j \in J$, and algebras $\mathbf{A}_i$ for $i \in I_j$ and a filter $F$ over $J$ and for $j \in J$ a filter $F_j$ over $I_j$ define

$$I = \bigcup_{j \in J} I_j$$

and let

$$\hat{F} = \{S \subseteq I : \{j \in J : S \cap I_j \in F_j\} \in F\}.$$

Then $\hat{F}$ is easily seen to be a filter over $I$, and we will show that

$$\prod_{j \in J}\left(\prod_{i \in I_j} \mathbf{A}_i/F_j\right)\Big/F \cong \prod_{i \in I} \mathbf{A}_i/\hat{F}.$$

For each $j \in J$ define

$$\alpha_j \colon \prod_{i \in I} A_i \to \prod_{i \in I_j} A_i$$

by

$$\alpha_j(a) = a \restriction I_j.$$

Then $\alpha_j$ is a surjective homomorphism from $\prod_{i \in I} \mathbf{A}_i$ to $\prod_{i \in I_j} \mathbf{A}_i$. Let

$$v_j \colon \prod_{i \in I_j} A_i \to \prod_{i \in I_j} A_i/F_j$$

be the natural mapping. Define

$$\beta \colon \prod_{i \in I} A_i \to \prod_{j \in J} (\prod_{i \in I_j} A_i/F_j)$$

to be the natural mapping derived from the $v_j$'s, i.e.,

$$\beta(a)(j) = v_j(a\restriction_{I_j}).$$

Let

$$v \colon \prod_{j \in J}\left(\prod_{i \in I_j} A_i/F_j\right) \to \prod_{j \in J}\left(\prod_{i \in I_j} A_i/F_j\right)\Big/F$$

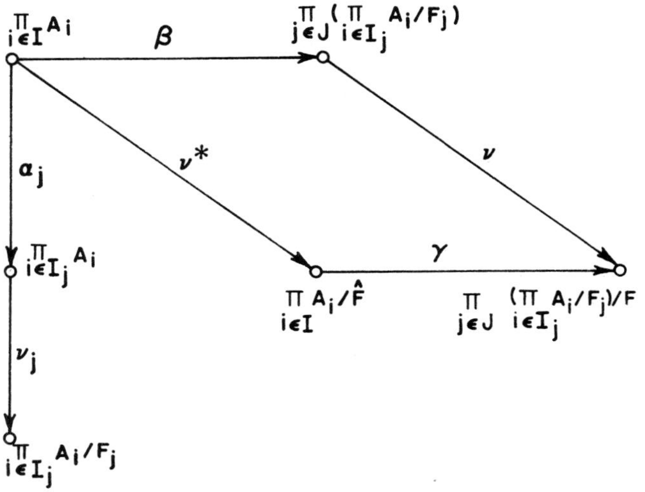


Figure 33

be the natural map (see Figure 33). The mapping $\nu \circ \beta$ is surjective as each of $\nu$ and $\beta$ is surjective. Also let

$$\nu^*: \prod_{i\in I} A_i \to \prod_{i\in I} A_i/\hat{F}$$

be the natural map. Let us show that

$$\ker(\nu \circ \beta) = \theta_F.$$

We have

$$\begin{aligned} \langle a,b\rangle \in \ker(\nu \circ \beta) &\Leftrightarrow \langle \beta a,\beta b\rangle \in \ker \nu = \theta_{\hat{F}} \\ &\Leftrightarrow [\![\beta a = \beta b]\!] \in F \\ &\Leftrightarrow \{j \in J: \nu_j(a\restriction_{I_j}) = \nu_j(b\restriction_{I_j})\} \in F \\ &\Leftrightarrow \{j \in J: [\![a = b]\!] \cap I_j \in F_j\} \in F \\ &\Leftrightarrow [\![a = b]\!] \in \hat{F}. \end{aligned}$$

Thus we have a bijection

$$\gamma: \prod_{i\in I} A_i/\hat{F} \to \prod_{j\in J}\left(\prod_{i\in I_j} A_i/F_j\right)\bigg/F$$

such that $\gamma \circ \nu^* = \nu \circ \beta$. If we were working in a language of algebras we could use the first isomorphism theorem to show $\gamma$ is an isomorphism. We will leave the details of showing that $\gamma$ preserves fundamental relations to the reader.

(c) If $F$ is a filter over $I$ let $J$ be the set of ultrafilters over $I$ containing $F$. Given $\mathbf{A}_i$, $i \in I$, define, for $U \in J$,

$$\alpha_U: \prod_{i\in I} A_i/F \to \prod_{i\in I} A_i/U$$

by

$$\alpha_U(a/F) = a/U,$$

and then let

$$\alpha\colon \prod_{i\in I} A_i/F \to \prod_{U\in J} (\prod_{i\in I} A_i/U)$$

be the natural map. We claim that since one clearly has

$$F = \bigcap J$$

we must have an injective map $\alpha$. For if

$$a/F \neq b/F$$

then

$$[\![a = b]\!] \notin F$$

so we can find an ultrafilter $U$ extending $F$ with

$$[\![a = b]\!] \notin U.$$

Thus

$$\alpha_U(a) \neq \alpha_U(b)$$

so $\alpha$ is injective. If we were working with algebras we would clearly have an embedding, and we again leave the details concerning fundamental relations to the reader. □

**Theorem 2.23.** *Let $K$ be a class of structures of type $\mathscr{L}$. Then the following are equivalent*:

(a) *$K$ is a universal Horn class.*
(b) *$K$ is closed under $I$, $S$, and $P_R$.*
(c) *$K$ is closed under $I$, $S$, $P$, and $P_U$.*
(d) *$K = ISP_R(K^*)$, for some $K^*$.*
(e) *$K = ISPP_U(K^*)$, for some $K^*$.*

Proof. (a) ⇒ (b) is easily checked using 2.7, and (b) ⇒ (c), (b) ⇒ (d) and (c) ⇒ (e) are clear. For (d) ⇒ (a) and (e) ⇒ (a) let $\mathrm{Th}_{\forall H}(K^*)$ be the set of universal Horn sentences of type $\mathscr{L}$ which are true of $K^*$. Certainly $K \models \mathrm{Th}_{\forall H}(K^*)$. Suppose

$$\mathbf{A} \models \mathrm{Th}_{\forall H}(K^*).$$

Let $\mathrm{Th}_0^*(\mathbf{A})$ be the set of atomic or negated atomic sentences true of $\mathbf{A}$ in $\mathscr{L}_A$. (This is called the *open diagram* of $\mathbf{A}$.) If we are given

$$\{\Phi_1(a_1, \ldots, a_n), \ldots, \Phi_k(a_1, \ldots, a_n)\} \subseteq \mathrm{Th}_0^*(\mathbf{A})$$

then

$$\mathbf{A} \models \exists x_1 \ldots \exists x_n[\Phi_1(x_1, \ldots, x_n) \,\&\, \cdots \,\&\, \Phi_k(x_1, \ldots, x_n)].$$

We want to show some member of $P(K^*)$ satisfies this sentence as well. For this purpose it suffices to show

$$P(K^*) \not\models \forall x_1 \ldots \forall x_n[\neg\Phi_1(x_1, \ldots, x_n) \vee \ldots \vee \neg\Phi_k(x_1, \ldots, x_n)].$$

If at most one $\Phi_i$ is negated atomic then the universal sentence above would be logically equivalent to a universal Horn sentence which is not true of **A**, hence not of $K^*$. So let us suppose at least two of the $\Phi_i$ are negated atomic, say $\Phi_i$ is negated atomic for $1 \leq i \leq t$ (where $2 \leq t \leq k$), and atomic for $t+1 \leq i \leq k$. Then, for $1 \leq i \leq t$, one can argue as above that

$$K^* \not\models \forall x_1 \ldots \forall x_n[\neg \Phi_i(x_1,\ldots,x_n) \vee \neg \Phi_{t+1}(x_1,\ldots,x_n) \vee \cdots \vee \neg \Phi_k(x_1,\ldots,x_n)];$$

hence for some $\mathbf{A}_i \in \mathrm{K}^*$,

$$\mathbf{A}_i \models \exists x_1 \ldots \exists x_n[\Phi_i(x_1,\ldots,x_n) \,\&\, \Phi_{t+1}(x_1,\ldots,x_n) \,\&\, \cdots \,\&\, \Phi_k(x_1,\ldots,x_n)].$$

For $1 \leq i \leq t$, $1 \leq j \leq n$, choose $a_j(i) \in A_i$ such that

$$\mathbf{A}_i \models \Phi_i(a_1(i),\ldots,a_n(i)) \,\&\, \Phi_{t+1}(a_1(i),\ldots,a_n(i)) \,\&\cdots\&\, \Phi_k(a_1(i),\ldots,a_n(i)).$$

Then

$$\prod_{1 \leq i \leq t} \mathbf{A}_i \models \mathop{\&}_{1 \leq i \leq k} \Phi_i(a_1,\ldots,a_n)$$

and

$$\prod_{1 \leq i \leq t} \mathbf{A}_i \in P(K^*).$$

Let $I$ be the collection of finite subsets of $\mathrm{Th}_0^*(\mathbf{A})$, and proceed as in the proof of 2.16, replacing $\mathrm{Th}^*(\mathbf{A})$ by $\mathrm{Th}_0^*(\mathbf{A})$, to obtain

$$\mathbf{A} \in ISP_U P(K^*).$$

From 2.22,

$$ISP_R \leq ISPP_U \leq ISP_R P_R \leq ISP_R;$$

hence

$$ISP_R = ISPP_U.$$

Now

$$ISP_U P \leq ISP_R P_R = ISP_R;$$

hence

$$\mathbf{A} \in ISP_U P(K^*) \subseteq ISP_R(K^*) = ISPP_U(K^*) = K. \qquad \square$$

Let us now turn to algebras.

**Definition 2.24.** A *quasi-identity* is an identity or a formula of the form $(p_1 \approx q_1 \,\&\cdots\&\, p_n \approx q_n) \rightarrow p \approx q$. A *quasivariety* is a class of algebras closed under $I$,$S$, and $P_R$, and containing the one-element algebras.

**Theorem 2.25.** *Let K be a class of algebras. Then the following are equivalent*:

(a) *K can be axiomatized by quasi-identities,*
(b) *K is a quasivariety,*
(c) *K is closed under I,S,P, and* $P_U$ *and contains a trivial algebra,*
(d) *K is closed under* $ISP_R$ *and contains a trivial algebra, and*
(e) *K is closed under* $ISPP_U$ *and contains a trivial algebra.*

Proof. As quasi-identities are logically equivalent to universal Horn formulas, and as trivial algebras satisfy any quasi-identity, we have (a) $\Rightarrow$ (b).

(b) ⇒ (c), (b) ⇒ (d) and (c) ⇒ (e) are obvious. If (d) or (e) holds then $K$ can be axiomatized by universal Horn formulas by 2.23 which we may assume to be of the form $\forall x_1 \ldots \forall x_n(\Psi_1 \vee \cdots \vee \Psi_k)$ with each $\Psi_i$ an atomic or negated atomic formula. (Why?) As a trivial algebra cannot satisfy a negated atomic formula, exactly one of $\Psi_1, \ldots, \Psi_k$ is atomic. Such an axiom is logically equivalent to a quasi-identity. □

For us the study of universal algebra has been almost synonomous with the study of varieties, but the Russian mathematicians under the leadership of Mal'cev have vigorously pursued the subject of quasivarieties as well.

EXAMPLE. The *cancellation law*

$$x \cdot y \approx x \cdot z \rightarrow y \approx z$$

is a quasi-identity.

## REFERENCES


1. J. L. Bell and A. B. Slomson [2]
2. C. C. Chang and H. J. Keisler [8]
3. A. I. Mal'cev [24]


## EXERCISES §2

1. If $\mathbf{R}$ is the ordered field of real numbers show that $\mathbf{R}^\omega/U$ is a non-Archimedian ordered field if $U$ is a nonprincipal ultrafilter on $\omega$. Show that the class of Archimedian ordered fields is not an elementary class.
2. With $P$ the set of prime numbers show that $\prod_{p \in P} \mathbf{Z}/(p)$ is a ring of characteristic zero. Hence show that "being a field of finite characteristic" is not a first-order property.
3. Show that "being a finite structure of type $\mathscr{L}$" is not a first-order property.
4. Show that "being isomorphic to the ring of integers" is not a first-order property. [Hint: Use IV§6 Ex. 7].
5. Prove that the following hold: (a) $P_U S \leq ISP_U$; (b) $P_R S \leq ISP_R$.
6. Prove that a graph is $n$-colorable iff each finite subgraph is $n$-colorable.

Given two languages $\mathscr{L}, \mathscr{L}'$ with $\mathscr{L} \subseteq \mathscr{L}'$ and a structure $\mathbf{A}$ of type $\mathscr{L}'$, let $\mathbf{A}\restriction_{\mathscr{L}}$ denote the *reduct* of $\mathbf{A}$ to $\mathscr{L}$, i.e., retain only those fundamental operations and relations of $\mathbf{A}$ which correspond to symbols in $\mathscr{L}$. Then define $K\restriction_{\mathscr{L}} = \{\mathbf{A}\restriction_{\mathscr{L}} : \mathbf{A} \in K\}$.

7. Let $K$ be an elementary class of type $\mathscr{L}'$, and let $\mathbf{A}$ be a structure of type $\mathscr{L}$, $\mathscr{L} \subseteq \mathscr{L}'$. Show that $\mathbf{A} \in IS(K\restriction_{\mathscr{L}})$ iff $\mathbf{A} \models \mathrm{Th}_\forall(K\restriction_{\mathscr{L}})$.
8. Prove that a group $\mathbf{G}$ can be linearly ordered iff each of its finitely generated subgroups can be linearly ordered.

9. If $\Phi$ is a sentence such that $\mathbf{A} \models \Phi \Rightarrow S(\mathbf{A}) \models \Phi$, then show that $\Phi$ is logically equivalent to a universal sentence.

10. If $\Phi$ is a sentence such that $K \models \Phi \Rightarrow SP_R(K) \models \Phi$ then show that $\Phi$ is logically equivalent to a universal Horn sentence.

11. Given a language $\mathscr{L}$ let $K$ be an elementary class and let $\Phi$ be a sentence such that for $\mathbf{A},\mathbf{B} \in K$ with $\mathbf{B} \leq \mathbf{A}$, if $\mathbf{A} \models \Phi$ then $\mathbf{B} \models \Phi$. Show that there is a universal sentence $\Psi$ such that $K \models \Phi \leftrightarrow \Psi$. [Hint: Make appropriate changes in the proof of 2.20.]

12. Given a first-order structure $\mathbf{A}$ of type $\mathscr{L}$ let $D^+(\mathbf{A})$ be the set of atomic sentences in the language $\mathscr{L}_A$ true of $\mathbf{A}$. Given a set of sentences $\Sigma$ of type $\mathscr{L}$, show that there is a homomorphism from $\mathbf{A}$ to some $\mathbf{B}$ with $\mathbf{B} \models \Sigma$ iff there is a $\mathbf{C}$ with $\mathbf{C} \models D^+(\mathbf{A}) \cup \Sigma$.

## §3. Principal Congruence Formulas

Principal congruence formulas are the obvious first-order formulas for describing principal congruences. We give two applications of principal congruence formulas, namely McKenzie's theorem on definable principal congruences, and Taylor's theorem on the number of subdirectly irreducible algebras in a variety. Throughout this section we are working with a fixed language $\mathscr{F}$ of algebras. First we look at how to construct principal congruences using unary polynomials.

**Lemma 3.1** (Mal'cev). *Let* $\mathbf{A}$ *be an algebra of type* $\mathscr{F}$ *and suppose* $a,b,c,d \in A$. *Then*

$$\langle a,b\rangle \in \Theta(c,d)$$

*iff there are terms*

$$p_i(x,y_1,\ldots,y_k),$$

$1 \leq i \leq m$, *and elements* $e_1,\ldots,e_k \in A$ *such that*

$$\begin{aligned} a &= p_1(s_1,\vec{e}),\\ p_i(t_i,\vec{e}) &= p_{i+1}(s_{i+1},\vec{e}) \quad \text{for } 1 \leq i < m,\\ p_m(t_m,\vec{e}) &= b, \end{aligned}$$

*where*

$$\{s_i,t_i\} = \{c,d\}$$

*for* $1 \leq i \leq m$.

Proof. Let $p_i(x,y_1,\ldots,y_k)$ be any terms of type $\mathscr{F}$ and let $e_1,\ldots,e_k$ be any elements of $A$. Then clearly

$$\langle p_i(c,\vec{e}),\ p_i(d,\vec{e})\rangle \in \Theta(c,d);$$

hence if

$$\{s_i,t_i\} = \{c,d\}$$

and

$$p_i(t_i,\vec{e}) = p_{i+1}(s_{i+1},\vec{e})$$

then by the transitivity of $\Theta(c,d)$,

$$\langle p_1(s_1,\vec{e}),\ p_m(t_m,\vec{e})\rangle \in \Theta(c,d).$$

Thus the collection $\theta^*$ of pairs $\langle a,b\rangle$ such that there are $p_i$'s and $e_j$'s as above form a subset of $\Theta(c,d)$. Now note that $\theta^*$ is an equivalence relation, and indeed a congruence. For if

$$\langle a_j,b_j\rangle \in \theta^*,$$

$1 \le j \le n$, and if $f$ is a fundamental $n$-ary operation, let

$$a_j = p_{j1}(s_{j1},\vec{e}_j),$$
$$p_{ji}(t_{ji},\vec{e}_j) = p_{j\,i+1}(s_{j\,i+1},\vec{e}_j),$$

and

$$p_{jm_j}(t_{jm_j},\vec{e}_j) = b_j.$$

Then

$$f(b_1,\ldots,b_{j-1},a_j,\ldots,a_n) = f(b_1,\ldots,b_{j-1},p_{j\,i+1}(s_{j\,i+1},\vec{e}_j),a_{j+1},\ldots,a_n),$$
$$f(b_1,\ldots,b_{j-1},p_{ji}(t_{ji},\vec{e}_j),a_{j+1},\ldots,a_n) = f(b_1,\ldots,b_{j-1},p_{j\,i+1}(s_{j\,i+1},\vec{e}_j),a_{j+1},\ldots,a_n),$$

$1 \le i < m_j$, and

$$f(b_1,\ldots,b_{j-1},p_{jm_j}(t_{jm_j},e_j),a_{j+1},\ldots,a_n) = f(b_1,\ldots,b_{j-1},b_j,a_{j+1},\ldots,a_n);$$

hence

$$\langle f(b_1,\ldots,b_{j-1},a_j,\ldots,a_n),\ f(b_1,\ldots,b_j,a_{j+1},\ldots,a_n)\rangle \in \theta^*,$$

so by transitivity

$$\langle f(a_1,\ldots,a_n),\ f(b_1,\ldots,b_n)\rangle \in \theta^*.$$

As

$$\langle c,d\rangle \in \theta^* \subseteq \Theta(c,d)$$

we must have

$$\Theta(c,d) = \theta^*,$$

since $\Theta(c,d)$ is the smallest congruence containing $\langle c,d\rangle$. □

**Definition 3.2.** A *principal congruence formula* (of type $\mathscr{F}$) is a formula

$$\pi(x,y,u,v)$$

of the form

$$\exists\vec{w}\{x \approx p_1(z_1,\vec{w})\ \&\left[\mathop{\&}_{1\le i<n} p_i(z_i',\vec{w}) \approx p_{i+1}(z_{i+1},\vec{w})\right]\&\ p_n(z_n',\vec{w}) \approx y\}$$

where

$$\{z_i,z_i'\} = \{u,v\},$$

$1 \leq i \leq n$. Let $\Pi$ be the set of principal congruence formulas in $\mathscr{F}(X)$ where $X$ is an infinite set of variables.

**Theorem 3.3.** *For $a,b,c,d \in A$ and $\mathbf{A}$ an algebra of type $\mathscr{F}$ we have*

$$\langle a,b \rangle \in \Theta(c,d)$$

*iff*

$$\mathbf{A} \models \pi(a,b,c,d)$$

*for some $\pi \in \Pi$.*

Proof. This is just a restatement of 3.1. □

**Definition 3.4.** A variety $V$ has *definable principal congruences* if there is a finite subset $\Pi_0$ of $\Pi$ such that for $\mathbf{A} \in V$ and $a,b,c,d \in A$,

$$\langle a,b \rangle \in \Theta(c,d) \quad \text{iff} \quad \mathbf{A} \models \pi(a,b,c,d)$$

for some $\pi \in \Pi_0$.

**Theorem 3.5** (McKenzie). *If $V$ is a directly representable variety, then $V$ has definable principal congruences.*

Proof. Choose finite algebras $\mathbf{A}_1, \ldots, \mathbf{A}_k \in V$ such that for any finite $\mathbf{B} \in V$,

$$\mathbf{B} \in IP(\{\mathbf{A}_1, \ldots, \mathbf{A}_k\}),$$

and let

$$m_i = |A_i|.$$

Now let

$$K = \{\mathbf{A}_1^{j_1} \times \cdots \times \mathbf{A}_k^{j_k} : j_i \leq m_i^4, 1 \leq i \leq k\}.$$

As $K$ is a finite set of finite algebras it is clear that there is a finite $\Pi_0 \subseteq \Pi$ such that for $\mathbf{A} \in K$ and $a,b,c,d \in A$,

$$\langle a,b \rangle \in \Theta(c,d)$$

iff

$$\mathbf{A} \models \pi(a,b,c,d)$$

for some $\pi \in \Pi_0$. Now suppose $\mathbf{B}$ is any finite member of $P(\{\mathbf{A}_1, \ldots, \mathbf{A}_k\})$ and $a,b,c,d \in B$ with

$$\langle a,b \rangle \in \Theta(c,d).$$

Let

$$\mathbf{B} = \mathbf{A}_1^{s_1} \times \cdots \times \mathbf{A}_k^{s_k}.$$

Let us rewrite the latter as

$$\mathbf{B}_{11} \times \cdots \times \mathbf{B}_{1s_1} \times \cdots \times \mathbf{B}_{k1} \times \cdots \times \mathbf{B}_{ks_k},$$

with $\mathbf{B}_{ij} = \mathbf{A}_i$. For some $\pi \in \Pi$ we have

$$\mathbf{B} \models \pi(a,b,c,d).$$

Let $\pi(x,y,u,v)$ be

$$\exists w_1 \ldots \exists w_r \Phi(x,y,u,v,w_1,\ldots,w_r),$$

where $\Phi$ is open. Let $e_1,\ldots,e_r \in B$ be such that

$$\mathbf{B} \models \Phi(a,b,c,d,e_1,\ldots,e_r).$$

As there are at most $m_i^4$ possible 4-tuples

$$\langle a(i,j), b(i,j), c(i,j), d(i,j)\rangle$$

for $1 \leq j \leq s_i$ we can partition the indices $i1,\ldots,is_i$ into sets $J_{i1},\ldots,J_{it_i}$ with $t_i \leq m_i^4$ such that on each $J_{ij}$ the elements $a,b,c,d$ are all constant. Thus in view of the description of congruence formulas we can assume the $e$'s are all constant on $J_{ij}$. The set of elements of $\mathbf{B}$ which are constant on each $J_{ij}$ form a subuniverse $C$ of $\mathbf{B}$, and let $\mathbf{C}$ be the corresponding subalgebra. Then $\mathbf{C} \in I(K)$, for if we select one index $(ij)^*$ from each $J_{ij}$ then the map

$$\alpha: C \to \prod B_{(ij)^*}$$

defined by

$$\alpha(c)(ij)^* = c((ij)^*)$$

is easily seen to be an isomorphism. As $\alpha\mathbf{C} \in H(\mathbf{B})$,

$$\alpha\mathbf{C} \models \pi(\alpha a,\alpha b,\alpha c,\alpha d);$$

hence

$$\mathbf{C} \models \pi(a,b,c,d).$$

It follows that for some $\pi^* \in \Pi_0$ (as $\mathbf{C} \in I(K)$),

$$\mathbf{C} \models \pi^*(a,b,c,d).$$

But then

$$\mathbf{B} \models \pi^*(a,b,c,d).$$

Hence for any finite member $\mathbf{B}$ of $V$, the principal congruences of $\mathbf{B}$ can be described just by using the formulas in $\Pi_0$.

Finally, if $\mathbf{B}$ is any member of $V$ and $a,b,c,d \in B$ with

$$\langle a,b\rangle \in \Theta(c,d)$$

then for some $\pi \in \Pi$ we have

$$\mathbf{B} \models \pi(a,b,c,d).$$

If $\pi$ is

$$\exists w_1 \ldots \exists w_r \Phi(x,y,u,v,w_1,\ldots,w_r)$$

with $\Phi$ open, choose $e_1,\ldots,e_r \in B$ such that

$$\mathbf{B} \models \Phi(a,b,c,d,e_1,\ldots,e_r).$$

Let $\mathbf{C}$ be the subalgebra of $\mathbf{B}$ generated by $\{a,b,c,d,e_1,\ldots,e_r\}$. Then

$$\mathbf{C} \models \Phi(a,b,c,d,e_1,\ldots,e_r)$$

so

$$\mathbf{C} \models \pi(a,b,c,d);$$

hence for some $\pi^* \in \Pi_0$,

$$\mathbf{C} \models \pi^*(a,b,c,d),$$

so

$$\mathbf{B} \models \pi^*(a,b,c,d).$$

Thus $V$ has definable principal congruences. □

Before proving Taylor's Theorem we need a combinatorial lemma, a proof of which can be found in (3).

**Lemma 3.6** (Erdös). *Let $\kappa$ be an infinite cardinal and let $A$ be a set with $|A| > 2^\kappa$, $C$ a set with $|C| \leq \kappa$. Let $A^{(2)}$ be the set of doubletons $\{c,d\}$ contained in $A$ with $c \neq d$. If $\alpha$ is a map from $A^{(2)}$ to $C$ then for some infinite subset $B$ of $A$,*

$$\alpha(B^{(2)}) = \{e\}$$

*for some $e \in C$.*

**Theorem 3.7** (Taylor). *Let $V$ be a variety of type $\mathscr{F}$, and let $\kappa = \max(\omega,|\mathscr{F}|)$. If $V$ has a subdirectly irreducible algebra $\mathbf{A}$ with $|A| > 2^\kappa$ then $V$ has arbitrarily large subdirectly irreducible algebras.*

Proof. If $\mathbf{A} \in V$ is subdirectly irreducible and $|A| > 2^\kappa$ then let $a,b \in A$ be such that $\Theta(a,b)$ is the smallest congruence not equal to $\Delta$. As there are only $\kappa$ many formulas in $\Pi$, and as

$$\mathbf{A} \models \pi(a,b,c,d)$$

for some $\pi \in \Pi$, if $c \neq d$, it follows from 3.3 and 3.6 that for some infinite subset $B$ of $A$ there is a $\pi^* \in \Pi$ such that for $c,d \in B$, if $c \neq d$ then

$$\mathbf{A} \models \pi^*(a,b,c,d).$$

Given an infinite set $\mathscr{I}$ of new nullary function symbols with $|\mathscr{I}| = m$ and an infinite set of variables $X$, let $\Sigma$ be

$$\{i \not\approx j : i,j \in \mathscr{I} \text{ and } i \neq j\} \cup (\mathrm{Id}_V(X)) \cup \{\pi^*(a,b,i,j) : i,j \in \mathscr{I} \text{ and } i \neq j\}.$$

Then for each finite $\Sigma_0 \subseteq \Sigma$ we see that by interpreting the $i$'s as suitable members of $B$ it is possible to find an algebra (essentially $\mathbf{A}$) satisfying $\Sigma_0$. Thus $\Sigma$ is satisfied by some algebra $\mathbf{A}^*$ of type $\mathscr{F} \cup \mathscr{I} \cup \{a,b\}$. Let $I \subseteq A^*$ be the elements of $A^*$ corresponding to $\mathscr{I}$, and let $a,b$ again denote appropriate elements of $A^*$. Then $|I| = m$, and $a \neq b$. Choose $\theta$ to be a maximal congruence on $\mathbf{A}^*$ among the congruences on $\mathbf{A}^*$ which do not identify $a$ and $b$. Then $i,j \in I$ and $i \neq j$ imply

$$\langle i,j \rangle \notin \theta,$$

as

$$\mathbf{A}^* \models \pi^*(a,b,i,j).$$

Consequently $\mathbf{A}^*/\theta$ is subdirectly irreducible and

$$|A^*/\theta| \geq |I| = m.$$

This shows that $V$ has arbitrarily large subdirectly irreducible members. □

The next result does not depend on principal congruence formulas, but does indeed nicely complement the previous theorem.

**Theorem 3.8** (Quackenbush). *If $V$ is a locally finite variety with, up to isomorphism, only finitely many finite subdirectly irreducible members, then $V$ has no infinite subdirectly irreducible members.*

PROOF. Let $V^*$ be the class of finite subdirectly irreducible members of $V$. If $\mathbf{A} \in V$ then let $K$ be the set of finitely generated subalgebras of $\mathbf{A}$. By 2.14 we have

$$\mathbf{A} \in ISP_U(K),$$

and from local finiteness

$$K \subseteq IP_S(V^*) \subseteq ISP(V^*);$$

hence

$$\mathbf{A} \in ISP_USP(V^*),$$

so

$$\mathbf{A} \in ISPP_U(V^*)$$

by 2.23. As an ultraproduct of finitely many finite algebras is isomorphic to one of the algebras we have

$$\mathbf{A} \in ISP(V^*);$$

hence

$$\mathbf{A} \in IP_S(V^*),$$

so $\mathbf{A}$ cannot be both infinite and subdirectly irreducible. □

REFERENCES

1. K. A. Baker [a]
2. J. T. Baldwin and J. Berman [1975]
3. P. Erdös [1942]
4. R. Freese and R. McKenzie [a]
5. W. Taylor [1972]

EXERCISES §3

1. Show that commutative rings with identity have definable principal congruences.
2. Show that abelian groups of exponent $n$ have definable principal congruences.
3. Show that discriminator varieties have definable principal congruences.
4. Show that distributive lattices have definable principal congruences.

5. Suppose $V$ is a variety such that there is a first-order formula $\phi(x,y,u,v)$ with

$$\langle a,b\rangle \in \Theta(c,d) \Leftrightarrow \mathbf{A} \models \phi(a,b,c,d)$$

for $a,b,c,d \in A$, $\mathbf{A} \in V$. Show that $V$ has definable principal congruences.

6. Show that a finitely generated semisimple arithmetical variety has definable principal congruences.

7. Are elementary substructures of subdirectly irreducible [simple] algebras also subdirectly irreducible [simple]? What about ultrapowers?

8. (Baldwin and Berman). If $V$ is a finitely generated variety with the CEP (see II§5 Ex. 10) show that $V$ has definable principal congruences.

## §4. Three Finite Basis Theorems

One of the older questions of universal algebra was whether or not the identities of a finite algebra of finite type $\mathscr{F}$ could be derived from finitely many of the identities. Birkhoff proved that this was true if a finite bound is placed on the number of variables, but in 1954 Lyndon constructed a seven-element algebra with one binary and one nullary operation such that the identities were not finitely based. Murskiĭ constructed a three-element algebra whose identities are not finitely based in 1965, and Perkins constructed a six-element semigroup whose identities are not finitely based in 1969. An example of a finite nonassociative ring whose identities are not finitely based was constructed by Polin in 1976. On the positive side we know that finite algebras of the following kinds have a finitely based set of identities: two-element algebras (Lyndon, 1951), groups (Oates–Powell, 1965), rings (Kruse; Lvov, 1973), algebras determining a congruence-distributive variety (Baker, 1977), and algebras determining a variety with finitely many finite subdirectly irreducibles and definable principal congruences (McKenzie, 1978). We will prove the theorems of Baker, Birkhoff, and McKenzie in this section.

**Definition 4.1.** Let $X$ be a set of variables and $K$ a class of algebras. We say that $\mathrm{Id}_K(X)$ is *finitely based* if there is a finite subset $\Sigma$ of $\mathrm{Id}_K(X)$ such that

$$\Sigma \models \mathrm{Id}_K(X)$$

and we say that the *identities of K are finitely based* if there is a finite set of identities $\Sigma$ such that for any $X$,

$$\Sigma \models \mathrm{Id}_K(X).$$

**Theorem 4.2** (Birkhoff). *Let* $\mathbf{A}$ *be a finite algebra of finite type* $\mathscr{F}$ *and let* $X$ *be a finite set of variables. Then* $\mathrm{Id}_{\mathbf{A}}(X)$ *is finitely based.*

PROOF. Let $\theta$ be the congruence on $\mathbf{T}(X)$ defined by

$$\langle p,q\rangle \in \theta$$

iff

$$\mathbf{A} \models p \approx q.$$

(This, of course, is the congruence used to define $\mathbf{F}_{V(\mathbf{A})}(\bar{X})$.) As $A$ is finite there are only finitely many equivalence classes of $\theta$. From each equivalence class of $\theta$ choose one term. Let this set of representatives be $Q = \{q_1, \ldots, q_n\}$. Now let $\Sigma$ be the set of equations consisting of

$$x \approx y \quad \text{if } x,y \in X \text{ and } \langle x,y\rangle \in \theta,$$
$$q_i \approx x \quad \text{if } x \in X \text{ and } \langle x,q_i\rangle \in \theta,$$
$$f(q_{i_1}, \ldots, q_{i_n}) \approx q_{i_{n+1}} \text{ if } f \in \mathscr{F}_n \text{ and } \langle f(q_{i_1}, \ldots, q_{i_n}), q_{i_{n+1}}\rangle \in \theta.$$

Then a proof by induction on the number of function symbols in a term $p \in T(X)$ shows that if

$$\langle p,q_i\rangle \in \theta$$

then

$$\Sigma \models p \approx q_i.$$

But then

$$\Sigma \models p \approx q$$

if

$$\mathbf{A} \models p \approx q,$$

and as

$$\mathbf{A} \models \Sigma,$$

$\mathrm{Id}_K(X)$ is indeed finitely based. □

**Theorem 4.3** (McKenzie). *If $V$ is a locally finite variety of finite type $\mathscr{F}$ with finitely many finite subdirectly irreducible members and if $V$ has definable principal congruences then the identities of $V$ are finitely based.*

PROOF. Let $\Pi_0 \subseteq \Pi$ be a finite set of principal congruence formulas which show that $V$ has definable principal congruences. Let $\Pi_0$ be $\{\pi_1, \ldots, \pi_n\}$, and define $\Phi$ to be

$$\pi_1 \vee \cdots \vee \pi_n.$$

Then for $\mathbf{A} \in V$ and $a,b,c,d \in A$,

$$\langle a,b\rangle \in \Theta(c,d) \Leftrightarrow \mathbf{A} \models \Phi(a,b,c,d).$$

Let $\mathbf{S}_1, \ldots, \mathbf{S}_n$ be finite subdirectly irreducible members of $V$ such that every finite subdirectly irreducible member of $V$ is isomorphic to one of the $\mathbf{S}_i$'s. By 3.8 they are, up to isomorphism, the only subdirectly irreducible algebras in $V$. Let $\Psi_1$ be a sentence which asserts "the collection of $\langle a,b\rangle$ such that

$\Phi(a,b,c,d)$ holds is $\Theta(c,d)$", i.e., $\Psi_1$ can be

$$\forall u \forall v \Big\{ \Phi(u,v,u,v) \,\&\, \forall x \Phi(x,x,u,v) \,\&\, \forall x \forall y [\Phi(x,y,u,v) \to \Phi(y,x,u,v)]$$

$$\&\, \forall x \forall y \forall z [\Phi(x,y,u,v) \,\&\, \Phi(y,z,u,v) \to \Phi(x,z,u,v)]$$

$$\&\, \mathop{\&}_{\mathscr{F}_n \neq \varnothing} \mathop{\&}_{f \in \mathscr{F}_n} \forall x_1 \forall y_1 \ldots \forall x_n \forall y_n \Big[ \mathop{\&}_{1 \leq i \leq n} \Phi(x_i,y_i,u,v) \to \Phi(f(\vec{x}), f(\vec{y}),u,v) \Big] \Big\}.$$

Thus for $\mathbf{A}$ any algebra of type $\mathscr{F}$, $\mathbf{A} \models \Psi_1$ iff for all $a,b,c,d \in A$,

$$\langle a,b \rangle \in \Theta(c,d) \Leftrightarrow \mathbf{A} \models \Phi(a,b,c,d).$$

Next let $\Psi_2$ be a sentence which says

"an algebra is isomorphic to one of $\mathbf{S}_1, \ldots, \mathbf{S}_n$"

(see §1 Exercise 3). Then let $\Psi_3$ be a sentence which says

"an algebra satisfies $\Psi_1$, and if it is subdirectly irreducible then it is isomorphic to one of $\mathbf{S}_1, \ldots, \mathbf{S}_n$".

For example $\Psi_3$ could be

$$\Psi_1 \,\&\, (\{\exists x \exists y [x \not\approx y \,\&\, \forall u \forall v (u \not\approx v \to \Phi(x,y,u,v))] \to \Psi_2\} \vee \forall x \forall y (x \approx y)).$$

Let $\Sigma$ be the set of identities of $V$ over an infinite set of variables $X$. As

$$\Sigma \models \Psi_3,$$

there must be a finite subset $\Sigma_0$ of $\Sigma$ such that

$$\Sigma_0 \models \Psi_3$$

by 2.13. But then the subdirectly irreducible algebras satisfying $\Sigma_0$ will satisfy $\Psi_3$; hence they will be in $V$. Thus the variety defined by $\Sigma_0$ must be $V$. □

Now we turn to the proof of Baker's finite basis theorem. From this paragraph until the statement of Corollary 4.18 we will assume that our finite language of algebras is $\mathscr{F}$, and that we are working with a congruence-distributive variety $V$. Let $p_0(x,y,z), \ldots, p_n(x,y,z)$ be ternary terms which satisfy Jónsson's conditions II§12.6.

**Lemma 4.4.** $V \models p_i(x,u,x) \approx p_i(x,v,x), \quad 1 \leq i \leq n-1$

$$V \models x \not\approx y \to [p_1(x,x,y) \not\approx p_1(x,y,y) \vee \cdots \vee p_{n-1}(x,x,y) \not\approx p_{n-1}(x,y,y)].$$

Proof. These are both immediate from II§12.6. □

The proof of Baker's theorem is much easier to write out if we can assume that the $p_i$'s are function symbols.

**Definition 4.5.** Let $\mathscr{F}^*$ be the language obtained by adjoining new ternary operation symbols $t_1, \ldots, t_{n-1}$ to $\mathscr{F}$, and let $V^*$ be the variety defined by the identities $\Sigma$ of type $\mathscr{F}$ over some infinite set $X$ of variables true of $V$ plus the identities

$$t_i(x,y,z) \approx p_i(x,y,z),$$

$1 \leq i \leq n-1$.

**Lemma 4.6.** *If the identities $\Sigma^*$ of $V^*$ are finitely based, then so are the identities $\Sigma$ of $V$.*

PROOF. Let $\Sigma^{**}$ be

$$\Sigma \cup \{t_i(x,y,z) \approx p_i(x,y,z) : 1 \leq i \leq n-1\},$$

and let $\Sigma_0^*$ be a finite basis for $\Sigma^*$. Then

$$\Sigma^{**} \models \Sigma_0^*;$$

hence by 2.13 there is a finite subset $\Sigma_0^{**}$ of $\Sigma^{**}$ such that

$$\Sigma_0^{**} \models \Sigma_0^*.$$

Thus $\Sigma_0^{**}$ is a set of axioms for $V^*$; hence there is a finite $\Sigma_0 \subseteq \Sigma$ such that

$$\Sigma_0 \cup \{t_i(x,y,z) \approx p_i(x,y,z) : 1 \leq i \leq n-1\}$$

axiomatizes $V^*$. Hence it is clear that

$$\Sigma_0 \models \Sigma$$

as one can add new functions $t_i$ to any $\mathbf{A}$ satisfying $\Sigma_0$ to obtain $\mathbf{A}^*$ with

$$\mathbf{A}^* \models \Sigma_0^{**},$$

so

$$\mathbf{A} \models \Sigma.$$

□

**Definition 4.7.** Let $T^*$ be the set of all terms $p(x,\vec{y})$ of type $\mathscr{F}^*$ such that (i) no variable occurs twice in $p$, and (ii) the variable $x$ occurs in every nonvariable subterm of $p$ (as defined in II§14.13).

**Lemma 4.8.** *For $\mathbf{A} \in V^*$ and $a,b,a',b' \in A$ we have*

$$\Theta(a,b) \cap \Theta(a',b') \neq \Delta$$

*iff*

$$\mathbf{A} \models \exists\vec{z}\exists\vec{w}[t_i(p(a,\vec{z}), q(a',\vec{w}), p(b,\vec{z})) \not\approx t_i(p(a,\vec{z}), q(b',\vec{w}), p(b,\vec{z}))]$$

*for some $p(x,\vec{z}), q(x,\vec{z}) \in T^*$ and some $i$, $1 \leq i \leq n-1$.*

PROOF. ($\Rightarrow$) Suppose $c \neq d$ and

$$\langle c,d \rangle \in \Theta(a,b) \cap \Theta(a',b').$$

Then we claim that for some $p(x,\vec{y}) \in T^*$, for some $j$, and for some $\vec{g}$ from $A$ we have

$$t_j(c,\hat{p}(a,\vec{g}),d) \neq t_j(c,\hat{p}(b,\vec{g}),d).$$

To see this first note that the equivalence relation on $A$ generated by

$$\{\langle \hat{p}(a,\vec{g}),\ \hat{p}(b,\vec{g})\rangle : \hat{p} \in T^*, \vec{g} \text{ from } A\}$$

is $\Theta$(a,b) (one can argue this in a manner similar to the proof of 3.1). As

$$\langle c,d \rangle \in \Theta(a,b)$$

we see that for each $i$,

$$\langle t_i(c,c,d), t_i(c,d,d)\rangle$$

is in the equivalence relation generated by

$$\{\langle t_i(c,\hat{p}(a,\vec{g}),d), t_i(c,\hat{p}(b,\vec{g}),d)\rangle : \hat{p} \in T^*, \vec{g} \text{ from } A\}.$$

As $c \neq d$, for some $j$ we know

$$t_j(c,c,d) \neq t_j(c,d,d)$$

by 4.4; hence for some $\hat{p}$, some $\vec{g}$, and the same $j$,

$$t_j(c,\hat{p}(a,\vec{g}),d) \neq t_j(c,\hat{p}(b,\vec{g}),d),$$

proving the claim. By incorporating $c,d$ into the parameters we have a $p \in T^*$ and parameters $\vec{e}$ such that

$$p(a,\vec{e}) \neq p(b,\vec{e}),$$

and furthermore

$$\langle p(a,\vec{e}),p(b,\vec{e})\rangle \in \Theta(a',b')$$

as

$$\langle p(a,\vec{e}),p(b,\vec{e})\rangle \in \Theta(c,d)$$

because of 4.4. Now starting with $\langle p(a,\vec{e}),\ p(b,\vec{e})\rangle$ instead of $\langle c,d \rangle$, we can repeat the above argument to find $q \in T^*$, $t_i$, and $\vec{f}$ from $A$ such that

$$t_i(p(a,\vec{e}), q(a',\vec{f}), p(b,\vec{e})) \neq t_i(p(a,\vec{e}), q(b',\vec{f}), p(b,\vec{e})),$$

as desired.

($\Leftarrow$) If for some $i$

$$t_i(p(a,\vec{e}), q(a',\vec{f}), p(b,\vec{e})) \neq t_i(p(a,\vec{e}), q(b',\vec{f}), p(b,\vec{e}))$$

then, as the ordered pair consisting of these two distinct elements is in both $\Theta(a,b)$ and $\Theta(a',b')$ by 4.4, we have

$$\Theta(a,b) \cap \Theta(a',b') \neq \Delta. \qquad \square$$

**Definition 4.9.** Suppose the operation symbols in $\mathscr{F}^*$ have arity at most $r$, with $r$ finite. For $m < \omega$ let $T_m^*$ be the subset of $T^*$ consisting of those terms $p$ in $T^*$ with no more than $m$ occurrences of function symbols. Then define

$\delta_m(x,y,u,v)$ to be

$$\bigvee_{\substack{1 \le i \le n-1 \\ p,q \in T_m^*}} \exists \vec{z} \exists \vec{w} [t_i(p(x,\vec{z}), q(u,\vec{w}), p(y,\vec{z})) \not\approx t_i(p(x,\vec{z}), q(v,\vec{w}), p(y,\vec{z}))]$$

where the $z$'s come from $\{z_1, \ldots, z_{mr}\}$, and the $w$'s come from $\{w_1, \ldots, w_{mr}\}$.

The next lemma is just a restatement of Lemma 4.8.

**Lemma 4.10.** *For* $\mathbf{A} \in V^*$ *and* $a,b,a',b' \in A$, *we have*

$$\Theta(a,b) \cap \Theta(a',b') \neq \Delta$$

*iff*

$$\mathbf{A} \models \delta_m(a,b,a',b')$$

*for some* $m < \omega$.

**Definition 4.11.** Let $\delta_m^*$ be the sentence

$$\forall x \forall y \forall u \forall v [\delta_{m+1}(x,y,u,v) \to \delta_m(x,y,u,v)].$$

**Lemma 4.12.** (a)
$$V^* \models \delta_m^* \to \delta_{m+1}^*$$

*for* $m < \omega$, *and*

(b) *for* $\mathbf{A} \in V^*$, *if*

$$\mathbf{A} \models \delta_m^*$$

*and*

$$\mathbf{A} \models \delta_k(a,b,c,d),$$

*then*

$$\mathbf{A} \models \delta_m(a,b,c,d)$$

*for* $k,m < \omega$.

Proof. To prove (a) suppose, for $\mathbf{A} \in V^*$,

$$\mathbf{A} \models \delta_m^*$$

and, for some $a,b,c,d \in A$,

$$\mathbf{A} \models \delta_{m+2}(a,b,c,d).$$

We want to show

$$\mathbf{A} \models \delta_{m+1}(a,b,c,d).$$

Choose $p,q \in T_{m+2}^*$, $\vec{f},\vec{g} \in A$, and $i$ such that

$$t_i(p(a,\vec{f}), q(c,\vec{g}), p(b,\vec{f})) \neq t_i(p(a,\vec{f}), q(d,\vec{g}), p(b,\vec{f})).$$

Then one can find $p',q' \in T_1^*$, $p'',q'' \in T_{m+1}^*$ with

$$p(x,\vec{z}) = p''(p'(x,\vec{z}_{(1)}),\vec{z}_{(2)})$$
$$q(x,\vec{w}) = q''(q'(x,\vec{w}_{(1)}),\vec{w}_{(2)})$$

where $\vec{z}_{(1)},\vec{z}_{(2)},\vec{w}_{(1)},\vec{w}_{(2)}$ are subsequences of $\vec{z}$, respectively $\vec{w}$. Let

$$\begin{aligned} a' &= p'(a,\vec{f}_{(1)}),\\ b' &= p'(b,\vec{f}_{(1)}),\\ c' &= q'(c,\vec{g}_{(1)}),\\ d' &= q'(d,\vec{g}_{(1)}). \end{aligned}$$

Then

$$t_i(p''(a',\vec{f}_{(2)}), q''(c',\vec{g}_{(2)}), p''(b',\vec{f}_{(2)})) \neq t_i(p''(a',\vec{f}_{(2)}), q''(d',\vec{g}_{(2)}), p''(b',\vec{f}_{(2)}));$$

hence

$$\mathbf{A} \models \delta_{m+1}(a',b',c',d').$$

As $\mathbf{A} \models \delta_m^*$ it follows that

$$\mathbf{A} \models \delta_m(a',b',c',d'),$$

so there are $\hat{p},\hat{q} \in \mathrm{T}_m^*$, and $\vec{h},\vec{k} \in A$, and $j$ such that

$$t_j(\hat{p}(a',\vec{h}), \hat{q}(c',\vec{k}), \hat{p}(b',\vec{h})) \neq t_j(\hat{p}(a',\vec{h}), \hat{q}(d',\vec{k}), \hat{p}(b',\vec{h})),$$

i.e.,

$$\begin{aligned} &t_j(\hat{p}(p'(a,\vec{f}_{(1)}),\vec{h}), \hat{q}(q'(c,\vec{g}_{(1)}),\vec{k}), \hat{p}(p'(b,\vec{f}_{(1)}),\vec{h}))\\ &\neq t_j(\hat{p}(p'(a,\vec{f}_{(1)}),\vec{h}), \hat{q}(q'(d,\vec{g}_{(1)}),\vec{k}), \hat{p}(p'(b,\vec{f}_{(1)}),\vec{h})). \end{aligned}$$

Now

$$\hat{p}(p'(x,\vec{z}_{(1)}),\vec{u}) \in T_{m+1}^*$$

for suitable $\vec{u}$, and likewise

$$\hat{q}(q'(x,\vec{w}_{(1)}),\vec{v}) \in T_{m+1}^*$$

for suitable $\vec{v}$, so

$$\mathbf{A} \models \delta_{m+1}(a,b,c,d),$$

as was to be shown.

Combining (a) with the fact that

$$V^* \models \delta_k \rightarrow \delta_{k+1},$$

$k < \omega$, we can easily show (b). □

**Definition 4.13.** An algebra $\mathbf{A}$ is *finitely subdirectly irreducible* if for $a,b,a',b' \in A$ with $a \neq b$, $a' \neq b'$ we always have

$$\Theta(a,b) \cap \Theta(a',b') \neq \Delta.$$

(Any subdirectly irreducible algebra is finitely subdirectly irreducible.) If $V$ is a variety then $V_{FSI}$ denotes *the class of finitely subdirectly irreducible algebras in* $V$.

**Lemma 4.14.** *If* $V_{FSI}^*$ *is a strictly elementary class then, for some* $n_0 < \omega$,

$$V_{FSI}^* \models (x \not\approx y \,\&\, u \not\approx v) \rightarrow \delta_{n_0}(x,y,u,v)$$

*and*

$$V^* \models \delta_{n_0}^* .$$

Proof. Let $\Phi$ axiomatize $V^*_{FSI}$. Then the set of formulas

$$\{\Phi \,\&\, (a \not\approx b \,\&\, c \not\approx d) \,\&\, \neg\delta_m(a,b,c,d)\}_{m<\omega}$$

cannot be satisfied by any algebra $\mathbf{A}$ and elements $a,b,c,d \in A$ in view of 4.10. Hence by the compactness theorem there is an $n_0 < \omega$ such that

$$\{\Phi \,\&\, (x \not\approx y \,\&\, u \not\approx v) \,\&\, \neg\delta_m(x,y,u,v)\}_{m \le n_0}$$

cannot be satisfied. By taking negations we see that every algebra of type $\mathscr{F}^*$ satisfies one of

$$\{\Phi \to [(x \not\approx y \,\&\, u \not\approx v) \to \delta_m(x,y,u,v)]\}_{m \le n_0};$$

hence if $\mathbf{A} \in V^*_{FSI}$ and $a,b,c,d \in A$ we have

$$\mathbf{A} \models \bigvee_{m \le n_0} (a \not\approx b \,\&\, c \not\approx d) \to \delta_m(a,b,c,d)$$

so

$$\mathbf{A} \models (a \not\approx b \,\&\, c \not\approx b) \to \bigvee_{m \le n_0} \delta_m(a,b,c,d)$$

and as

$$V^*_{FSI} \models \delta_m \to \delta_{m+1},$$

we have

$$\mathbf{A} \models (a \not\approx b \,\&\, c \not\approx d) \to \delta_{n_0}(a,b,c,d).$$

Thus

$$V^*_{FSI} \models (x \not\approx y \,\&\, u \not\approx v) \to \delta_{n_0}(x,y,u,v).$$

Again if

$$\mathbf{A} \in V^*_{FSI}$$

and $a,b,c,d \in A$ and

$$\mathbf{A} \models \delta_{n_0+1}(a,b,c,d)$$

then

$$\Theta(a,b) \cap \Theta(c,d) \neq \Delta$$

by 4.10 so $a \neq b$ and $c \neq d$. From the first part of this lemma we have

$$\mathbf{A} \models \delta_{n_0}(a,b,c,d).$$

Thus

$$V^*_{FSI} \models \delta^*_{n_0}.$$

Now if

$$\mathbf{A} \in P_s(V^*_{FSI}),$$

say

$$\mathbf{A} \le \prod_{i \in I} \mathbf{A}_i \quad \text{(as a subdirect product)}$$

where $\mathbf{A}_i \in V^*_{FSI}$, and if $a,b,c,d \in A$ and

$$\mathbf{A} \models \delta_{n_0+1}(a,b,c,d)$$

then for some $p,q \in T^*_{n_0+1}$, for some $\vec{e},\vec{f}$ from $A$, and for some $j$, we have

$$t_j(p(a,\vec{e}), q(c,\vec{f}), p(b,\vec{e})) \neq t_j(p(a,\vec{e}), q(d,\vec{f}), p(b,\vec{e}));$$

hence for some $i \in \mathrm{I}$,

$$t_j(p(a,\vec{e}), q(c,\vec{f}), p(b,\vec{e}))(i) \neq t_j(p(a,\vec{e}), q(d,\vec{f}), p(b,\vec{e}))(i).$$

Thus

$$\mathbf{A}_i \vDash \delta_{n_0+1}(a(i),b(i),c(i),d(i)).$$

As $V^*_{FSI} \vDash \delta^*_{n_0}$ it follows that

$$\mathbf{A}_i \vDash \delta_{n_0}(a(i),b(i),c(i),d(i)).$$

We leave it to the reader to see that the above steps can be reversed to show

$$\mathbf{A} \vDash \delta_{n_0}(a,b,c,d).$$

Consequently,

$$V^* \vDash \delta^*_{n_0}. \qquad \square$$

**Definition 4.15.** If $V^*_{FSI}$ is a strictly elementary class let $\Phi_1$ axiomatize $V^*_{FSI}$. Let $\Phi_2$ be the sentence

$$\forall x \forall u \forall v \left[ \mathop{\&}_{1 \le i \le n-1} t_i(x,u,x) \approx t_i(x,v,x) \right]$$

$$\& \, \forall x \forall y \left[ x \not\approx y \to \bigvee_{1 \le i \le n-1} t_i(x,x,y) \not\approx t_i(x,y,y) \right]$$

and let $\Phi_3$ be the sentence

$$\forall x \forall y \forall u \forall v [(x \not\approx y \; \& \; u \not\approx v) \to \delta_{n_0}(x,y,u,v)],$$

where $n_0$ is as in 4.14.

**Lemma 4.16.** *If $V^*_{FSI}$ is a strictly elementary class then*

$$V^* \vDash \delta^*_{n_0} \,\&\, \Phi_2 \,\&\, (\Phi_3 \to \Phi_1),$$

*where $n_0$ and the $\Phi_i$ are as in* 4.15.

Proof. We have

$$V^* \vDash \delta^*_{n_0}$$

from 4.14 and

$$V^* \vDash \Phi_2$$

follows from 4.4. Finally, the assertions

$$\mathbf{A} \vDash \Phi_3, \qquad \mathbf{A} \in V^*$$

imply

$$\mathbf{A} \in V^*_{FSI}$$

in view of 4.10; hence

$$V^* \models \Phi_3 \to \Phi_1. \qquad \square$$

The following improvement of Baker's theorem (4.18) was pointed out by Jónsson.

**Theorem 4.17.** *Suppose $V$ is a congruence-distributive variety of finite type such that $V_{FSI}$ is a strictly elementary class. Then $V$ has a finitely based equational theory.*

PROOF. Let $p_1, \ldots, p_{n-1}$ be the terms used in 4.4, and let $V^*$ be as defined in 4.5. Let $\Phi$ axiomatize $V_{FSI}$. Then

$$\Phi \& \left[ \mathop{\&}_{1 \le i \le n-1} t_i(x,y,z) \approx p_i(x,y,z) \right]$$

axiomatizes $V^*_{FSI}$, so $V^*_{FSI}$ is also a strictly elementary class. Now let $\Phi_1, \Phi_2, \Phi_3$ and $n_0$ be as in 4.15. If $\Sigma^*$ is the set of identities true of $V^*$ over some infinite set of variables then

$$\Sigma^* \models \delta^*_{n_0} \& \Phi_2 \& (\Phi_3 \to \Phi_1)$$

by 4.16. By 2.13 it follows that there is a finite subset $\Sigma^*_0$ of $\Sigma^*$ such that

$$\Sigma_0 \models \delta^*_{n_0} \& \Phi_2 \& (\Phi_3 \to \Phi_1).$$

We want to show that $\Sigma^*_0$ axiomatizes $V^*$, so suppose $\mathbf{A}$ is finitely subdirectly irreducible and $\mathbf{A} \models \Sigma^*_0$. The only time we have made use of congruence-distributivity was to obtain terms for 4.4. All of the subsequent results have depended only on 4.4 (this is not surprising in view of Exercise 3). As $\Phi_2$ holds in the variety defined by $\Sigma^*_0$ we can use these subsequent results. Hence if $a,b,c,d \in A$ and $a \neq b$, $c \neq d$ then

$$\mathbf{A} \models \delta_m(a,b,c,d)$$

for some $m < \omega$ by 4.10. As $\mathbf{A} \models \delta^*_{n_0}$ we know

$$\mathbf{A} \models \delta_{n_0}(a,b,c,d)$$

by 4.12. Thus

$$\mathbf{A} \models \Phi_3,$$

and as

$$\mathbf{A} \models \Phi_3 \to \Phi_1,$$

it follows that

$$\mathbf{A} \models \Phi_1.$$

This means

$$\mathbf{A} \in V^*_{FSI};$$

hence every subdirectly irreducible algebra satisfying $\Sigma^*_0$ also satisfies $\Sigma^*$. In view of Birkhoff's theorem (II§8.6), $\Sigma^*_0$ is a set of axioms for $V^*$. From 4.6 it is clear that $V$ has a finitely based set of identities. $\square$

**Corollary 4.18** (Baker). *If V is a finitely generated congruence-distributive variety of finite type then V has a finitely based equational theory.*

REFERENCES

1. K. A. Baker [1977]
2. G. Birkhoff [1935]
3. S. Burris [1979]
4. R. McKenzie [1978]

EXERCISES §4

1. Given a finite algebra **A** of finite type and a finite set of variables $X$ show that there is an algorithm to find a finite basis for $\mathrm{Id}_{\mathbf{A}}(X)$.

2. Show that the identities of a variety are finitely based iff the variety is a strictly elementary class.

3. (Baker). If $V$ is a variety with ternary terms $p_1, \ldots, p_{n-1}$ which satisfy the statements in Lemma 4.4 show that $V$ is congruence-distributive.

# §5. Semantic Embeddings and Undecidability

In this section we will see that by assuming a few basic results about undecidability we will be able to prove that a large number of familiar theories are undecidable. The fundamental work on undecidability was developed by Church, Gödel, Rosser, and Turing in the 1930's. Rosser proved that the theory of the natural numbers is undecidable, and Turing constructed a Turing machine with an undecidable halting problem. These results were subsequently encoded into many problems to show that the latter were also undecidable—some of the early contributors were Church, Novikov, Post, and Tarski. Popular new techniques of encoding were developed in the 1960's by Ershov and Rabin.

We will look at two methods, the embedding of the natural numbers used by Tarski, and the embedding of finite graphs used by Ershov and Rabin.

The precise definition of decidability cannot be given here—however it suffices to think of a set of objects as being decidable if there is an "algorithm" to determine whether or not an object is in the set, and it is common to think of an algorithm as a computer program.

Let us recall the definition of the theory of a class of structures.

**Definition 5.1.** Let $K$ be a class of structures of type $\mathscr{L}$. The *theory* of $K$, written $\mathrm{Th}(K)$, is the set of all first-order sentences of type $\mathscr{L}$ (over some fixed "standard" countably infinite set of variables) which are satisfied by $K$.

**Definition 5.2.** Let **A** be a structure of type $\mathscr{L}$ and let **B** be a structure of type $\mathscr{L}^*$. Suppose we can find formulas

$$\begin{aligned}&\Delta(x)\\&\Phi_f(x_1,\dots,x_n,y)\quad \text{for } f\in\mathscr{F}_n,\ n\geq 1\\&\Phi_r(x_1,\dots,x_n)\qquad \text{for } r\in\mathscr{R}_n, n\geq 1\end{aligned}$$

of type $\mathscr{L}^*$ such that if we let

$$B_0=\{b\in B:\mathbf{B}\models\Delta(b)\}$$

then the set

$$\{\langle\langle b_1,\dots,b_n\rangle,b\rangle\in B_0^n\times B_0:\mathbf{B}\models\Phi_f(b_1,\dots,b_n,b)\}$$

defines an $n$-ary function $f$ on $B_0$, for $f\in\mathscr{F}_n$, $n\geq 1$, and the set

$$\{\langle b_1,\dots,b_n\rangle\in B_0^n:\mathbf{B}\models\Phi_r(b_1,\dots,b_n)\}$$

defines an $n$-ary relation $r$ on $B_0$ for $r\in\mathscr{R}_n$, $n\geq 1$, such that by suitably interpreting the constant symbols of $\mathscr{L}$ in $B_0$ we have a structure $\mathbf{B}_0$ of type $\mathscr{L}$ isomorphic to **A**. Then we say **A** can be *semantically embedded* in **B**, written $\mathbf{A}\xrightarrow[\text{sem}]{}\mathbf{B}$. The notation $\mathbf{A}\xrightarrow[\text{sem}]{}K$ means **A** can be semantically embedded in *some* member of $K$, and the notation $H\xrightarrow[\text{sem}]{}K$ means every member of $H$ can be semantically embedded in at least one member of $K$, using the *same* formulas $\Delta$, $\Phi_f$, $\Phi_r$.

**Lemma 5.3.** *If* $G\xrightarrow[\text{sem}]{}H$ *and* $H\xrightarrow[\text{sem}]{}K$ *then* $G\xrightarrow[\text{sem}]{}K$, *i.e., the notion of semantic embedding is transitive.*

PROOF. (Exercise.) □

**Definition 5.4.** If $K$ is a class of structures of type $\mathscr{L}$ and $c_1,\dots,c_n$ are symbols not appearing in $\mathscr{L}$, then $K(c_1,\dots,c_n)$ denotes the class of *all* structures of type $\mathscr{L}\cup\{c_1,\dots,c_n\}$, where each $c_i$ is a constant symbol, obtained by taking the members **B** of $K$ and arbitrarily designating elements $c_1,\dots,c_n$ in $B$.

**Definition 5.5.** Let $N$ be the set of natural numbers, and let **N** be $\langle N,+,\cdot,1\rangle$.

We will state the following result without proof, and use it to prove that the theory of rings and the theory of groups are undecidable. (See [33].)

**Theorem 5.6** (Tarski). *Given* $K$, *if for some* $n<\omega$ *we have* $\mathbf{N}\xrightarrow[\text{sem}]{}K(c_1,\dots,c_n)$, *then* $\mathrm{Th}(K)$ *is undecidable.*

**Lemma 5.7** (Tarski). $\mathbf{N}\xrightarrow[\text{sem}]{}\mathbf{Z}=\langle Z,+,\cdot,1\rangle$, *$Z$ being the set of integers.*

PROOF. Let $\Delta(x)$ be

$$\exists y_1 \cdots \exists y_4[x \approx y_1 \cdot y_1 + \cdots + y_4 \cdot y_4 + 1].$$

By a well-known theorem of Lagrange,

$$\mathbf{Z} \models \Delta(n) \quad \text{iff } n \in N.$$

Let $\Phi_+(x_1,x_2,y)$ be

$$x_1 + x_2 \approx y,$$

and let $\Phi_\cdot(x_1,x_2,y)$ be

$$x_1 \cdot x_2 \approx y.$$

Then it is easy to see $\mathbf{N} \xrightarrow[\text{sem}]{} \mathbf{Z}$. □

**Theorem 5.8** (Tarski). *The theory of rings is undecidable.*

PROOF. $\mathbf{Z}$ is a ring, so 5.6 applies. □

*Remark.*

In the above theory of rings we can assume the language being used is any of the usual languages such as $\{+,\cdot\}$, $\{+,\cdot,1\}$, $\{+,\cdot,-,0,1\}$ in view of 5.6.

**Lemma 5.9** (Tarski).

$$\langle Z,+,\cdot,1\rangle \xrightarrow[\text{sem}]{} \langle Z,+,{}^2,1\rangle,$$

*where* $^2$ *denotes the function mapping a to* $a^2$.

PROOF. Let $\Delta(x)$ be

$$x \approx x,$$

let $\Phi_+(x_1,x_2,y)$ be

$$x_1 + x_2 \approx y,$$

and let $\Phi_\cdot(x_1,x_2,y)$ be

$$y + y + x_1^2 + x_2^2 \approx (x_1 + x_2)^2.$$

To see that the latter formula actually defines $\cdot$ in $\mathbf{Z}$, note that in $\mathbf{Z}$

$$a \cdot b = c \Leftrightarrow 2c + a^2 + b^2 = (a + b)^2.$$ □

**Lemma 5.10** (Tarski).

$$\langle Z,+,{}^2,1\rangle \xrightarrow[\text{sem}]{} \langle Z,+,|,1\rangle,$$

*where* $a|b$ *means a divides b.*

PROOF. Let $\Delta(x)$ be

$$x \approx x,$$

let $\Phi_+(x_1,x_2,y)$ be

$$x_1 + x_2 \approx y$$

and let $\Phi_2(x_1,y)$ be

$$\begin{aligned}\forall z[(x_1 + y)|z \leftrightarrow ((x_1|z) \,\&\, (x_1 + 1|z))]\\ \&\, \forall u \forall v \forall z[((u + x_1 \approx y) \,\&\, (v + 1 \approx x_1))\\ \to (u|z \leftrightarrow (x_1|z \,\&\, v|z))].\end{aligned}$$

Then $\Phi_2(a,b)$ holds for $a,b \in Z$ iff

$$\begin{aligned}a + b &= \pm a(a + 1)\\ b - a &= \pm a(a - 1)\end{aligned}$$

and thus iff

$$b = a^2.$$

□

**Lemma 5.11** (Tarski). *Let* Sym(*Z*) *be the set of bijections from Z to Z, let* $\circ$ *denote composition of bijections, and let* $\pi$ *be the bijection defined by* $\pi(a) = a + 1$, $a \in Z$. *Then*

$$\langle Z, +, |, 1\rangle \xrightarrow[\text{sem}]{} \langle \mathrm{Sym}(Z), \circ, \pi\rangle.$$

Proof. Let $\Delta(x)$ be

$$x \circ \pi \approx \pi \circ x,$$

let $\Phi_+(x_1,x_2,y)$ be

$$x_1 \circ x_2 \approx y,$$

and let $\Phi_|(x_1,x_2)$ be

$$\forall z(x_1 \circ z \approx z \circ x_1 \to x_2 \circ z \approx z \circ x_2).$$

For $\sigma \in \mathrm{Sym}(Z)$ note that

$$\sigma \circ \pi = \pi \circ \sigma$$

iff for $a \in Z$,

$$\sigma(a + 1) = \sigma(a) + 1;$$

hence if

$$\sigma \circ \pi = \pi \circ \sigma$$

then

$$\sigma(a) = \sigma(0) + a,$$

i.e.,

$$\sigma = \pi^{\sigma(0)}.$$

Thus

$$\langle \mathrm{Sym}(Z), \circ, \pi\rangle \models \Delta(\sigma)$$

iff

$$\sigma \in \{\pi^n : n \in Z\}.$$

Clearly $\Phi_+$ defines a function on this set, and indeed

$$\langle \mathrm{Sym}(Z), \circ, \pi\rangle \models \Phi_+(\pi^a, \pi^b, \pi^c)$$

iff

$$a + b = c.$$

Next we wish to show

$$\langle \mathrm{Sym}(Z), \circ, \pi \rangle \models \Phi_{|}(\pi^a, \pi^b) \quad \text{iff} \quad a \mid b,$$

in which case the mapping

$$a \mapsto \pi^a$$

for $a \in Z$ gives the desired isomorphism to show

$$\langle Z, +, |, 1 \rangle \xrightarrow[\text{sem}]{} \langle \mathrm{Sym}(Z), \circ, \pi \rangle.$$

So suppose $a \mid b$ in $Z$. If $\sigma \in \mathrm{Sym}(Z)$ and

$$\sigma \circ \pi^a = \pi^a \circ \sigma$$

we have

$$\sigma(c + a) = \sigma(c) + a$$

for $c \in Z$, hence

$$\sigma(c + d \cdot a) = \sigma(c) + d \cdot a$$

for $c,d \in Z$, so in particular

$$\sigma(c + b) = \sigma(c) + b;$$

hence

$$\sigma \circ \pi^b = \pi^b \circ \sigma.$$

Thus

$$a \mid b \Rightarrow \langle \mathrm{Sym}(Z), \circ, \pi \rangle \models \Phi_{|}(\pi^a, \pi^b).$$

Conversely suppose

$$\langle \mathrm{Sym}(Z), \circ, \pi \rangle \models \Phi_{|}(\pi^a, \pi^b)$$

for some $a,b \in Z$. If $b = 0$ then $a \mid b$, so suppose $b \neq 0$. Let

$$\rho(c) = \begin{cases} c + a & \text{if } a \mid c \\ c & \text{if } a \nmid c \end{cases}$$

for $c \in Z$. Clearly $\rho \in \mathrm{Sym}(Z)$. An easy calculation shows

$$\rho \circ \pi^a = \pi^a \circ \rho;$$

hence looking at $\Phi_{|}$ we must have

$$\rho \circ \pi^b = \pi^b \circ \rho.$$

Now

$$\pi^b \circ \rho(c) = \begin{cases} c + a + b & \text{if } a \mid c \\ c + b & \text{if } a \nmid c \end{cases}$$

and

$$\rho \circ \pi^b(c) = \begin{cases} c + b + a & \text{if } a \mid c + b \\ c + b & \text{if } a \nmid c + b. \end{cases}$$

Thus

$$a \mid c \quad \text{iff} \quad a \mid c + b,$$

for $c \in Z$; hence $a \mid b$. □

**Corollary 5.12** (Tarski). *The theory of groups is undecidable.*

PROOF. From 5.3, 5.7, 5.9, 5.10, and 5.11 we have $\mathbf{N} \xrightarrow[\text{sem}]{} \langle \mathrm{Sym}(Z),\circ,\pi\rangle$. If $K$ is the class of groups (in the language $\{\cdot\}$) then $\langle \mathrm{Sym}(Z),\circ,\pi\rangle \in K(c_1)$; hence by 5.6 it follows that $\mathrm{Th}(K)$ is undecidable. □

A major result of J. Robinson was to show $\langle N,+,\cdot,1\rangle \xrightarrow[\text{sem}]{} \langle Q,+,\cdot,1\rangle$; hence the theory of fields is undecidable.

Now we turn to our second technique for proving undecidability. Recall that a *graph* is a structure $\langle G,r\rangle$ where $r$ is an irreflexive and symmetric binary relation.

**Definition 5.13.** $G_{\mathrm{fin}}$ will denote the class of *finite graphs.*

The following result we state without proof. (See [13]; Rabin [1965].)

**Theorem 5.14** (Ershov, Rabin). *If we are given K, and for some* $n < \omega$ we have

$$G_{\mathrm{fin}} \xrightarrow[\text{sem}]{} K(c_1, \ldots, c_n),$$

*then* $\mathrm{Th}(K)$ *is undecidable.*

**Corollary 5.15** (Grzegorczyk). *The theory of distributive lattices is undecidable.*

PROOF. If $\mathbf{P} = \langle P,\leq\rangle$ is a poset recall that a lower segment of $\mathbf{P}$ means a subset $S$ of $P$ such that $a\in P$, $b\in S$ and $a \leq b$ imply $a\in S$. In I§3 Ex. 4 it was stated that a finite distributive lattice is isomorphic to the lattice of nonempty lower segments (under $\subseteq$) of the poset of join irreducible elements of the lattice; and if we are given any poset with 0 then the nonempty lower segments form a distributive lattice, with the poset corresponding to the join irreducibles.

Thus given a finite graph $\langle G,r\rangle$, let us define a poset $\mathbf{P} = \langle P,\leq\rangle$ by

$$P = G \cup \{\{a,b\} \subseteq G\colon arb \text{ holds}\} \cup \{0\},$$

and require $p \leq q$ to hold iff $p = q$, $p = 0$, or $p\in G$ and $q$ is of the form $\{p,b\}$. Then in the lattice $\mathbf{L}$ of lower segments of $\mathbf{P}$ the minimal join irreducible elements are precisely the lower segments of the form $\{a,0\}$ for $a\in G$; and $arb$ holds in $G$ iff there is a join irreducible element above $\{a,0\}$ and $\{b,0\}$ in $L$. (See Figure 34 for the poset corresponding to the graph in Figure 30.) Hence if we let $Irr(x)$ be

$$\forall y \forall z(y \vee z \approx x \rightarrow (y \approx x \vee z \approx x))$$

and then let $\Delta(x)$ be

$$Irr(x) \,\&\, \forall y[(y \leq x \,\&\, Irr(y)) \rightarrow (y \approx 0 \vee y \approx x)] \,\&\, (x \not\approx 0)$$

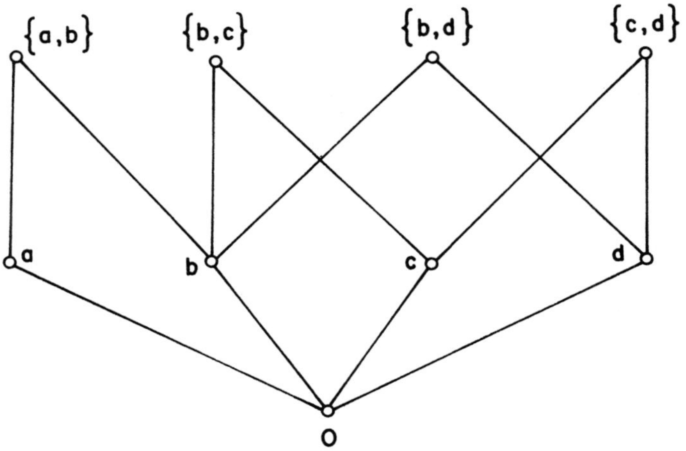


Figure 34

and let $\Phi_r(x_1,x_2)$ be

$$(x_1 \not\approx x_2) \,\&\, \exists y[Irr(y) \,\&\, x_1 \leq y \,\&\, x_2 \leq y],$$

where in the above formulas $u \leq v$ is to be replaced by $u \wedge v = u$, then we see that $\langle G,r\rangle$ is semantically embedded in $\langle L, \vee, \wedge, 0\rangle$. □

**Corollary 5.16** (Rogers). *The theory of two equivalence relations is undecidable, i.e., if $K$ is the class of structures $\langle A,r_1,r_2\rangle$ where $r_1$ and $r_2$ are both equivalence relations on $A$, then* Th$(K)$ *is undecidable.*

PROOF. Given a finite graph $\langle G,r\rangle$ let $\leq$ be a linear order on $G$. Then let $S$ be the set

$$G \cup \{\langle a,b\rangle : arb\}.$$

Let the equivalence class $a/r_1$ be

$$\{a\} \;\cup\; \{\langle a,b\rangle : arb, a < b\} \;\cup\; \{\langle b,a\rangle : arb, a < b\}$$

and let the equivalence class $b/r_2$ be

$$\{b\} \cup \{\langle a,b\rangle : arb, a < b\} \cup \{\langle b,a\rangle : arb, a < b\}.$$

(See Figure 35 for the structure $\langle S,r_1,r_2\rangle$ corresponding to the graph in Figure 30 with $a < b < c < d$. The rows give the equivalence classes of $r_1$, the columns the equivalence classes of $r_2$.) Then $\langle S,r_1,r_2\rangle$ is a set with two equivalence relations. Let

$$r_0 = r_1 \cap r_2.$$

Then the elements of $G$ are precisely those $s \in S$ such that

$$s/r_0 = \{s\},$$

| | | | |
|---|---|---|---|
| $\circ\, a$ | $\circ\langle a,b\rangle$<br>$\circ\langle b,a\rangle$ | | |
| | $\circ\, b$ | $\circ\langle b,c\rangle$<br>$\circ\langle c,b\rangle$ | $\circ\langle b,d\rangle$<br>$\circ\langle d,b\rangle$ |
| | | $\circ\, c$ | $\circ\langle c,d\rangle$<br>$\circ\langle d,c\rangle$ |
| | | | $\circ\, d$ |

Figure 35

and for $a,b \in G$, $arb$ holds iff

$$|\{c \in S : ar_1c \text{ and } cr_2b\} \cup \{c \in S : br_1c \text{ and } cr_2a\}| = 2.$$

Thus the formulas

$$\Delta(x) = \forall y[(xr_1y \,\&\, xr_2y) \rightarrow x \approx y]$$

$$\Phi_r(x_1,x_2) = \exists y_1 \exists y_2 \left\{ y_1 \not\approx y_2 \,\&\, \left[ \left( \underset{i=1,2}{\&} x_1r_1y_i \,\&\, y_ir_2x_2 \right) \vee \left( \underset{i=1,2}{\&} x_2r_1y_i \,\&\, y_ir_2x_1 \right) \right] \right\}$$

suffice to show

$$\langle G,r\rangle \xrightarrow[\text{sem}]{} \langle S,r_1,r_2\rangle. \qquad \square$$

A more general notion of a semantic embedding of a structure **A** into a structure **B** is required for some of the more subtle undecidability results, namely the interpretation of the elements of **A** as equivalence classes of $n$-tuples of elements of **B**. Of course this must all be done in a first-order fashion. For notational convenience we will define this only for the case of **A** a binary structure, but it should be obvious how to formulate it for other structures.

**Definition 5.17.** Let $\mathbf{A} = \langle A,r\rangle$ be a binary structure, and **B** a structure of type $\mathscr{L}$. **A** can be semantically embedded in **B**, written $\mathbf{A} \xrightarrow{\text{sem}} \mathbf{B}$, if there are $\mathscr{L}$-formulas, for some $n < \omega$,

$$\Delta(x_1, \ldots, x_n)$$
$$\Phi_r(x_1, \ldots, x_n; y_1, \ldots, y_n)$$
$$\mathrm{Eq}(x_1, \ldots, x_n; y_1, \ldots, y_n)$$

such that if we let

$$D = \{\langle b_1, \ldots, b_n\rangle \in B^n : \mathbf{B} \models \Delta(b_1, \ldots, b_n)\}$$

and if $r^D$ is the binary relation

$$r^D = \{\langle \vec{b},\vec{c}\rangle \in D \times D : \mathbf{B} \models \Phi_r(\vec{b},\vec{c})\}$$

and $\equiv$ is the binary relation

$$\equiv \;=\; \{\langle \vec{b},\vec{c}\rangle \in D \times D : \mathbf{B} \models \mathrm{Eq}(\vec{b},\vec{c})\}$$

then $\equiv$ is an equivalence relation on $D$ and we have

$$\langle A,r\rangle \cong \langle D,r^D\rangle/\equiv$$

where

$$r^D/\equiv \;=\; \{\langle \vec{b}/\equiv,\vec{c}/\equiv\rangle \in D/\equiv \times D/\equiv : r^D \cap (\vec{b}/\equiv \times \vec{c}/\equiv) \neq \varnothing\}.$$

A class $H$ of binary structures can be semantically embedded into a class $K$ of structures of type $\mathscr{L}$, written $H \xrightarrow{\text{sem}} K$, if there are formulas $\Delta,\Phi_r,\mathrm{Eq}$ as above such that for each structure $\mathbf{A}$ in the class $H$ there is a member $\mathbf{B}$ of $K$ such that $\Delta,\Phi_r,\mathrm{Eq}$ provide a semantic embedding of $\mathbf{A}$ into $\mathbf{B}$.

Using our more general notion of semantic embedding we still have the general results from before, two of which we repeat here for convenience.

**Theorem 5.18.** (a) *The semantic embeddability relation* $\xrightarrow{\text{sem}}$ *is transitive.*

(b) (*Ershov, Rabin*). *If finite graphs can be semantically embedded into a class* $K(c_1,\ldots,c_n)$ *then the first-order theory of* $K$ *is undecidable.*

For the last part of this section we will look at results on Boolean pairs.

**Definition 5.19.** A *Boolean pair* is a structure $\langle B,B_0,\leq\rangle$ where $\langle B,\leq\rangle$ is a Boolean algebra (i.e., this is a complemented distributive lattice) and $B_0$ is a unary relation which gives a subalgebra $\langle B_0,\leq\rangle$. The class of all Boolean pairs is called $BP$.

The Boolean pairs $\langle B,B_0,\leq\rangle$ such that $\langle B,\leq\rangle$ is *atomic* (i.e., every element is a sup of atoms) and $B_0$ contains all the atoms of $\langle B,\leq\rangle$ form the class $BP^1$.

The Boolean pairs $\langle B,B_0,\leq\rangle$ such that for every element $b \in B$ there is a least element $\bar{b} \in B_0$ with $b \leq \bar{b}$ constitute the class $BP^M$.

The Boolean pairs $\langle B,B_0,\leq\rangle$ in $BP^M$ such that $\langle B,\leq\rangle$, $\langle B_0,\leq\rangle$ are atomic form the class $BP^2$.

**Definition 5.20.** Let $G^*_{\text{fin}}$ be the class of finite graphs $\langle G,r\rangle$ such that $r \neq \varnothing$.

**Lemma 5.21.** $G_{\text{fin}} \xrightarrow{\text{sem}} G^*_{\text{fin}}(c)$.

Proof. (Exercise.) □

Adapting a technique of Rubin, McKenzie proved the following.

**Theorem 5.22** (McKenzie). *The theory of $BP^2$ is undecidable.*

PROOF. Given a member $\mathbf{G} = \langle G,r \rangle$ of $G^*_{\text{fin}}$ let $X = G \times \omega$. Two sets $Y$ and $Z$ are said to be "almost equal", written $Y \stackrel{a}{=} Z$, if $Y$ and $Z$ differ by only finitely many points. For $g \in G$, let $C_g = \{\langle g,j \rangle : j \in \omega\} \subseteq X$, a "cylinder" of $X$. Let $B$ be all subsets of $X$ which are almost equal to a union of cylinders, i.e., all $Y$ such that for some $S \subseteq G$, $Y \stackrel{a}{=} \bigcup_{g \in S} C_g$. Note that $\langle B, \subseteq \rangle$ is a Boolean algebra containing all the finite subsets of $X$.

To define $B_0$ first let

$$E = \{\{a,b\} : \langle a,b \rangle \in r\},$$

the set of unordered edges of $\mathbf{G}$, and then for each $g \in G$ choose a surjective map

$$\alpha_g : C_g \to E \times \omega$$

such that

$$|\alpha_g^{-1}(\langle e,j \rangle)| = \begin{cases} 2 & \text{if } g \in e \\ 3 & \text{if } g \notin e. \end{cases}$$

Then, for $\langle e,j \rangle \in E \times \omega$, let

$$D_{e,j} = \bigcup_{g \in G} \alpha_g^{-1}(\langle e,j \rangle).$$

This partitions $X$ into finite sets $D_{e,j}$ such that for $g \in G$,

$$|D_{e,j} \cap C_g| = \begin{cases} 2 & \text{if } g \in e \\ 3 & \text{if } g \notin e. \end{cases}$$

Let $B_0$ be the set of finite and cofinite unions of $D_{e,j}$'s. Note that $\langle B_0, \subseteq \rangle$ is a subalgebra of $\langle B, \subseteq \rangle$ as a Boolean algebra.

Now we want to show $\langle G,r \rangle \xrightarrow{\text{sem}} \langle B, B_0, \subseteq \rangle$:

$\Delta(x)$ is "for all atoms $y$ of $B_0$ there are exactly two or three atoms of $B$ below $x \wedge y$"

$\text{Eq}(x,y)$ is $\forall z \exists u[u$ is an atom of $B_0$ and there are exactly two or three atoms of $B$ below $x \wedge y \wedge u$ and $(x \wedge y \wedge z \wedge u \approx 0$ or $x \wedge y \wedge z' \wedge u \approx 0)]$

$\Phi_r(x,y)$ is $x \not\approx y \;\&\; \forall u \forall v[\text{Eq}(u,x) \;\&\; \text{Eq}(v,y) \to$ (for some atom $z$ of $B_0$ there are exactly two atoms of $B$ below each of $u \wedge z$ and $v \wedge z$)].

To see that $\mathbf{G} \xrightarrow{\text{sem}} \langle B, B_0, \leq \rangle$ it suffices to check the following claims:

(a) $\langle B, B_0, \leq \rangle \models \Delta(Z)$ implies $Z \stackrel{a}{=} C_g$ for some $g \in G$ (just recall the description of the elements of $B$),
(b) $\langle B, B_0, \leq \rangle \models \Delta(C_g)$ for $g \in G$,
(c) for $X, Y$ such that $\Delta(X), \Delta(Y)$ hold we have $\text{Eq}(X, Y)$ iff $X \stackrel{a}{=} Y$,
(d) for $X, Y$ such that $\Delta(X), \Delta(Y)$ hold we have $\Phi_r(X, Y)$ iff $X \stackrel{a}{=} C_g$, $Y \stackrel{a}{=} C_{g'}$ for some $g, g' \in G$ with $\langle g, g' \rangle \in r$,

(e) the mapping $g \mapsto C_g/\equiv$ establishes $\mathbf{G} \cong \langle D, r^D \rangle/\equiv$.

Thus we have proved

$$G^*_{\text{fin}} \xrightarrow{\text{sem}} BP^2;$$

hence

$$G^*_{\text{fin}}(c) \xrightarrow{\text{sem}} BP^2(c);$$

thus by Lemma 5.21

$$G_{\text{fin}} \xrightarrow{\text{sem}} BP^2(c). \qquad \square$$

**Theorem 5.23** (Rubin). *The theory of $CA_1$, the variety of monadic algebras, is undecidable.*

Proof. It suffices to show $BP^2 \xrightarrow{\text{sem}} CA_1$ as we have $G_{\text{fin}} \xrightarrow{\text{sem}} BP^2(c_1)$. Given $\langle B, B_0, \leq \rangle \in BP^2$, let $c$ be the unary function defined on the Boolean algebra $\langle B, \leq \rangle$ by

$$c(b) = \text{the least member of } B_0 \text{ above } b.$$

Then $\langle B, \vee, \wedge, ', c, 0, 1 \rangle$ is a monadic algebra, and with

$$\begin{array}{ll} \Delta(x) & \text{defined as } x \approx x \\ \Phi_{B_0}(x) & \text{defined as } x \approx c(x) \end{array}$$

we have, using the old definition of semantic embedding, $\langle B, B_0, \leq \rangle \xrightarrow[\text{sem}]{} \langle B, \vee, \wedge, ', c, 0, 1 \rangle$. $\square$

Actually the class $BP^M$ defined above is just an alternate description of monadic algebras, and $BP^2 \subseteq BP^M$.

Finally we turn to the class $BP^1$, a class which has played a remarkable role in the classification of decidable locally finite congruence modular varieties.

**Theorem 5.24** (McKenzie). *The theory of $BP^1$ is undecidable.*

Proof. Given a finite graph $\langle G, r \rangle$ with $r \neq \varnothing$ first construct $\langle B, B_0, \subseteq \rangle$ as in 5.22, so $\mathbf{B}$ is a field of subsets of $X = G \times \omega$. Let $\pi_1$ be the first projection map from $X \times \omega$ to $X$, and define

$$\begin{aligned} B^* &= \{\pi_1^{-1}(Y) : Y \in B\} \\ B_0^* &= \{\pi_1^{-1}(Y) : Y \in B_0\}. \end{aligned}$$

Then

$$\langle B^*, B_0^*, \subseteq \rangle \cong \langle B, B_0, \subseteq \rangle,$$

and each nonzero member of $B^*$ contains infinitely many points from $X \times \omega$. Now let

$$\begin{aligned} B^{**} &= \{Y \subseteq X \times \omega : Y \stackrel{a}{=} Z \text{ for some } Z \in B^*\} \\ B_0^{**} &= \{Y \subseteq X \times \omega : Y \stackrel{a}{=} Z \text{ for some } Z \in B_0^*\}. \end{aligned}$$

Then $\langle B^{**}, B_0^{**}, \subseteq \rangle \in BP^1$ as all finite subsets of $X \times \omega$ belong to both $B^{**}$ and $B_0^{**}$, and furthermore

$$\langle B^{**}, B_0^{**}, \subseteq \rangle / \overset{a}{=} \cong \langle B^*, B_0^*, \subseteq \rangle.$$

Now "$Y$ is finite" can be expressed for $Y \in B^{**}$ by

$$\forall x[x \leq Y \rightarrow x \in B_0^{**}]$$

as every nonzero element $b_0$ of $B_0$ has an element $b \in B - B_0$ below it. Thus

$$\langle B^*, B_0^*, \subseteq \rangle \xrightarrow{\text{sem}} \langle B^{**}, B_0^{**}, \subseteq \rangle;$$

hence

$$\langle B, B_0, \subseteq \rangle \xrightarrow{\text{sem}} \langle B^{**}, B_0^{**}, \subseteq \rangle.$$

This shows

$$BP^2 \xrightarrow{\text{sem}} BP^1;$$

hence

$$G_{\text{fin}} \xrightarrow{\text{sem}} BP^1(c_1).$$ □

REFERENCES

1. S. Burris and R. McKenzie [1981]
2. Ju. L. Ershov, I. A. Lavrov, A. D. Taĭmanov, and M. A. Taĭclin [1965]
3. H. P. Sankappanavar [31]
4. A. Tarski, A. Mostowski, and R. M. Robinson [33]

# Recent Developments and Open Problems

At several points in the text we have come very close to some of the most exciting areas of current research. Now that the reader has had a substantial introduction to universal algebra, we will survey the current situation in these areas and list a few of the problems being considered. (This is not a comprehensive survey of recent developments in universal algebra—the reader will have a good idea of the breadth of the subject if he reads Taylor's survey article [35], Jónsson's report [20], and the appendices to Grätzer's book [16].)

## §1. The Commutator and the Center

One of the most promising developments has been the creation of the commutator by Smith [1976]. He showed that, for any algebra **A** in a congruence-permutable variety, there is a unique function $[-,-]$, called the *commutator*, from (Con **A**) × (Con **A**) to Con **A** with certain properties. In the case of groups this is just the familiar commutator (when one considers the corresponding normal subgroups). Rather abruptly, several concepts one had previously considered to belong exclusively to the study of groups have become available on a grand scale: viz., solvability, nilpotence, and the center. Hagemann and Herrmann [1979] subsequently extended the commutator to any algebra in a congruence-modular variety. Freese and McKenzie [b] have given another definition of the commutator, and of course we used their (first-order) definition of the center (of an arbitrary algebra) in II§13. These new concepts have already played key roles in Burris and McKenzie [1981] and Freese and McKenzie [a], [b].

Problem 1. For which varieties can we define a commutator?

Problem 2. Find a description of all **A** (parallel to II§13.4) such that $Z(\mathbf{A}) = \nabla_A$.

# §2. The Classification of Varieties

Birkhoff's suggestion in the 1930's that congruence lattices should be considered as fundamental associated structures has proved to be remarkably farsighted. An important early result was the connection between modular congruence lattices and the unique factorization property due to Ore [1936]. A major turning point in showing the usefulness of classifying a variety by the behavior of the congruence lattices was Jónsson's theorem [1967] that if $V(K)$ is congruence-distributive then $V(K) = IP_S HSP_U(K)$.

The role of a single congruence, the center, is rapidly gaining attention. Let us call a variety *modular abelian* if it is congruence-modular and, for any algebra $\mathbf{A}$ in the variety, $Z(\mathbf{A}) = \nabla_A$. Such varieties are essentially varieties of unitary left $\mathbf{R}$-modules. A variety $V$ is said to be (discriminator) $\otimes$ (modular abelian) if it is congruence modular and there are two subvarieties $V_1, V_2$ such that $V_1$ is a discriminator variety, $V_2$ is a modular abelian variety, and $V = V_1 \vee V_2$. For such a variety $V$ (see Burris and McKenzie [1981]) each algebra in $V$ is, up to isomorphism, uniquely decomposable as a product of an algebra from $V_1$ and an algebra from $V_2$. The importance of this class of varieties is discussed in §3 and §5 below. The following Hasse diagram (Figure 36) shows some of the most useful classes of varieties in research.

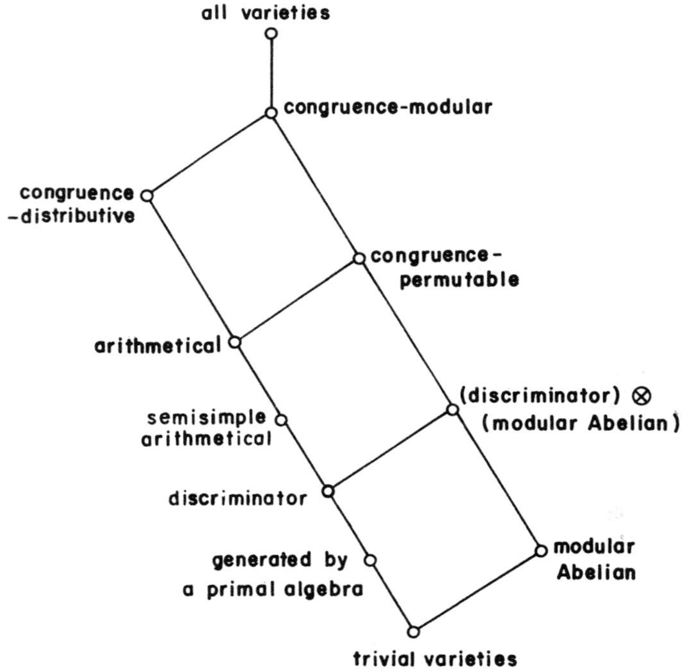


Figure 36

# §3. Decidability Questions

Decidability problems have been a popular area of investigation in universal algebra, thanks to the fascinating work of Mal'cev [24] and Tarski [33]. Let us look at several different types of decidability questions being studied.

(a) First-order Theories. In V§5 we discussed the semantic embedding technique for proving that theories are undecidable. There has been a longstanding conviction among researchers in this area that positive decidability and nice structure theory go hand in hand. The combined efforts of Szmielew [1954], Ershov [1972] and Zamjatin [1978a] show that a variety of groups is decidable iff it is abelian. This has recently been strengthened by McKenzie [d] as follows: any class of groups containing $P_S(\mathbf{G})$, where $\mathbf{G}$ can be any nonabelian group, has an undecidable theory. In Burris and Werner [1979] techniques of Comer [1975] for cylindric algebras have been extended to prove that every finitely generated discriminator variety of finite type has a decidable theory. Zamjatin [1976] showed that a variety of rings has a decidable theory iff it is generated by a zero-ring and finitely many finite fields. Recently Burris and McKenzie [1981] have applied the center and commutator to prove the following: if a locally finite congruence modular variety has a decidable theory then it must be of the form (discriminator) $\otimes$ (modular abelian). Indeed there is an algorithm such that, given a finite set $K$ of finite algebras of finite type, one can decide if $V(K)$ is of this form, and if so, one can construct a finite ring $\mathbf{R}$ with 1 such that $V(K)$ has a decidable theory iff the variety of unitary left $\mathbf{R}$-modules has a decidable theory. This leads to an obvious question.

Problem 3. Which locally finite varieties of finite type have a decidable theory?

Zamjatin [1976] has examined the following question for varieties of rings.

Problem 4. For which varieties of finite type is the theory of the finite algebras in the variety decidable?

Actually we know very little about this question, so let us pose two rather special problems.

Problem 5. Do the finite algebras in any finitely generated arithmetical variety of finite type have a decidable theory?

Problem 6. Do the finite algebras in any finitely generated congruence-distributive, but not congruence-permutable, variety of finite type have an undecidable theory?

(b) Equational Theories. Tarski [1953] proved that there is no algorithm for deciding if an equation holds in all relation algebras (hence the first-order theory is certainly undecidable). Mal'cev [24] showed the same for unary algebras. Murskiĭ [1968] gave an example of a finitely based variety of semigroups with an undecidable equational theory. R. Freese [1979] has proved that there is no algorithm to decide which equations in at most 5 variables hold in the variety of modular lattices. From Dedekind's description (see

[3]) of the 28 element modular lattice freely generated by 3 elements it is clear that one can decide which equations in at most 3 variables hold in the variety of modular lattices.

Problem 7. Is there an algorithm to decide which equations in at most 4 variables hold in a modular lattice?

(c) Word Problems. Given a variety $V$ of type $\mathscr{F}$, a *presentation* (of an algebra **A**) *in* $V$ is an ordered pair $\langle G,R \rangle$ of *generators* $G$ and *defining relations* $R$ such that the following hold.

(i) $R$ is a set of equations $p(g_1, \ldots, g_n) \approx q(g_1, \ldots, g_n)$ of type $\mathscr{F} \cup G$ (we assume $\mathscr{F} \cap G = \emptyset$) with $g_1, \ldots, g_n \in G$.
(ii) If $\hat{V}$ is the variety of type $\mathscr{F} \cup G$ defined by $\Sigma \cup R$, where $\Sigma$ is a set of equations defining $V$, then **A** is the reduct (see II§1 Exercise 1) of $\mathbf{F}_{\hat{V}}(\emptyset)$ to the type of $V$.

When the above holds we write $\mathscr{P}_V(G,R)$ for **A**, and say "$\mathscr{P}_V(G,R)$ is the algebra in $V$ freely generated by $G$ subject to the relations $R$." If $R = \emptyset$ we just obtain $\mathbf{F}_V(G)$. A presentation $\langle G, R \rangle$ is *finite* if both $G$ and $R$ are finite, and in such case $\mathscr{P}_V(G,R)$ is said to be *finitely presented*.

The *word problem* for a given presentation $\langle G,R \rangle$ in $V$ asks if there is an algorithm to determine, for any pair of "words", i.e., terms $r(g_1, \ldots, g_n)$, $s(g_1, \ldots, g_n)$, whether or not

$$\mathbf{F}_V(\emptyset) \models r(g_1, \ldots, g_n) \approx s(g_1, \ldots, g_n).$$

If so, the word problem for $\langle G,R \rangle$ is *decidable* (or *solvable*); otherwise it is *undecidable* (or *unsolvable*). The question encountered in (b) above of "which equations in the set of variables $X$ hold in a variety $V$" is often called the *word problem* for the free algebra $\mathbf{F}_V(X)$. The *word problem* for a given variety $V$ asks if every finite presentation $\langle G,R \rangle$ in $V$ has a decidable word problem. If so, the word problem for $V$ is decidable; otherwise it is undecidable.

Markov [1947] and Post [1947] proved that the word problem for semigroups is undecidable. (A fascinating introduction to decidability and word problems is given in Trakhtenbrot [36].) Perhaps the most celebrated result is the undecidability of the word problem for groups (Novikov [1955]). A beautiful algebraic characterization of finitely presented groups $\mathscr{P}_V(G,R)$ with solvable word problems is due to Boone and Higman [1974], namely $\mathscr{P}_V(G,R)$ has a solvable word problem iff it can be embedded in a simple group **S** which in turn can be embedded in a finitely presented group **T**. This idea has been generalized by Evans [1978] to the variety of all algebras of an arbitrary type. Other varieties where word problems have been investigated include loops (Evans [1951]) and modular lattices (Hutchinson [1973], Freese [1979] and Lipschitz [1974]). The survey article of Evans [14] is recommended.

Problem 8. Is the word problem for orthomodular lattices decidable?

(d) Base Undecidability. This topic has been extensively developed by McNulty [1976] and Murskiĭ [1971]. The following example suffices to explain the subject. Suppose one takes a finite set of equations which are true of Boolean algebras and asks: "Do these equations axiomatize Boolean algebras?" Surprisingly, there is no algorithm to decide this question.

Problem 9. Can one derive the Linial–Post theorem [1949] (that there is no algorithm to determine if a finite set of tautologies with modus ponens axiomatizes the propositional calculus) from the above result on Boolean algebras, or vice versa?

(e) Other Undecidable Properties. Markov [26] showed that a number of properties of finitely presented semigroups are undecidable, for example there is no algorithm to determine if the semigroup is trivial, commutative, etc. Parallel results for groups were obtained by Rabin [1958]; and McNulty [1976] investigates such questions for arbitrary types. In [1975] McKenzie shows that the question of whether or not a single groupoid equation has a nontrivial finite model is undecidable, and then he derives the delightful result that there is a certain groupoid equation which will have a nontrivial finite model iff Fermat's Last Theorem is actually false. For decidability questions concerning whether a quasivariety is actually a variety see Burris [b] and McNulty [1977]. A difficult question is the following.

Problem 10. (Tarski). Is there an algorithm to determine if $V(\mathbf{A})$ has a finitely based equational theory, given that $\mathbf{A}$ is a finite algebra of finite type?

## §4. Boolean Constructions

Comer's work [1971], [1974], [1975], and [1976] connected with sheaves has inspired a serious development of this construction in universal algebra. Comer was mainly interested in sheaves over Boolean spaces, and one might say that this construction, which we have formulated as a Boolean product, bears the same relation to the direct product that the variety of Boolean algebras bears to the class of power set algebras $\mathbf{Su}(I)$. Let us discuss the role of Boolean constructions in two major results.

The decidability of any finitely generated discriminator variety of finite type (Burris and Werner [1979]) is proved by semantically embedding the countable members of the variety into countable Boolean algebras with a fixed finite number of distinguished filters, and then applying a result of Rabin [1969]. The semantic embedding is achieved by first taking the Boolean product representation of Keimel and Werner [1974], and then converting this representation into a better behaved Boolean product called a *filtered Boolean power* (the filtered Boolean power is the construction introduced by Arens and Kaplansky in [1948]).

The newest additions to the family of Boolean constructions are the modified Boolean powers, introduced by Burris in the fall of 1978. Whereas

Boolean products of finitely many finite structures give a well-behaved class of algebras, the modified Boolean powers give a uniform method for constructing deviant algebras from a wide range of algebras. This construction is a highly specialized subdirect power, but not a Boolean product.

The construction is quite easy. Given a field $\mathbf{B}$ of subsets of a set $I$, a subfield $\mathbf{B}_0$ of $\mathbf{B}$, an algebra $\mathbf{A}$, and a congruence $\theta$ on $\mathbf{A}$, let

$$A[\mathbf{B},\mathbf{B}_0,\theta]^* = \{f \in A^I : f^{-1}(a) \in B, f^{-1}(a/\theta) \in B_0, \text{ for all } a \in A\}.$$

This is a subuniverse of $\mathbf{A}^I$, and the corresponding subalgebra is what we call the modified Boolean power $\mathbf{A}[\mathbf{B},\mathbf{B}_0,\theta]^*$. McKenzie developed a subtle generalization of this construction in the fall of 1979 for the decidability result of Burris and McKenzie mentioned in §3(a) above. His variation proceeds as follows: let $\mathbf{B},\mathbf{B}_0,\mathbf{A}$ and $\theta$ be as above, and suppose $\mathbf{A} \leq \mathbf{S}$. Furthermore assume that $B_0$ contains all singletons $\{i\}$, for $i \in I$. Then the set

$$A[\mathbf{B},\mathbf{B}_0,\theta,\mathbf{S}]^* = \{f \in S^I : \exists g \in A[\mathbf{B},\mathbf{B}_0,\theta]^* \text{ with } [\![f \neq g]\!] \text{ finite}\}$$

is a subuniverse of $\mathbf{S}^I$. The corresponding subalgebra $\mathbf{A}[\mathbf{B},\mathbf{B}_0,\theta,\mathbf{S}]^*$ is the algebra we want.

## §5. Structure Theory

We have seen two beautiful results on the subject of structure theory, namely the Bulman–Fleming, Keimel, and Werner theorem (IV§9.4) that every discriminator variety can be represented by Boolean products of simple algebras, and McKenzie's proof [c] that every directly representable variety is congruence-permutable. McKenzie goes on to show that in a directly representable variety every directly indecomposable algebra is modular abelian or functionally complete.

The definition of a Boolean product was introduced in Burris and Werner [1979] as a simplification of a construction sometimes called a Boolean sheaf. Subsequently Krauss and Clark [1979] showed that the general sheaf construction could be described in purely algebraic terms, reviewed much of the literature on the subject, and posed a number of interesting problems. Recently Burris and McKenzie [1981] have proved that if a variety $V$ can be written in the form $I\Gamma^a(K)$, with $K$ consisting of finitely many finite algebras, then $V$ is of the form (discriminator) $\otimes$ (modular abelian); and then they discuss in detail the possibility of Boolean powers, or filtered Boolean powers, of finitely many finite algebras representing a variety. The paper concludes with an internal characterization of all quasiprimal algebras $\mathbf{A}$ such that the [countable] members of $V(\mathbf{A})$ can be represented as filtered Boolean powers of $\mathbf{A}$, generalizing the work of Arens and Kaplansky [1948] on finite fields.

Let us try to further crystallize the mathematically imprecise question of "which varieties admit a nice structure theory" by posing some specific questions.

Problem 11. For which varieties does there exist a bound on the size of the directly indecomposable members?

Problem 12. For which varieties $V$ is every algebra in $V$ a Boolean product of directly indecomposable algebras? (Krauss and Clark [1979]) of subdirectly irreducible algebras? of simple algebras?

Problem 13. For which finite rings **R** with 1 is the variety of unitary left **R**-modules directly representable?

## §6. Applications to Computer Science

Following Kleene's beautiful characterization [1956] of languages accepted by finite state acceptors and Myhill's study [1957] of the monoid of a language, considerable work has been devoted to relating various subclasses of regular languages and the associated class of monoids. For example Schützenberger [1965] showed that the class of star-free languages corresponds to the class of groupfree monoids. For this direction see the books [11], [12] of Eilenberg, and the problem set and survey of Brzozowski [7], [7a].

## §7. Applications to Model Theory

Comer [1974] formulated a version of the Feferman–Vaught theorem (on first-order properties of direct products) for certain Boolean products, and in Burris and Werner [1979] it is shown that all of the known variations on the Feferman–Vaught theorem can be derived from Comer's version.

Macintyre [1973/74] used sheaf constructions to describe the model companions of certain classes of rings, and this was generalized somewhat by Comer [1976] and applied to varieties of monadic algebras. In Burris and Werner [1979] a detailed study is made of model companions of discriminator varieties, and then the concept of a discriminator formula is introduced to show that the theorems of Macintyre and Comer are easy consequences of the results on discriminator varieties. A formula $\tau(x,y,u,v)$ is a *discriminator formula* for a class $K$ of algebras if it is an existential formula in prenex form such that the matrix is a conjunction of atomic formulas, and we have

$$K \models \tau(x, y, u, v) \leftrightarrow [(x \approx y \,\&\, u \approx v) \vee (x \not\approx y \,\&\, x \approx v)].$$

Problem 14. For which varieties $V$ can one find a discriminator formula for the subclass of subdirectly irreducible members?

Problem 15. Which finite algebras have a discriminator formula?

## §8. Finite Basis Theorems

In V§4 we looked at the three known general results on the existence of a finite basis for a variety (i.e., the variety is finitely axiomatizable). For many years universal algebraists hoped to amalgamate the Oates–Powell theorem [1965] (that a variety generated by a finite group has a finite basis) with Baker's theorem (that a finitely generated congruence-distributive variety of finite type has a finite basis) into one theorem saying that a finitely generated congruence-modular variety of finite type would have a finite basis. This was shown impossible by Polin [1976] who gave an example of a finitely generated but not finitely based (congruence-permutable) variety of non-associative rings.

16. Find a common generalization of the Oates–Powell theorem and Baker's theorem.

## §9. Subdirectly Irreducible Algebras

Let $\mathscr{F}$ be a type of algebras, and let $\kappa = |\mathscr{F}| + \omega$. As we have seen in V§3, Taylor [1972] proved that if a variety $V$ of type $\mathscr{F}$ has a subdirectly irreducible algebra of size greater than $2^{\kappa}$ then $V$ has arbitrarily large subdirectly irreducible members. Later McKenzie and Shelah [1974] proved a parallel result for simple algebras. In V§3 we proved Quackenbush's result [1971] that if $\mathbf{A}$ is finite and $V(\mathbf{A})$ has only finitely many finite subdirectly irreducible members (up to isomorphism) then $V(\mathbf{A})$ contains no infinite subdirectly irreducible members. Using the commutator Freese and McKenzie [a] proved that a finitely generated congruence-modular variety with no infinite subdirectly irreducible members has only finitely many finite subdirectly irreducible members.

Problem 17. (Quackenbush). If a finitely generated variety has no infinite subdirectly irreducible members, must it have only finitely many finite subdirectly irreducible algebras?

# Bibliography

## §1. Books and Survey Articles


1. R. Balbes and P. Dwinger, *Distributive Lattices*, Univ. of Missouri Press, Columbia, 1974.
2. J. L. Bell and A. B. Slomson, *Models and Ultraproducts*, North-Holland, Amsterdam, 1969.
3. G. Birkhoff, *Lattice Theory*, 3rd edition. Colloq. Publ., vol. **25**, Amer. Math. Soc., Providence, 1967.
4. ———, The rise of modern algebra to 1936, in J. D. Tarwater, J. T. White, and J. D. Miller, eds., *Men and Institutions in American Mathematics*, Graduate Studies, Texas Tech Press, Lubbock, 41–63.
5. ———, The rise of modern algebra, 1936 to 1950, in J. D. Tarwater, J. T. White, and J. D. Miller, eds., *Men and Institutions in American Mathematics*, Graduate Studies, Texas Tech Press, Lubbock, 65–85.
6. R. H. Bruck, *A Survey of Binary Systems*, Ergebnisse der Mathematik und ihrer Grenzgebiete, Band **20**, Springer-Verlag, Berlin-Göttingen-Heidelberg.
7. J. Brzozowski, *Open Problems about Regular Languages*, Dept. of Computer Science, Univ. of Waterloo, 1980.

7a. ———, *Developments in the Theory of Regular Languages*, Dept. of Computer Science, Univ. of Waterloo, 1980.

8. C. C. Chang and H. J. Keisler, *Model Theory*, Studies in Logic and the Foundations of Mathematics, vol. **73**, North-Holland, Amsterdam-London, 1973.
9. P. M. Cohn, *Universal Algebra*, Harper and Row, New York, 1965.
10. P. Crawley and R. P. Dilworth, *Algebraic Theory of Lattices*, Prentice-Hall, Englewood Cliffs, 1973.
11. S. Eilenberg, *Automata, Languages, and Machines*, A, Pure and Applied Mathematics, vol. **59-A**, Academic Press, New York, 1974.
12. ———, *Automata, Languages, and Machines*, B, Pure and Applied Mathematics, vol. **59-B**, Academic Press, New York, 1976.
13. Ju. L. Ershov, I. A. Lavrov, A. D. Taĭmanov, and M. A. Taĭclin, Elementary Theories, *Russian Math. Surveys* **20** (1965), 35-105.
14. T. Evans, Word Problems, *Bull. Amer. Math. Soc.* **84** (1978), 789-802.

15. G. Grätzer, *General Lattice Theory*, Pure and Applied Mathematics, vol. **75**, Academic Press, New York, 1978; Mathematische Reihe, Band **52**, Birkhäuser Verlag, Basel; Akademie Verlag, Berlin, 1978.
16. ———, *Universal Algebra*, 2nd edition, Springer-Verlag, New York, 1979.
17. P. R. Halmos, *Algebraic Logic*, Chelsea, New York, 1962.
18. ———, *Lectures on Boolean Algebra*, van Nostrand, Princeton, 1963.
19. L. Henkin, J. D. Monk, and A. Tarski, *Cylindric Algebras*, Part 1, North-Holland, Amsterdam, 1971.
20. B. Jónsson, *Topics in Universal Algebra*, Lecture Notes in Mathematics, vol. **250**, Springer–Verlag, Berlin-New York, 1972.
21. ———, Congruence varieties, *Algebra Universalis* **10** (1980), 355–394.
22. A. G. Kurosh, *Lectures on General Algebra*, Chelsea, New York, 1963.
23. H. Lausch and W. Nöbauer, *Algebra of Polynomials*, Mathematics Studies, vol. **5**, North-Holland, Amsterdam, 1973.
24. A. I. Mal'cev, *The Metamathematics of Algebraic Systems*, Collected papers 1936–1967, translated and edited by B. F. Wells III, North-Holland, Amsterdam, 1971.
25. ———, *Algebraic Systems*, Grundlehren der mathematischen Wissenschaften, vol. **192**, Springer-Verlag, New York, 1973.
26. A. A. Markov, *Theory of Algorithms*, Academy of Science USSR, Works of the Mathematical Institute Steklov. (Transl. by NSF).
27. B. H. Neumann, *Universal Algebra*, Lecture notes, Courant Institute of Math. Sci., New York University, New York, 1962.
28. R. S. Pierce, *Introduction to the Theory of Abstract Algebra*, Holt, Rinehart and Winston, New York, 1968.
29. H. Rasiowa, *An Algebraic Approach to Non-Classical Logics*, North-Holland, Amsterdam, 1974.
30. H. Rasiowa and R. Sikorski, *The Mathematics of Metamathematics*, Panstwowe Wydawnictwo Naukowe, Warszawa, 1963.
31. H. P. Sankappanavar, Decision Problems: History and Methods, in A. I. Arruda, N. C. A. da Costa, and R. Chuaqui, eds., *Mathematical Logic: Proceedings of the First Brazilian Conference*, Marcel Dekker, New York, 1978, 241–291.
32. R. Sikorski, *Boolean Algebras*, Ergebnisse der Mathematik und ihrer Grenzgebiete, Band **25**, Springer-Verlag, Berlin, 1964.
33. A. Tarski, with A. Mostowski and R. Robinson, *Undecidable Theories*, North-Holland, Amsterdam, 1953.
34. ———, Equational logic and equational theories of algebras, in K. Schütte, ed., *Contributions to Mathematical Logic*, North-Holland, Amsterdam, 1968, pp. 275–288.
35. W. Taylor, *Equational Logic*, Survey 1979, Houston J. of Math., 1979.
36. B. A. Trakhtenbrot, *Algorithms and Automatic Computing Machines*, Heath, Boston, 1963.
37. J. von Neumann, *The Computer and the Brain*, Yale University Press, New Haven, 1958.
38. R. Wille, *Kongruenzklassengeometrien*, Lecture Notes in Mathematics, vol. **113**, Springer-Verlag, Berlin, 1970.

## §2. Research Papers and Monographs

R. F. Arens and I. Kaplansky
[1948] Topological representation of algebras, *Trans. Amer. Math. Soc.* **63**, 457–481.
K. A. Baker
[1977] Finite equational bases for finite algebras in a congruence-distributive equa-

tional class, *Adv. in Math.* **24**, 207–243. [a] Definable normal closures in locally finite varieties of groups, (manuscript).

K. A. Baker and A. F. Pixley

[1975] Polynomial interpolation and the Chinese remainder theorem for algebraic systems, *Math. Z.* **143**, 165–174.

J. T. Baldwin and J. Berman

[1975] The number of subdirectly irreducible algebras in a variety, *Algebra Universalis* **5**, 379–389.

[1977] A model-theoretic approach to Mal'cev conditions, *J. Symbolic Logic* **42**, 277–288.

B. Banaschewski and E. Nelson

[1980] Boolean powers as algebras of continuous functions, *Dissertationes Mathematicae*, **179**, Warsaw.

G. M. Bergman

[1972] Boolean rings of projection maps, *J. London Math. Soc.* **4**, 593–598.

J. Berman

(See J. Baldwin and J. Berman)

G. Birkhoff

[1933] On the combination of subalgebras, *Proc. Camb. Philos. Soc.* **29**, 441–464.

[1935] On the structure of abstract algebras, *Proc. Camb. Philos. Soc.* **31**, 433–454.

[1944] Subdirect unions in universal algebra, *Bull. Amer. Math. Soc.* **50**, 764–768.

G. Birkhoff and O. Frink

[1948] Representations of lattices by sets, *Trans. Amer. Math. Soc.* **64**, 299–316.

G. Birkhoff and J. D. Lipson

[1970] Heterogeneous algebras, *J. Combin. Theory* **8**, 115–133.

[1974] Universal algebra and automata, pp. 41–51 in Proc. Tarski Symposium (1971), Vol. *25* of Symposia in Pure Math., Amer. Math. Soc., Providence, R.I.

W. W. Boone and G. Higman

[1974] An algebraic characterization of groups with soluble word problem, *J. Austral. Math. Soc.* **81**, 41–53.

R. H. Bruck

[1963] What is a loop? *Studies in Modern Algebra*, MAA Studies in Mathematics, vol. **2**, Math. Assoc. of America, 59–99.

S. Bulman–Fleming and H. Werner

[1977] Equational compactness in quasi-primal varieties, *Algebra Universalis* **7**, 33–46.

S. Burris

[1975a] Separating sets in modular lattices with applications to congruence lattices, *Algebra Universalis* **5**, 213–223.

[1975b] Boolean powers, *Algebra Universalis* **5**, 341–360.

[1979] On Baker's finite basis theorem for congruence distributive varieties, *Proc. Amer. Math. Soc.* **73**, 141–148.

[a] A note on directly indecomposable algebras, (to appear in *Algebra Universalis*)

[b] Remarks on reducts of varieties, (to appear in *Proceedings of the Conference on Universal Algebra*, Esztergom

S. Burris and J. Lawrence

[1979] Definable principal congruences in varieties of groups and rings, *Algebra Universalis* **9**, 152–164.

[a] A correction to "Definable principal congruence in varieties of groups and rings", (to appear in *Algebra Universalis*).

[b] Two undecidability results using modified Boolean powers, (manuscript, 1980).

S. Burris and R. McKenzie

[1981] Decidability and Boolean Representations, *Mem. Amer. Math. Soc.* no. 246.

S. Burris and H. P. Sankappanavar

[1975] Lattice-theoretic decision problems in universal algebra, *Algebra Universalis* **5**, 163–177.

S. Burris and H. Werner

[1979] Sheaf constructions and their elementary properties, *Trans. Amer. Math. Soc.* **248**, 269–309.

[1980] Remarks on Boolean products, *Algebra Universalis* **10**, 333–344.

A. Church

[1936] A note on the Entscheidungs problem, *J. Symbolic Logic* **1**, 40–41, 101–102.

D. M. Clark

(see P. H. Krauss and D. M. Clark)

S. D. Comer

[1971] Representations by algebras of sections over Boolean spaces, *Pacific J. Math.* **38**, 29–38.

[1974] Elementary properties of structures of sections, *Bol. Soc. Mat. Mexicana* **19**, 78–85.

[1975] Monadic algebras with finite degree, *Algebra Universalis* **5**, 315–327.

[1976] Complete and model-complete theories of monadic algebras, *Colloq. Math.* **34**, 183–190.

J. Dauns and K. H. Hofmann

[1966] The representation of biregular rings by sheaves, *Math. Z.* **91**, 103–123.

A. Day

[1969] A characterization of modularity for congruence lattices of algebras, *Canad. Math. Bull.* **12**, 167–173.

[1971] A note on the congruence extension property, *Algebra Universalis* **1**, 234–235.

P. Erdös

[1942] Some set-theoretical properties of graphs. *Revista, Ser. A., Matematicas y Fisica Teorica, Universidad Nacional de Tucuman* **3**, 363–367.

Ju. L. Ershov

[1972] Elementary group theories, *Soviet Math. Dokl.* **13**, 528–532.

T. Evans

[1951] The word problem for abstract algebras, *J. London Math. Soc.* **26**, 64–71.

[1953] Embeddability and the word problem, *J. London Math. Soc.* **28**, 76–80.

[1978] An algebra has a solvable word problem if and only if it is embeddable in a finitely generated simple algebra, *Algebra Universalis* **8**, 197–204.

[1979] Universal algebra and Euler's officer problem, *Amer. Math. Monthly* **86**, 466–473.

I. Fleischer

[1955] A note on subdirect products, *Acta Math. Acad. Sci. Hungar.* **6**, 463–465.

A. L. Foster

[1953a] Generalized "Boolean" theory of universal algebras. Part I: Subdirect sums and normal representation theorem, *Math. Z.* **58**, 306–336.

[1953b] Generalized "Boolean" theory of universal algebras. Part II: Identities and subdirect sums in functionally complete algebras, *Math. Z.* **59**, 191–199.

[1969] Automorphisms and functional completeness in universal algebras, *Math. Ann.* **180**, 138–169.

[1971] Functional completeness and automorphisms I, *Monatsh. Math.* **75**, 303–315.

[1972] Functional completeness and automorphisms II, *Monatsh. Math.* **76**, 226–238.

A. L. Foster and A. F. Pixley

[1964a] Semi-categorical algebras I, *Math. Z.* **83**, 147–169.

[1964b] Semi-categorical algebras II, *Math. Z.* **85**, 169–184.

R. Freese

[1979] Free modular lattices (abstract), *Notices Amer. Math. Soc.* **26**, A-2.

R. Freese and R. McKenzie
[a] Residually small varieties with modular congruence lattices, (to appear in *Trans. Amer. Math. Soc.*).
[b] The modular commutator, an overview, (manuscript, 1981).
O. Frink
(See G. Birkhoff and O. Frink)
B. Ganter and H. Werner
[1975] Equational classes of Steiner systems, *Algebra Universalis* **5**, 125–140.
I. Gelfand
[1941] Normierte Ringe, *Rec. Math.* (*Math. Sbornik*) *N.S.* **9**, 1–23.
K. Gödel
[1931] Über formal unentscheidbare Sätze der Principia Mathematica und verwandter Systeme I. *Monatshefte für Mathematik und Physik*, **38**, 173–198.
G. Grätzer
[1967] On the spectra of classes of algebras, *Proc. Amer. Math. Soc.* **18**, 729–735.
G. Grätzer and E. T. Schmidt
[1963] Characterizations of congruence lattices of abstract algebras, *Acta. Sci. Math.* (Szeged) **24**, 34–59.
A. Grzegorczyk
[1951] Undecidability of some topological theories, *Fund. Math.* **38**, 137–152.
H. P. Gumm
[1979] Algebras in permutable varieties: Geometrical properties of affine algebras, *Algebra Universalis* **9**, 8–34.
[a] An easy way to the commutator in modular varieties, (to appear in *Arch. Math.* (Basel)).
[b] Congruence modularity is permutability composed with distributivity, (manuscript 1980).
J. Hagemann and C. Herrmann
[1979] A concrete ideal multiplication for algebraic systems and its relation to congruence distributivity, *Arch. Math.* (Basel) **32**, 234–245.
C. Herrmann
[1979] Affine algebras in congruence modular varieties, *Acta. Sci. Math.* (Szeged) **41**, 119–125.
(See J. Hagemann and C. Herrmann)
G. Higman
(See W. W. Boone and G. Higman)
K. H. Hofmann
(See J. Dauns and K. H. Hofmann)
G. Hutchinson
[1973] Recursively unsolvable word problems of modular lattices and diagram-chasing, *J. Algebra* **26**, 385–399.
B. Jónsson
[1953] On the representation of lattices, *Math. Scand.* **1**, 193–206.
[1967] Algebras whose congruence lattices are distributive, *Math. Scand.* **21**, 110–121.
I. Kaplansky
(See R. F. Arens and I. Kaplansky)
K. Keimel and H. Werner
[1974] Stone duality for varities generated by quasi primal algebras, *Mem. Amer. Math. Soc.*, no. 148, 59–85.
S. C. Kleene
[1956] Representation of events in nerve nets and finite automata, in C. E. Shannon and J. McCarthy, eds., *Automata Studies*, Annals of Math. Studies **34**, Princeton Univ. Press, Princeton, pp. 3–41.

S. R. Kogalovskiĭ
[1965] On Birkhoff's theorem (Russian), *Uspehi Mat. Nauk.* **20**, 206–207.
P. H. Krauss and D. M. Clark
[1979] Global subdirect products, *Mem. Amer. Math. Soc.* no. 210.
R. L. Kruse
[1973] Indentities satisfied by a finite ring, *J. Algebra* **26**, 298–318.
J. Lawrence
[a] A note on Boolean powers of groups, (to appear in *Proc. Amer. Math. Soc.*).
(See also S. Burris and J. Lawrence)
S. Linial and E. L. Post
[1949] Recursive unsolvability of the deducibility, Tarski's completeness and independence of axioms problems of propositional calculus (Abstract), *Bull. Amer. Math. Soc.* **55**, p. 50.
L. Lipschitz
[1974] The undecidability of the word problem for projective geometries and modular lattices, *Trans. Amer. Math. Soc.* **193**, 171–180.
J. D. Lipson
(See G. Birkhoff and J. D. Lipson)
J. Łoś
[1955] Quelques remarques théorèmes et problèmes sur les classes définissables d'algèbres, in T. Skolem et al., eds., *Mathematical Interpretation of Formal Systems*, Studies in Logic and the Foundations of Mathematics, North-Holland, Amsterdam, 98–113.
I. V. Lvov
[1973] Varieties of associative rings I, II, *Algebra and Logic* **12**, 150–167, 381–393.
R. Lyndon
[1951] Identities in two-valued calculi, *Trans. Amer. Math. Soc.* **71**, 457–465.
[1954] Identities in finite algebras, *Proc. Amer. Math. Soc.* **5**, 8–9.
A. Macintyre
[1973/74] Model-completeness for sheaves of structures, *Fund. Math.* **81**, 73–89.
R. Magari
[1969] Una dimonstrazione del fatto che ogni varietà ammette algebre semplici, *Ann. Univ. Ferrara Sez. VII (N.S.)* **14**, 1–4.
A. I. Mal'cev
[1954] On the general theory of algebraic systems, *Mat. Sb.* (*N.S.*) **35**, 3–20.
A. Markov
[1947] On the impossibility of certain algorithms in the theory of associative systems, *Dokl. Akad. Nauk SSSR* (N.S.) **55**, 583–586.
[1954] *Theory of Algorithms*, Academy of Science USSR, Works of the Mathematical Institute Steklov. (Translated by the NSF, Washington, D.C.)
R. McKenzie
[1970] Equational bases for lattice theories, *Math. Scand.* **27**, 24–38.
[1975] On spectra, and the negative solution of the decision problem for identities having a finite non-trivial model, *J. Symbolic Logic* **40**, 186–196.
[1978] Para-primal varieties: A study of finite axiomatizability and definable principal congruences in locally finite varieties, *Algebra Universalis* **8**, 336–348.
[a] On minimal locally finite varieties with permuting congruences, (manuscript).
[b] Residually small varieties of K-algebras, (to appear in *Algebra Universalis*).
[c] Narrowness implies uniformity, (to appear in *Algebra Universalis*).
[d] Subdirect powers of non-abelian groups, (manuscript, 1980).
(See also S. Burris and R. McKenzie, R. Freese and R. McKenzie)
R. McKenzie and S. Shelah
[1974] The cardinals of simple models for universal theories, in E. Adams et al.,

eds, *Tarski Symposium* (1971), Proc. Symposia in Pure Math., vol. **25**, Amer. Math. Soc., Providence, 53–74.
G. F. McNulty
[1976] The decision problem for equational bases of algebras, *Annals Math. Logic* **11**, 193–259.
[1976] Undecidable properties of finite sets of equations, *J. Symbolic Logic* **41**, 589–604.
[1977] Fragments of first-order logic I: universal Horn logic, *J. Symbolic Logic* **42**, 221–237.
G. F. McNulty and W. Taylor
[1975] Combinatory interpolation theorems, *Discrete Math.* **12**, 193–200.
A. Mostowski and A. Tarski
[1939] Boolesche Ringe mit geordneter Basis, *Fund. Math.* **32**, 69–86.
V. L. Murskiĭ
[1965] The existence in the three-valued logic of a closed class with a finite basis, not having a finite complete system of identities, *Soviet Math. Dokl.* **6**, 1020–1024.
[1968] Some examples of varieties of semigroups (Russian), *Mat. Zametki* **3**, 663–670.
[1971] Non-discernible properties of finite systems of identity relations, *Soviet Math. Dokl.* **12**, 183–186.
[1975] The existence of a finite basis, and some other properties, for "almost all" finite algebras (Russian), *Problemy Kibernet.* **50**, 43–56.
J. Myhill
[1957] Finite automata and representation of events, WADC Tech. Rept. 57-624.
E. Nelson
[1967] Finiteness of semigroups of operators in universal algebra, *Canad. J. Math.* **19**, 764–768. (See also B. Banaschewski and E. Nelson)
W. D. Neumann
[1974] On Mal'cev conditions, *J. Austral. Math. Soc.* **17**, 376–384.
P. S. Novikov
[1955] On the algorithmic unsolvability of the word problem in group theory, *Trudy Mat. Inst. Steklov* **44**; English translation, *Proc. Steklov Inst. Math.* (2) **9** (1958), 1–122.
S. Oates and M. B. Powell
[1965] Identical relations in finite groups, *J. Algebra* **1**, 11–39.
O. Ore
[1935] On the foundation of abstract algebra, I, *Ann. of Math.* **36**, 406–437.
[1936] On the foundation of abstract algebra, II, *Ann. of Math.* **37**, 265–292.
P. Perkins
[1966] Decision problems for equational theories of semigroups, Ph.D Thesis, Univ. of California, Berkeley.
[1969] Bases for equational theories of semigroups, *J. Algebra* **11**, 298–314.
R. S. Pierce
[1967] Modules over commutative regular rings, *Mem. Amer. Math. Soc.*, no. 70.
D. Pigozzi
[1972] On some operations on classes of algebras, *Algebra Universalis* **2**, 346–353.
A. F. Pixley
[1963] Distributivity and permutability of congruence relations in equational classes of algebras, *Proc. Amer. Math. Soc.* **14**, 105–109.
[1971] The ternary discriminator function in universal algebra, *Math. Ann.* **191**, 167–180.
(See also K. A. Baker and A. F. Pixley, A. L. Foster and A. F. Pixley)
S. V. Polin
[1976] Identities of finite algebras, *Siberian Math. J.* **17**, 992–999.

E. L. Post
[1947] Recursive unsolvability of a problem of Thue, *J. Symbolic Logic* **12**, 1–11.
(See S. Linial and E. L. Post)
M. B. Powell
(See S. Oates and M. B. Powell)
P. Pudlák
[1976] A new proof of the congruence lattice representation theorem, *Algebra Universalis* **6**, 269–275.
R. W. Quackenbush
[1971] Equational classes generated by finite algebras, *Algebra Univeralis* **1**, 265–266.
[1974a] Semi-simple equational classes with distributive congruence lattices, *Ann. Eötvös Lorand Univ.* **17**, 15–19.
[1974b] Near Boolean algebras I: Combinatorial aspects, *Discrete Math.* **10**, 301–308.
[1975] Near-vector spaces over $GF(q)$ and $(v,q + 1,1)$-BIBDS, *Linear Algebra Appl.* **10**, 259–266.
[1976] Varieties of Steiner loops and Steiner quasigroups. *Canad. J. Math.* **28**, 1187–1198.
M. O. Rabin
[1958] Recursive unsolvability of group theoretic problems, *Ann. Math.* **67**, 172–194.
[1965] A simple method for undecidability proofs and some applications, in Y. Bar-Hillel, ed., *Logic, Methodology, Philosophy of Science*, Studies in Logic and the Foundations of Mathematics, North-Holland, Amsterdam, 58–68.
[1969] Decidability of second-order theories and automata on infinite trees, *Trans. Amer. Math. Soc.* **141**, 1–35.
J. Robinson
[1949] Definability and decision problems in arithmetic, *J. Symbolic Logic* **14**, 98–114.
H. Rogers
[1956] Certain logical reduction and decision problems, *Ann. of Math.* **64**, 264–284.
P. C. Rosenbloom
[1942] Post algebras I. Postulates and general theory, *Amer. J. Math.* **64**, 167–188.
B. Rosser
[1936] Extensions of some theorems of Gödel and Church, *J. Symbolic Logic* **1**, 87–91.
M. Rubin
[1976] The theory of Boolean algebras with a distinguished subalgebra is undecidable, *Ann. Sci. Univ. Clermont* No. 13, 129–134.
H. P. Sankappanavar
(See S. Burris and H. P. Sankappanavar)
E. T. Schmidt
(See G. Grätzer and E. T. Schmidt)
J. Schmidt
[1952] Über die Rolle der transfiniten Schlussweisen in einer allgemeinen Idealtheorie, *Math. Nachr.* **7**, 165–182.
M. P. Schützenberger
[1965] On finite monoids having only trivial subgroups, *Inform. and Control* **8**, 190–194.
S. Shelah
(See R. McKenzie and S. Shelah)
J. D. H. Smith
[1976] *Mal'cev Varieties*. Lecture Notes in Mathematics, vol. **554**, Springer-Verlag, Berlin-New York.

M. H. Stone
[1936] The theory of representation for Boolean algebras, *Trans. Amer. Math. Soc.* **40**, 37–111.
[1937] Applications of the theory of Boolean rings to general topology, *Trans. Amer. Math. Soc.* **41**, 375–481.
W. Szmielew
[1954] Elementary properties of Abelian groups, *Fund. Math.* **41**, 203–271.
A. Tarski
[1930] Fundamentale Begriffe der Methodologie der deduktiven Wissenschaften. I. *Monatsh. Math. Phys.* **37**, 360–404.
[1946] A remark on functionally free algebras, *Ann. of Math.* **47**, 163–165.
[1953] (Abstracts), *J. Symbolic Logic* **18**, 188–189.
[1975] An interpolation theorem for irredundant bases of closure structures, *Discrete Math.* **12**, 185–192.
W. Taylor
[1972] Residually small varieties, *Algebra Universalis* **2**, 33–53.
[1973] Characterizing Mal'cev conditions, *Algebra Universalis* **3**, 351–397.
[a] Some applications of the term condition (manuscript).
(See G. McNulty and W. Taylor)
A. M. Turing
[1937] On computable numbers, with an application to the Entscheidungsproblem, *Proc. London Math. Soc.* **42**, 230–265 (Correction **43** (1937), 544–546).
H. Werner
[1974] Congruences on products of algebras and functionally complete algebras, *Algebra Universalis* **4**, 99–105.
[1978] *Discriminator Algebras*, Studien zur Algebra und ihre Anwendungen, Band **6**, Akademie-Verlag, Berlin.
(See also S. Burris and H. Werner, S. Bulman-Fleming and H. Werner, B. Davey and H. Werner, B. Ganter and H. Werner, K. Keimel and H. Werner)
A. P. Zamjatin
[1973] A prevariety of semigroups whose elementary theory is solvable, *Algebra and Logic* **12**, 233–241.
[1976] Varieties of associative rings whose elementary theory is decidable, *Soviet Math. Dokl.* **17**, 996–999.
[1978a] A non-Abelian variety of groups has an undecidable elementary theory, *Algebra and Logic* **17**, 13–17.
[1978b] Prevarieties of associative rings whose elementary theory is decidable, *Siberian Math. J.* **19**, 890–901.

# Author Index

# Subject Index

# Graduate Texts in Mathematics

Soft and hard cover editions are available for each volume up to Vol. 14, hard cover only from Vol. 15.

1 TAKEUTI/ZARING. Introduction to Axiomatic Set Theory.
2 OXTOBY. Measure and Category. 2nd ed.
3 SCHAEFFER. Topological Vector Spaces.
4 HILTON/STAMMBACH. A Course in Homological Algebra.
5 MACLANE. Categories for the Working Mathematician.
6 HUGHES/PIPER. Projective Planes.
7 SERRE. A Course in Arithmetic.
8 TAKEUTI/ZARING. Axiomatic Set Theory.
9 HUMPHREYS. Introduction to Lie Algebras and Representation Theory. 2nd printing, revised.
10 COHEN. A Course in Simple Homotopy Theory.
11 CONWAY. Functions of One Complex Variable. 2nd ed.
12 BEALS. Advanced Mathematical Analysis.
13 ANDERSON/FULLER. Rings and Categories of Modules.
14 GOLUBITSKY/GUILLEMIN. Stable Mappings and Their Singularities.
15 BERBERIAN. Lectures in Functional Analysis and Operator Theory.
16 WINTER. The Structure of Fields.
17 ROSENBLATT. Random Processes. 2nd ed.
18 HALMOS. Measure Theory.
19 HALMOS. A Hilbert Space Problem Book.
20 HUSEMOLLER. Fibre Bundles. 2nd ed.
21 HUMPHREYS. Linear Algebraic Groups.
22 BARNES/MACK. An Algebraic Introduction to Mathematical Logic.
23 GREUB. Linear Algebra. 4th ed.
24 HOLMES. Geometric Functional Analysis and Its Applications.
25 HEWITT/STROMBERG. Real and Abstract Analysis. 4th printing.
26 MANES. Algebraic Theories.
27 KELLEY. General Topology.
28 ZARISKI/SAMUEL. Commutative Algebra I.
29 ZARISKI/SAMUEL. Commutative Algebra II.
30 JACOBSON. Lectures in Abstract Algebra I: Basic Concepts.
31 JACOBSON. Lectures in Abstract Algebra II: Linear Algebra.
32 JACOBSON. Lectures in Abstract Algebra III: Theory of Fields and Galois Theory.
33 HIRSCH. Differential Topology.
34 SPITZER. Principles of Random Walk. 2nd ed.
35 WERMER. Banach Algebras and Several Complex Variables. 2nd ed.
36 KELLEY/NAMIOKA. Linear Topological Spaces.
37 MONK. Mathematical Logic.
38 GRAUERT/FRITZSCHE. Several Complex Variables.
39 ARVESON. An Invitation to $C^*$-Algebras.
40 KEMENY/SNELL/KNAPP. Denumerable Markov Chains. 2nd ed.
41 APOSTOL. Modular Functions and Dirichlet Series in Number Theory.

42 Serre. Linear Representations of Finite Groups.
43 Gillman/Jerison. Rings of Continuous Functions.
44 Kendig. Elementary Algebraic Geometry.
45 Loève. Probability Theory. 4th ed. Vol. 1.
46 Loève. Probability Theory. 4th ed. Vol. 2.
47 Moise. Geometric Topology in Dimensions 2 and 3.
48 Sachs/Wu. General Relativity for Mathematicians.
49 Gruenberg/Weir. Linear Geometry. 2nd ed.
50 Edwards. Fermat's Last Theorem.
51 Klingenberg. A Course in Differential Geometry.
52 Hartshorne. Algebraic Geometry.
53 Manin. A Course in Mathematical Logic.
54 Graver/Watkins. Combinatorics with Emphasis on the Theory of Graphs.
55 Brown/Pearcy. Introduction to Operator Theory. Vol. 1: Elements of Functional Analysis.
56 Massey. Algebraic Topology: An Introduction.
57 Crowell/Fox. Introduction to Knot Theory.
58 Koblitz. $p$-adic Numbers, $p$-adic Analysis, and Zeta-Functions.
59 Lang. Cyclotomic Fields.
60 Arnold. Mathematical Methods in Classical Mechanics.
61 Whitehead. Elements of Homotopy Theory.
62 Kargapolov/Merzijakov. Fundamentals of the Theory of Groups.
63 Bollobas. Graph Theory—An Introductory Course.
64 Edwards. Fourier Series. 2nd ed. Vol. 1.
65 Wells. Differential Analysis on Complex Manifolds.
66 Waterhouse. Introduction to Affine Group Schemes.
67 Serre. Local Fields.
68 Weidmann. Linear Operators in Hilbert Spaces.
69 Lang. Cyclotomic Fields II.
70 Massey. Singular Homology Theory.
71 Farkas/Kra. Riemann Surfaces.
72 Stillwell. Classical Topology and Combinatorial Group Theory.
73 Hungerford. Algebra.
74 Davenport. Multiplicative Number Theory.
75 Hochschild. Basic Theory of Algebraic Groups and Lie Algebras.
76 Iitaka. Algebraic Geometry.
77 Hecke. Lectures on the Theory of Algebraic Numbers.
78 Burris/Sankappanavar. A Course in Universal Algebra.
79 Walters. An Introduction to Ergodic Theory.
80 Robinson. A Course in the Theory of Groups.
81 Forster. Lectures on Riemann Surfaces.